20th International School on Condensed Matter Physics 2018

Physics and Applications of Advanced and Multifunctional Materials

Journal of Physics: Conference Series
Volume 1186

Varna, Bulgaria
3 – 6 September 2018

ISBN: 978-1-5108-8497-7
ISSN: 1742-6588

Printed from e-media with permission by:

Curran Associates, Inc.
57 Morehouse Lane
Red Hook, NY 12571

Some format issues inherent in the e-media version may also appear in this print version.

Institute of Physics
Dirac House, Temple Back
Bristol BS1 6BE UK

Phone: 44 1 17 929 7481
Fax: 44 1 17 920 0979

techtracking@iop.org

Additional copies of this publication are available from:

Curran Associates, Inc.
57 Morehouse Lane
Red Hook, NY 12571 USA
Phone: 845-758-0400
Fax: 845-758-2633
Email: curran@proceedings.com
Web: www.proceedings.com

Table of contents

Volume 1186

20th International School on Condensed Matter Physics

3–6 September 2018, Varna, Bulgaria

Accepted papers received: 8 February 2019
Published online: 3 April 2019

Preface

Papers

Invited Lectures

Oral Presentations

Poster Presentations

IOP Conf. Series: Journal of Physics: Conf. Series **1186** (2019) 011001 doi:10.1088/1742-6596/1186/1/011001

PREFACE

We are pleased to introduce the Proceedings of the 20th JUBILEE INTERNATIONAL SCHOOL ON CONDENSED MATTER PHYSICS "Physics and Applications of Advanced and Multifunctional Materials (20th ISCMP). The school was held from September 3rd to September 7th, 2018 in Varna, Bulgaria. It was organized by the Institute of Solid State Physics of the Bulgarian Academy of Sciences (ISSP-BAS), and took place at one of the fine resorts on the Bulgarian Black sea "Saints Constantine and Helena".

The aim of this international school is to bring together top experimentalists and theoreticians, with interests in interdisciplinary areas, with the younger generation of scientists, in order to discuss current research and to communicate new forefront ideas. This year special focus was given to advanced organic and inorganic materials and their application in biophysics and sensors. Traditionally covered areas, such as characterization of nanostructured materials and thin films were also discussed. Participants from 16 countries presented 26 invited lectures, 28 short oral presentations and 60 posters.

The hope of the organizing committee is that the 20th ISCMP provided enough opportunities for direct scientific contacts, interesting discussions and interactive exchange of ideas between the participants. The nice weather certainly helped a lot in this respect. The editors would like to thank all authors for their high-quality contributions and the members of the international program committee for their commitment.

The papers submitted for publication in the Proceedings were refereed according to the publishing standards of the Journal of Physics: Conference Series. The Editorial Committee members are very grateful to the Journal's staff for the continuous fruitful relations and for giving us the opportunity to present the work from the 20th ISCMP.

Prof. DSc Hassan Chamati, Prof. DSc Doriana Dimova-Malinovska, Assoc. Prof. Dr. Julia Genova Prof. DSc Diana Nesheva, and Prof. DSc Albena Paskaleva

IOP Conf. Series: Journal of Physics: Conf. Series **1186** (2019) 011002 doi:10.1088/1742-6596/1186/1/011002

Bulgarian Academy of Sciences

GEORGI NADJAKOV INSTITUTE OF SOLID STATE PHYSICS

SOFIA, BULGARIA

20th JUBILEE INTERNATIONAL SCHOOL ON CONDENSED MATTER PHYSICS

Physics and Applications of Advanced and Multifunctional Materials

September 3rd - 7th, 2018
Varna, Bulgaria

IOP Conf. Series: Journal of Physics: Conf. Series **1186** (2019) 011002 doi:10.1088/1742-6596/1186/1/011002

Organising Committee

Chairman

Prof. Hassan Chamati, Director, Georgi Nadjakov Institute of Solid State Physics, Bulgarian Academy of Sciences, Sofia, Bulgaria (director@issp.bas.bg)

Honorary Chairman

Academician Alexander G. Petrov, Georgi Nadjakov Institute of Solid State Physics, Bulgarian Academy of Sciences, Sofia, Bulgaria (agpetrov@issp.bas.bg)

Local Organising Committee

Julia Genova (scientific secretary, iscmp@issp.bas.bg), Radostina Kamburova, Temenuga Hristova-Vassileva, Albena Paskaleva, Zdravka Slavkova, Vasilka Steflekova, Stanislav Varbev, Ekaterina Yordanova, Krastyo Buchkov

International Programme Committee

Sergei Baranovski (Philipps-University Marburg, Germany), Isak Bivas (Sofia, Bulgaria), Vladan Celebonovic (Institute of Physics, Belgrade, Serbia), Hassan Chamati (ISSP-BAS, Sofia, Bulgaria), Alina Ciach (IPC PAS, Warsaw, Poland), Doriana Dimova-Malinovska (CLSENES-BAS, Sofia), Julia Genova (ISSP-BAS, Sofia, Bulgaria), Ales Iglic (University of Ljubljana, Slovenia), Nikola Malinovski (IOMT-BAS, Bulgaria), Paul Montgomery (Unistra-CNRS, Strasbourg, France), Diana Nesheva (ISSP-BAS, Sofia, Bulgaria), Albena Paskaleva (ISSP-BAS, Sofia, Bulgaria), Alexander Petrov (ISSP-BAS, Sofia, Bulgaria), Stephen Reynolds (University of Dundee, Scotland, UK), Philippe Vanderbemden (Institut Montefiore, Liège, Belgium), Viktoria Vitkova (ISSP-BAS, Sofia, Bulgaria), Cheong Kuan Yew (Universiti Sains Malaysia, Malaysia)

Local Scientific Council

Bulgarian Academy of Sciences: J. Revalski (President of BAS), O. Stoilova (Scientific Secretary of BAS)

Georgi Nadjakov Institute of Solid State Physics: H. Chamati, J. Genova, D. Nesheva, A. Paskaleva

Central Laboratory of Solar Energy and New Energy Sources: D. Dimova-Malinovska

Sofia University - Faculty of Physics: V. Donchev, E. Valcheva (Heads of Departments)

Plovdiv University Paisii Hilendarski - Faculty of Physics and Engineering Technologies: T. Yovcheva (Head of Department)

IOP Conf. Series: Journal of Physics: Conf. Series **1186** (2019) 011003 doi:10.1088/1742-6596/1186/1/011003

Peer review statement

All papers published in this volume of *Journal of Physics: Conference Series* have been peer reviewed through processes administered by the proceedings Editors. Reviews were conducted by expert referees to the professional and scientific standards expected of a proceedings journal published by IOP Publishing.

IOP Conf. Series: Journal of Physics: Conf. Series **1186** (2019) 012001 doi:10.1088/1742-6596/1186/1/012001

Theoretically choosing multifunctional materials

V Celebonovic and M G Nikolic

LEX Laboratory, Inst of Physics, Univ. of Belgrade, Pregrevica 118, 11080 Zemun-Belgrade, Serbia

E-mail: vladan@ipb.ac.rs

Abstract. The aim of this paper is to discuss the possibility of theoretically engineering multifunctional nanomaterials. The calculations were performed for 1D and 2D nanomaterials by using results obtained within the Hubbard model (HM). The main results of the HM are briefly reviewed. The conclusion is that the approach taken in this paper is a distinct improvement over those in the literature.In the present paper results of applications of the HM are directly used in examples of engineering nanomaterials. On the other hand, in the literature calculations were performed by ab initio methods and then fitted to the form of the Hamiltonian of the HM.

1. Introduction

This paper is devoted to the problem of theoretically predicting values of parameters of multifunctional materials. It seems appropriate to start by defining these materials. Multifunctional materials can be defined as the materials which perform multiple functions in a system to which they belong [1]. They can exist naturally, but they can also be artificially synthetized. The probably best known natural multifunctional material is the human skin.

Multifunctional nanomaterials offer interesting possibilities for applications.A recent list [1] shows as much as 11 areas of human activity in which multifunctional materials have important impact.

Just as an illustration,note that applications of multifunctional nanomaterials in medicine are particularly interesting. A recent example is a drug called Kadcyla produced by the chemical industry Roche. In it,two materials in the form of nanoparticles which are separately known to be cancer-toxic (called transtuzumab and DM1) are joined together and sent to find and destroy HER2 positive cancer cells. Details are avaliable at http://www.roche.com.

In the example just mentioned, and in principle in any atempt of applying nanomaterials, there arises a practical question: how does one choose parameters of a multifunctional material? Experimentally, one could proceed by the "trial and error" method until reaching a satisfactory result.There is also the theoretical approach of modeling a material within some particular model. The calculations to be presented in this paper will be performed within the Hubbard model (HM).

Before embarking on the calculations, there are two questions which should be tackled to some extent.In view of the fact that large calculational projects, such as NOMAD (Novel Materials Discovery) (http://nomad-coe.eu), exist is it useful to perform small scale calculations like those to be presented here? Another big material science project is the Materials Genome

Initiative (`http://www.mgi.gov`). Concerning the Hubbard model, is it applicable to modeling of materials?

Concerning NOMAD, the answer is positive. It is true that NOMAD garantees high precision in treatment of any problem in material science. However, attempting to achieve the same goal by using any particular model is certainly easier as it bypasses problems related to access to NOMAD. Therefore, calculations discussed here can be treated as a first approximation to work possible by using NOMAD. A similar conclusion applies to MGI, which also has a data repository.

A bibliographical search concerning the applicability of the Hubbard model to modelling materials gave encouraging results. A search with the keywords "hubbard model material" on `http://prola.aps.org` published "anytime" performed begining October 2018 gave 9218 results. The clear implication is that this model can be applied to modelling materials. Classified as "most relevant" in this list is [2]. In this paper, the authors used *ab initio* electronic structure methods to design a material whose Hamiltonian matches as closely as possible that of the single band Hubbard model. Another interesting paper from this list is [3]. That paper is devoted to the computation of the ground-state phase diagram of the Hubbard model.

The next two sections contain a brief reminder of the main results of the HM needed in this paper, while the rest of the paper is devoted to the application of the HM in the choice of parameters of multifunctional nanomaterials.

2. The Hubbard model

This is a reminder of the main equations of the HM, while the reader interested in details is referred to [4,5]. See also [6].

The basic point for the study of any physical system is its Hamiltonian. For a 1D system, within the second quantisation formalism, the Hamiltonian has the following form

$$H = -w \sum_{l,\sigma} (c_{l,\sigma}^{+} c_{l+1,\sigma} + c_{l+1,\sigma}^{+} c_{l,\sigma}) + U \sum_{j} n_{j,\uparrow} n_{j,\downarrow} \tag{1}$$

The following symbols have been used: w denotes the hopping and U is the interaction energy of a pair of electrons having opposite spins on a lattice site j. Symbols of the form $c_{l,\sigma}^{+}$ denote a creation operator for an electron having spin σ on lattice site l, while $c_{l+1,\sigma}$ is the annihilation operator for an electron at lattice site $l+1$ with spin σ. The current operator is

$$j = -iw \sum_{l,\sigma} (c_{l,\sigma}^{+} c_{l+1,\sigma} - c_{l+1,\sigma}^{+} c_{l,\sigma}) \tag{2}$$

Knowing the Hamiltonian of a system enables the calculation of its transport properties. One of the methods for such calculations is the so called "memory function" approximation (MFA), developed in the 1970s as a consequence of previous work of Kubo. The MFA is extensively discussed in the literature (some references are given in [4]). Without repeating this discussion, we shall just state the final result for the electrical conductivity of a 1D system [4].

The final result for the electrical conductivity is given by

$$\sigma_R(\omega_0) = (1/2\chi_0)(\omega_P^2/\pi)[\omega_0^2 - (bw)^2]^{-1}(Uw/N^2)^2 S \tag{3}$$

and the symbol S denotes the following function

$$\begin{aligned} S = {} & 42.49916(1 + \exp(\beta(-\mu - 2w)))^{-2} + 78.2557(1 + \exp(\beta(-\mu + 2w\cos(1+\pi))))^{-2} \\ & + \frac{bw}{\omega_0 + bw}(4.53316 \times (1 + \exp(\beta(-\mu - 2w)))^{-2} \\ & + 24.6448(1 + \exp(\beta(-\mu + 2w\cos(1+\pi)))))^{-2}). \end{aligned} \tag{4}$$

The symbol μ denotes the chemical potential of the electron gas on a 1D lattice [6], n is the filling factor of the lattice, N the number of sites in the lattice, b a numerical constant,

$$\mu = \frac{(\beta w)^6 (ns - 1)\,|w|}{1.1029 + 0.1694(\beta w)^2 + 0.0654(\beta w)^4} \tag{5}$$

3. Optics

Relations between the electrical conductivity and reflectivity of solids are known since the middle of the twenties of the last century. These are the so called Kramers-Kroning (KK) relations. Detailed considerations on the KK equations can be found, for example, in [6,10]. In the context of this paper, there is one logically arising question: can these relations be used in theoretically choosing a nano-material? The answer is, as the reader may suppose, positive.

Skipping the details of the calculation (see [6] for example) it can be shown that the reflectivity is given by

$$R = \frac{(2\pi\sigma_R)^2 - K\omega[4\pi\sigma_R - K\omega(1+K^2)]}{(2\pi\sigma_R)^2 + K\omega[4\pi\sigma_R + K\omega(1+K^2)]} \tag{6}$$

and

$$K_{1,2}^2 = \frac{4\pi\omega\sigma_I - \omega^2 \pm [(4\pi\omega\sigma_R)^2 + (\omega^2 - 4\pi\omega\sigma_I)^2]^{1/2}}{2\omega^2}, \tag{7}$$

where the symbols ω,σ_R and σ_I denote the frequency, the real and imaginary components of the conductivity. These two expressions are applicable to 1D materials. Their applicability can be extended to $2D$ materials by a change of variables: $\sigma = [\sigma_x^2 + \sigma_y^2]^{1/2}$ [9].

The KK relations can be applied in several ways in engineering a nanomaterial. For example, what parameters a nanomaterial must have in order to have reflectivity close to zero? Or, if a material has a particular value of t_x what value of t_y must it have in order to have a particular value of the reflectivity?

4. Choosing multifunctional materials

Every part of any system performs a certain task within the domain of possibilities of the system as a whole. If a part of a system, or the system as a whole can perform multiple tasks, it is called *a multifunctional system.* Going a step further,a nanomaterial is called a *multifunctional nanomaterial* if its behavior can be controlled by the application of various external fields [8].

4.1. 1D material under a thermal field

One of the parameters of a material is the chemical potential of the electron gas. In the 1D case,it is given by eq.(5),and discussed in more detail [4,6]. It was shown by Lieb and Wu [7] that the chemical potential is zero for half filling, regardless of the temperature. Speaking from the experimental point of view, the half-filled case means that the material is pure (not doped) and that it is not a conductor of electricity [6].

Taking into account the fact that in real experiments materials are at some non-zero temperature, the obvious question is: can the chemical potential become close to zero for the lattice filling factor n different from one? A positive answer to that question would have some interesting consequences in the applications of the HM.

Take equation (5) and arbitrarily normalize the chemical potential to $\mu = 1$ at the point $n = 1.1, T = 116K$ and $t = 0.02$. Repeating the same calculation for $n = 1.3$, and then plotting the two graphs, one gets figure 1. Figure 1. shows that for $T \geq 180\ K$, the chemical potential tends approximately to zero.

This is a difference with the standard results of the HM, which state that the chemical potential and the electrical conductivity are zero for $n = 1$. In fact, this means that the

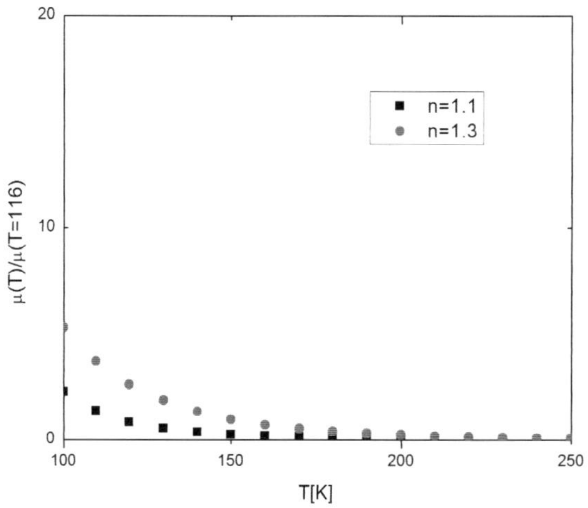

Figure 1. The chemical potential of 1D electron gas for $n = 1.1$ and $n = 1.3$

material becomes electrically neutral at values of temperature above some limiting value of the temperature. In the context of choosing a nano-material this can be taken to imply that designing an electrically neutral material can be achieved by heating it above a limiting value of the temperature.

4.2. the conductivity of a 1D material under a thermal field

As described in the preceding section, a prerequisite for the calculation of the electrical conductivity of any physical system is the knowledge of its Hamiltonian. A frequent choice for such calculations is the memory function method. Applying this approach to the 1D Hubbard model, and using the formalism of second quantisation, one gets expressions (3)-(4) for the electrical conductivity.

The chemical potential,which contains the dependence on the doping and temperature, is given by equation (5). It is generally agreed that equations (3)-(5) can be applied to 1D systems of correlated electrons, such as low dimensional organic conductors and/or superconductors. The following two figures illustrate the behavior of systems of 1D correlated electrons under the control of a thermal field (i.e.,under variable temperature).

Figure 2 shows the characteristic shape of the $\sigma(T)$ curve. As the position of the maximum is sharp, this result opens up (in principle) the possibility of theoretically choosing a nanomaterial which will have a predetermined position of the maximum conductivity in the (σ, T) plane.

Experimentally speaking, the filling factor n denotes the doping of a material. Figure 3 shows that exposing a material to a varying doping field changes its conductivity in a characteristic manner,and thus opens up the possibility of designing a material with a specified values of the couple σ, n.

Figure 4 is even more interesting, because it shows the behavior of the conductivity of a material under the influence of two variable fields: the thermal field, and (what could be called) the hopping field,under constant doping n. In the region $T < 180$ K the relative differences of the conductivities for the largest and smallest values of the hopping t amount to as much as 50 per cent. Using calculations like those which led to this figure, one can theoretically design

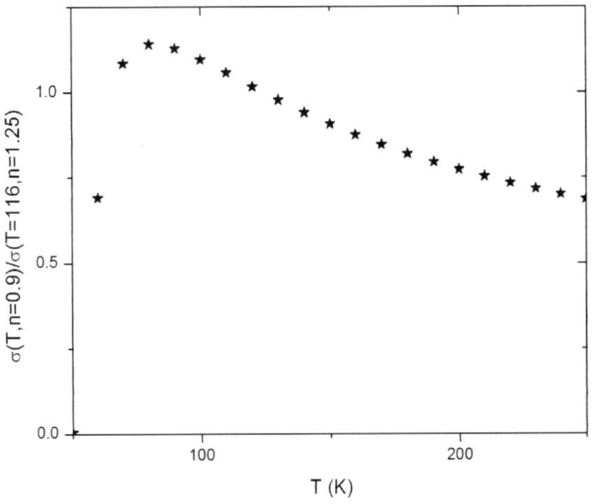

Figure 2. The normalized elect. conduct. of 1D electron gas for $n = 0.9$

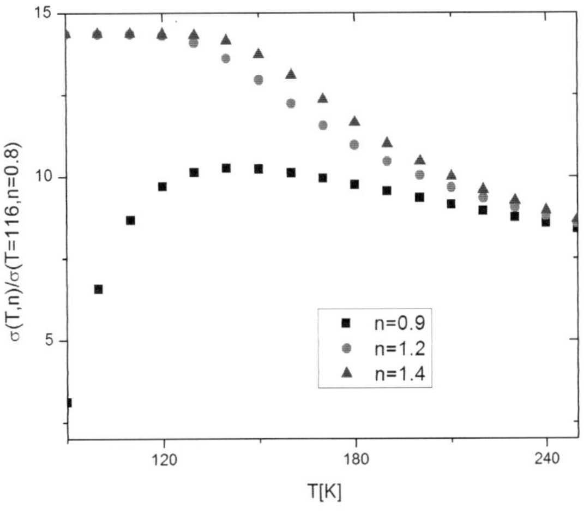

Figure 3. The normalized conductivity for three values of n for a 1D material

materials with predetermined values of the conductivity for certain values of the doping and/or hopping. Practically, one would have to choose pairs of (n, t) values, or fix n and vary t and then calculate the conductivity.

4.3. The two dimensional materials

The HM is at present solved only for the case of 1D [7]. In the special case of a 2D rectangular lattice where the conductivities along the axes are mutually independent and denoted by σ_x and

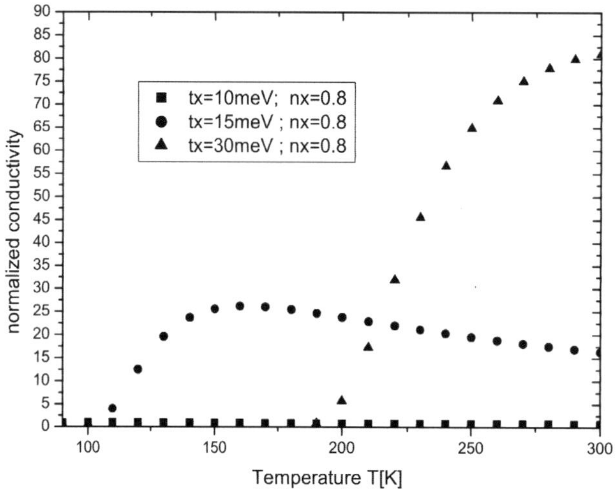

Figure 4. The normalized conduct. of a 2D rectang. lattice for 3 values of t

σ_y, the conductivity of the lattice can be expressed as

$$\sigma = [\sigma_x^2 + \sigma_y^2]^{1/2}. \tag{8}$$

Theoretically choosing the parameters of a system having such a link between the conductivities of its parts may seem complicated. In fact, this 2D problem reduces to two 1D problems along the two lattice axes.

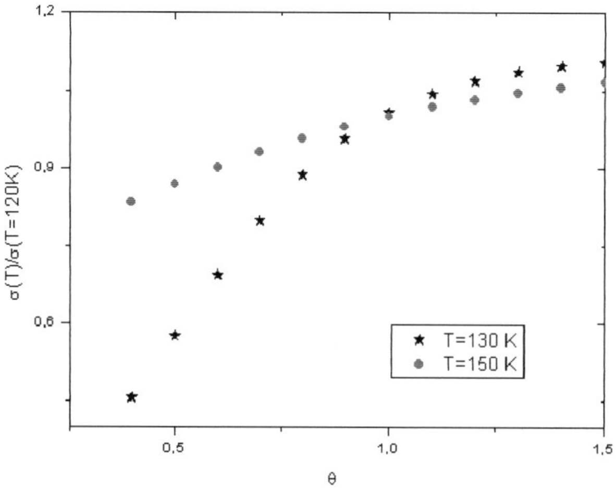

Figure 5. The normalized reflectivity of a 2D rectangular lattice for 2 values of T

Figure 5 taken from [6] is an example of the influence of external fields on a rectangular 2D lattice which is moderately useful for theoretically choosing a nanomaterial. It represents the dependence of the normalized electrical conductivity of a 2D lattice on the ratio of the filling factors along the two lattice axes. This ratio is symbolized by $n_y/n_x = \theta$ and the parameters for which this figure was calculated are avaliable in [6]. The point which could be useful in choosing a nanomaterial is here the turning point of one of the curves.

4.4. designing the optical properties of nano materials

An interesting, but complicated way of engineering a nano material is to predict theoretically the value of its reflectivity and link it to the values of various material parameters. It can be shown ([6] and references given there) that the reflectivity can be expressed as the following sum

$$R \cong 1 - 2(\frac{K\omega_0}{\pi\sigma_R}) + 2(\frac{K\omega_0}{\pi\sigma_R})^2 - .. \tag{9}$$

This expression for the reflectivity implicitly depends on the dimensionality of the system through σ_R and K. An illustrative example of the temperature dependence of the reflectivity is presented on fig.6.

Note an interesting detail on this figure, useful for theoretically choosing a nano material. The derivative dR/dT drastically changes at the temperature of approximately $T \cong 70K$. The position of this point can be theoretically predicted, and it depends on the values of the hopping t and the filling factor n. This means that fixing the temperature and which R changes, one can theoretically determine the values of t and n.

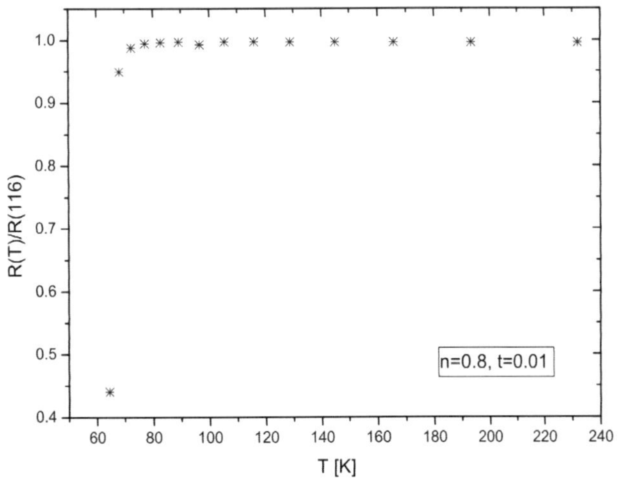

Figure 6. The reflectivity of 1D HM for $n = 0.8$ $t = 0.01$

The first step was the calculation of the electrical conductivity. It was normalized to $\sigma_R = 1$ for $n = 0.8$, $t = 0.005$ eV and $T = 116$ K. In the case of a 2D material, two important ratios influence the results of calculations discussed here. These are the ratio of the conductivities,denoted by $\sigma_{Ry}/\sigma_{Rx} = \Gamma$, and the ratio of the filling factors along the two lattice axes, symbolized by $n_y/n_x = \theta$. Note that all the 2D materials discussed here imply materials with a square lattice.

The dependence of the reflectivity under constant temperature of a 2D lattice on the parameter θ is shown on the following figure, taken from [6]. Note that the decrease of θ leads to a decrease of the value of R. In practical terms, this means that decreasing the ratio of filling factors along the two lattice axes leads to increasing difficulties in seeing the specimen. This conclusion could have implications in work on invisibility cloaks.

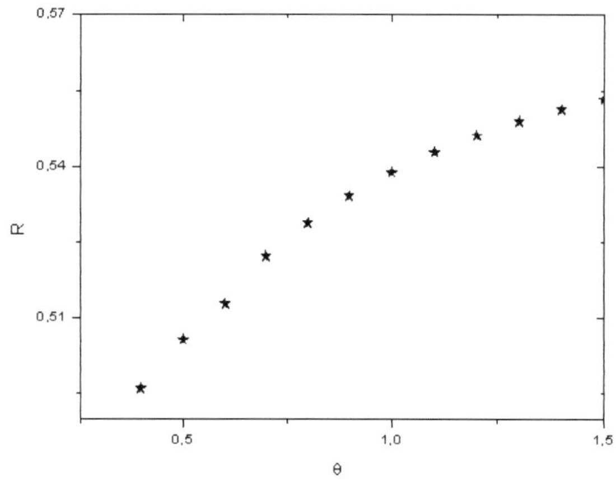

Figure 7. The reflectivity as a function of θ for $T = 150K$

5. Piezoresistivity under strain

In various situations, both in laboratory work and practical applications, the need may arise for a material which becomes better conductor of electricity when exposed to a strain ϵ. Such a material can be chosen theoretically, but the situation is slightly more complex that in the cases discussed so far.

Namely, in such a situation one has to take into account the dependence of the hopping t on the strain. Using results on the overlap of atomic wave functions,in [11], the dependence of the hopping t on the interatomic distance r was approximated as

$$t = t_0[1 + r + (1/3)r^2]\exp(-r) \tag{10}$$

where r is measured in the units of Bohr radius.

As a practical application, in which theoretical choice of a material is needed,let us consider an instrument based on piezoresistivity [4].In this case the problem we encounter can be stated as "At a given temperature T what value of the hopping t must the material have, in order to achieve some chosen value of the sensitivity G?"

The sensitivity of a sensor based on piezoresistivity is defined as

$$G = -\frac{1}{\sigma}(\frac{\partial\sigma}{\partial\epsilon}) \tag{11}$$

where σ is the conductivity and ϵ the strain. The dependence of G on the strain has been studied in [4]. Clearly, the lower the temperature the more pronounced this dependence becomes.

6. Compressibilities

What would be he result of atempting to engineer a nanomaterial under the influence of high pressure? To render the problem more complex, immagine that the nanomaterial in question is a compound, consisting of two specimen of nanomaterials with compressibilities of different signs. This combination is in principle possible [12]. Imagine an experiment in which two materials having different coefficients of thermal expansion are fixed together. If such a combination is heated,it will bend to one side. A similar situation will occur if the two nanomaterials are subdued to high pressure and have compressibilities of different signs [12].

The isothermal compressibility is defined as

$$\beta_T = -\frac{1}{V}(\frac{\partial V}{\partial p})_T \tag{12}$$

where all the symbols have their standard meanings. The pressure P and volume V of a material are linked through the equation of state. Choosing an appropriate equation of state for a given nanomaterial is a delicate research subject. See,for example,[13].

In the case of a nanomaterial distributed along a straight line, the problem returns to those discussed earlier. However, if the materials are distributed along a curved line, the solution is strongly dependent upon the compressibilities and their signs. A detailed consideration of this problem is in preparation.

7. Conclusions

This paper attempted to answer an apparently simple question - is it possible to choose parameters of a nano-material theoretically so that they have certain pre-selected values? This choice of parameters is dependent on the theoretical model chosen to describe the material under study.

The calculations discussed in the present paper were performed for 1D and 2D materials within the Hubbard model.Calculations reported here are more direct than those in [2,3]. For example, in [2] the calculations were performed by *ab initio* methods, and then the Hamiltonian was matched to the Hamiltonian of the HM. On the other hand, in the present paper results of direct applications of the HM and the memory function method were used to demonstrate possibilities of engineering nano-materials, which is a distinct advantage. It was shown on several examples that it is possible to engineer multifunctional nano-materials by calculating their conductivity, reflectivity and sensitivity for piezoresistivity, making calculations within the Hubbard model a viable way in theoretically modelling nanomaterials.

Acknowledgments

Prepared within the research projects 171005 and 171038 of the Ministry of Education, Research and Technological Development of Serbia. V C is grateful to ICTP (Trieste,Italy) for their hospitality and use of the libraries.

References

[1] `https://www.e-education.psu.edu/eme807/node/698`
[2] Griffin S M Staar Schulthess et al 2016 *Phys. Rev.* B **93** 075115
[3] B-Xiao Zheng and Chan G K L 2016,*Phys. Rev.* B **93** 035126
[4] Celebonovic V and Nikolic M G 2018 *J. Low Temp. Phys.* **190** 191
[5] Giammarchi T 2017 *Nature* **545** 414
[6] Celebonovic V 2016 in *Advanced 2D Materials, edited.by A Tiwari and M Syvajarvi* Advanced Materials Series, (Scrivener Publishing LLC) chap. 4
[7] Lieb E H and Wu F Y 1968 *Phys. Rev. Lett.* **20** 1445
[8] Suzdalev I P 2009 *Russ. Chem. Rev.* **78** 249
[9] Celebonovic V 2010 *J. Phys. Conf. Ser.* **253** 012004.

IOP Conf. Series: Journal of Physics: Conf. Series **1186** (2019) 012001 doi:10.1088/1742-6596/1186/1/012001

[10] Peiponen K E and Saarinen J J 2009 *Rep. Progr. Phys.* **72** 056401
[11] Celebonovic V 2014 *J. Phys. Conf. Ser* **558** 012006
[12] Gatt R and Grima J N 2008 *Phys. Stat. Sol. RRL* **2** 236
[13] Chandra J and Kholiya K 2016 *J. Taibah Univ. Science* **10** 386

IOP Conf. Series: Journal of Physics: Conf. Series **1186** (2019) 012002 doi:10.1088/1742-6596/1186/1/012002

Ferrimagnetism in a system of two antiferromagnetically coupled Heisenberg models

H Chamati and D V Shopova

Institute of Solid State Physics, Bulgarian Academy of Sciences, 1784 Sofia, Bulgaria

E-mail: chamati@issp.bas.bg

Abstract. We phenomenologically investigate the phase diagram of some ferrimagnetic materials on the basis of two bilinearly coupled Heisenberg models sitting on two interpenetrating simple magnetic cubic lattices. Computations were performed with the help of the Landau free energy obtained through applying the Hubbard-Stratonovich transformation to the initial microscopic Heisenberg Hamiltonian. The transitions to the magnetically ordered phases within the model are found to be of second order with the occurrence of a compensation point at lower temperatures for some values of the system's parameters. It is found that the main stable phase is a two-sublattice collinear ferrimagnet in addition to a metastable non-collinear phase. The numerical results give a detailed description of temperature dependence of magnetization on the strength of intersublattice interaction and the difference between the effective exchanges of two ferromagnetically ordered sublattices. Our results are in qualitative agreement with the experimental results on $Y_3Fe_5O_{12}$ in the strong coupling limit and the compounds $ErFe_2$ and $GdCo_{12}B_6$ for weak couplings.

1. Introduction

Ferrimagnets are substances made up of various components possessing different magnetic properties. The difference in magnetic moments leads to geometric frustration that may arise because either different elements occupy the lattice sites or the same element occupies nonequivalent crystallographic sites surrounded by a different number or type of non-magnetic ions, which effectively results in different magnetic properties. For complex alloys a combination of both may take place (For an extensive review see reference [1] and references therein). Within the mean-field approach it is generally accepted that ferrimagnets can be modeled with the help of several interpenetrating sublattices each ordered ferromagnetically with an effective antiferromagnetic coupling between them. In the pioneering works of Néel on ferrimagnetism within the molecular field approach (see e.g. [2]), a two sublattice model is used to compute the thermal magnetization behaviour of ferrimagnets, and six possible magnetization curves were derived. There, special attention was paid to iron garnets, where the spontaneous magnetization in comparison with experiment can be interpreted by applying a three-sublattice model. In view of experimental study of magnetocaloric effect of rare-earth based ferrimagnets [3] which has great potential for technological applications in environmentally-friendly refrigeration, the theoretical mean-field description of such alloys with three sublattices, is further elaborated [4]. Such studies are based on considering microscopic classical Heisenberg models with different exchange and spin-orbit interactions depending on the crystal structure, and chemical composition of the particular alloy under study. There is another theoretical mean field approach based on considering mixed spin Ising model for description of ferrimagnets, see for example [5–7], where a very detailed review of the literature on this approach is presented.

In the present paper we consider ferrimagnets that can be described by different magnetic ions sitting on two interpenetrating simple cubic sublattices arranged in a body-centred-cubic structure. The magnetic interaction between ions on each sublattice is supposed ferromagnetic, while ions on the different sublattices are coupled antiferromagnetically. The magnetic properties will be investigated on the basis of bilinearly coupled Heisenberg classical model in a mean-field approximation which is performed using the Hubbard-Stratonovich transformation for obtaining the respective Landau free energy and its analysis.

We aim at clarifying two aspects, namely, the influence of the magnitude of inter-sublattice interaction on the behaviour of sublattice magnetizations as a function of the temperature, as well as the dependence of the sublattice magnetizations on the difference between the magnitudes of exchange interactions in the sublattices. For this reason we will not consider any anisotropy effects and will adhere to high-symmetry crystal structure, for which such anisotropic effects can be neglected as small. Let us anticipate that despite the simplifications introduced in the model considered here our theoretical results compare qualitatively well with experimental measurements on the compounds $Y_3Fe_5O_{12}$ [4] in the strong interaction limit, as well as, $ErFe_2$ [8] and $GdCo_{12}B_6$ [9] in the weak interaction limit.

The rest of the paper is organized as follows: in Section 2 we describe in detail the derivation of the expression of the Landau free energy, within the mean field approach, from the classical Heisenberg model with competing interactions following closely the approach of Ref. [10]. In Section 3 the solutions of the equations of state obtained after minimization of the Landau free energy derived in Section 2 are analyzed. Section 4 discusses the analytical and the numerical results depending on the model's parameters both in strong- and weak-coupling limits for the ferrimagnetic substances under consideration. Section 5 generalizes the conclusions and possible further development of our study.

2. The model and derivation of Landau free energy

The microscopic Heisenberg Hamiltonian that describes two bilinearly coupled magnetic sublattices reads

$$H = -\frac{1}{2}\sum_{ij}^{2N}\left[\mathscr{J}_{ij}^{(1)}\mathbf{S}_i^{(1)}\cdot\mathbf{S}_j^{(1)} + \mathscr{J}_{ij}^{(2)}\mathbf{S}_i^{(2)}\cdot\mathbf{S}_j^{(2)} + 2\mathscr{K}_{ij}\mathbf{S}_i^{(1)}\cdot\mathbf{S}_j^{(2)}\right]. \tag{1}$$

Here $\mathbf{S}_j^{(1,2)}$, are n-component classical Heisenberg spins whose magnitude is normalized on the unit sphere in spin space through the condition $|\mathbf{S}_j^{(1,2)}| = 1$. The exchange parameters $\mathscr{J}_{ij}^{(1,2)}$, $\mathscr{K}_{ij}$ in the general case are $N \times N$ symmetric matrices with N – the number of lattice sites, assumed equal for both subsystems. This condition simplifies the consideration as it makes the system symmetric with respect to the interchange of the subsystems. The exchange matrices $\mathscr{J}_{ij}^{(1,2)}$ denote the interaction between ions on the same sublattice and $\mathscr{K}_{ij}$ – between magnetic ions on the different sublattices.

Hamiltonian (1) may be applied to the description of magnetic systems which consist of two distinct magnetic materials. Regardless of the microscopic origin of mechanisms underlying the magnetic properties such systems can be effectively described by different exchange interactions within the two subsystems. There may be other situations when the substance is made only of one type of magnetic ions that occupy two different crystallographic positions in the Bravais lattice, separated by some non-magnetic atoms. Such substance may also be considered as built of two magnetic subsystems with different exchange interactions within them.

In order to analyse the behaviour of the magnetization and the phase behaviour in systems that can be described by the microscopic Hamiltonian (1) we have to find the associated mean-field free energy. This is achieved by calculating the partition function which in this case is represented by functional integral in n-dimensional spin space, where n - is the number of spin components. To this end we apply the Hubabrd-Stratonovich transformation; see e.g. ref. [11] and references therein. This approach was described in detail and used in reference [10] to explore the magnetic properties of a system of two ferromagnetically coupled subsystems. Here we will just outline the main steps for the derivation of the Landau free energy, especially in respect to the antiferromagnetic coupling between the subsystems.

We consider the two interacting magnetic subsystems sitting on a body-centered crystal lattice, for which the corners of elementary cube are occupied by one kind of magnetic atoms, and at the centre of the cube the atoms of the other kind. Thus, the nearest neighbours belong to different magnetic subsystems and the next-nearest neighbours to same subsystem. Hence, the system may be described as two interpenetrating sublattices, consisting of different magnetic atoms and we assume that the interaction within the sublattices $\mathscr{J}_{ij}^{1,2}$ is ferromagnetic and between them, $\mathscr{K}_{ij}$, it is antiferromagnetic.

The Hubbard-Stratonovich transformation expresses the initial microscopic Hamiltonian in new n-component variables $\Psi_i^{(1)}, \Psi_i^{(2)}$ defined in real space, directly connected with the initial spins (see [10]), namely:

$$\begin{aligned}\mathscr{H} =& \frac{1}{2}\sum_{ij}^{2N}\left(J_{ij}^{(1)}\Psi_i^{(1)}\cdot\Psi_j^{(1)} + J_{ij}^{(2)}\Psi_i^{(2)}\cdot\Psi_j^{(2)} + 2K_{ij}\Psi_i^{(1)}\cdot\Psi_j^{(2)}\right) - \ln\left[\sum_i^{2N} I_{\frac{n}{2}-1}\left(x_i^{(1)}\right)\left(\frac{x_i^{(1)}}{2}\right)^{-\frac{n}{2}}\Gamma\left(\frac{n}{2}\right)\right] \\ & - \ln\left[\sum_i^{2N} I_{\frac{n}{2}-1}\left(x_i^{(2)}\right)\left(\frac{x_i^2}{2}\right)^{-\frac{n}{2}}\Gamma\left(\frac{n}{2}\right)\right]. \end{aligned} \tag{2}$$

Here $I_{\frac{n}{2}-1}\left(x_i^{(1,2)}\right)$ is the modified Bessel function of the first kind, and $\Gamma\left(\frac{n}{2}\right)$ is the Gamma function. In the above expression the exchange parameters $J_{ij}^{(1,2)}$ and K_{ij} are connected to those in the initial Hamiltonian (1) through the relations:

$$J_{ij}^{(1,2)} = \frac{\mathscr{J}_{ij}^{(1,2)}}{T}, \qquad K_{ij} = \frac{\mathscr{K}_{ij}}{T}, \tag{3}$$

with T - the temperature. We have denoted by $x_i^{(1)}$ and $x_i^{(2)}$ in (2) the following expressions:

$$x_i^{(1)} = \left|J_{ij}^{(1)}\Psi_j^{(1)} + K_{ij}\Psi_j^{(2)}\right|; \qquad x_i^{(2)} = \left|J_{ij}^{(2)}\Psi_j^{(2)} + K_{ij}\Psi_j^{(1)}\right|.$$

The terms in (2) containing the Bessel functions will be further used only in the form of an expansion with respect to $x_i^{(1,2)}$ up to forth order by using the relation:

$$\Gamma\left(\frac{n}{2}\right)\left(\frac{2}{x}\right)^{\frac{n}{2}-1} I_{\frac{n}{2}-1}(x) = 1 + \sum_{k=1}^{\infty}\frac{\left(\frac{x}{2}\right)^{2k}}{k!(k+\frac{n}{2}+1)(k+\frac{n}{2}-2)\cdots\frac{n}{2}}.$$

The next step is to perform a Fourier transformation to the **k**-space, and pass to the continuum limit in **k** as the finite size effects will not be considered at this stage. The quadratic part of the obtained Hamiltonian again contains a bilinear term with respect to $\Psi^{(1,2)}(\mathbf{k})$ and we have to diagonalise it. This is achieved with the help of the unitary matrix $\hat{S}$:

$$\hat{S} = \begin{pmatrix} S_0(\mathbf{k}) & -S_1(\mathbf{k}) \\ S_1^*(\mathbf{k}) & S_0(\mathbf{k}) \end{pmatrix}. \tag{4}$$

The eigenvalues of the matrix $\hat{S}$ read:

$$\lambda_{1,2}(\mathbf{k}) = \frac{1}{2}\left[J_1(\mathbf{k}) + J_2(\mathbf{k}) \pm \sqrt{(J_1(\mathbf{k}) - J_2(\mathbf{k}))^2 + 4K(\mathbf{k})^2}\right], \tag{5}$$

where $J_{1,2}(\mathbf{k})$ and $K(\mathbf{k})$ are the Fourier transforms of $J_{ij}^{(1,2)}$ and K_{ij} from (3), respectively.

In order to compute the integral in the partition function we use the steepest-descent method, *i.e.* the integration contour is taken around the maxima of the eigenvalues (5). For the bcc structure, if we take the nearest neighbour interaction between atoms of the same sort and the nearest neighbour interaction between the atoms of different sorts, the Fourier transforms of exchange interactions read:

$$J_{1,2}(\mathbf{k}) = \frac{J_{1,2}}{3}\sum_{i=1}^{3}\cos\mathbf{k}_i a$$

and

$$|K(\mathbf{k})| = \frac{K^2}{8}\left[1+\sum_{i=1}^{3}\cos\mathbf{k}_i a+\sum_{i<j}^{3}\cos\mathbf{k}_i a\cos\mathbf{k}_j a+\cos\mathbf{k}_1 a\cos\mathbf{k}_2 a\cos\mathbf{k}_3 a\right];$$

here $J_{1,2}$ are the exchange interactions between the nearest neighbour atoms of the same sort on sublattices 1 and 2, respectively, and K is the exchange interaction between nearest neighbour atoms of different sort; a is the lattice constant. The calculation using the above expressions shows that the eigenvalues $\lambda_{1,2}(\mathbf{k})$ have a maximum in the centre of the Brillouin zone that gives ferromagnetic ordering for the sublattices with antiferromagnetic $K<0$ interaction between them, and this is the focus of our consideration below. We should note that $\lambda_{1,2}(\mathbf{k})$ have maxima also at the border of the Brillouin zone $\mathbf{k}=\pi/a$ which presupposes antiferromagnetically ordered sublattices. There may be also some local maxima inside the Brillouin zone, which give some incommensurate ordering within the sublattices, but this case is beyond the scope of the present study.

After performing the inverse Fourier transform to real space, we obtain the dimensionless Landau free energy in the following form:

$$f \equiv \frac{F}{T} = \frac{t_1}{2}\vec{\psi}_1^2+\frac{t_2}{2}\vec{\psi}_2^2+\frac{g}{4}\left[(\vec{\psi}_1^2)^2+(\vec{\psi}_2^2)^2\right]+\frac{b}{2}\vec{\psi}_1^2\vec{\psi}_2^2+b(\vec{\psi}_1\cdot\vec{\psi}_2)^2+w(\vec{\psi}_2^2-\vec{\psi}_1^2)(\vec{\psi}_1\cdot\vec{\psi}_2), \quad (6)$$

where the coefficient

$$t_{1,2}=\frac{1}{\lambda_{1,2}}-\frac{1}{n},\quad g=\frac{u}{2}(S_0^4+S_1^4),\quad b=n^2(n+2)S_1^2S_0^2,\quad w=\frac{u}{2}S_0S_1(S_0^2-S_1^2). \quad (7)$$

are expressed in terms of the elements of the matrix (4) and its eigenvalues (5) with $\mathbf{k}=0$, *i.e.*

$$S_0=\frac{1}{D}\left(J_1-J_2+\sqrt{(J_1-J_2)^2+4K^2}\right),\quad S_1=\frac{2K}{D}, \quad (8)$$

where D is introduced to satisfy the condition $\|\hat{S}\|=1$, namely $S_0^2+S_1^2=1$:

$$D=\sqrt{2}\left[(J_1-J_2)^2+4K^2\right]^{1/4}\left[J_1-J_2+\sqrt{(J_1-J_2)^2+4K^2}\right]^{1/2}. \quad (9)$$

The real vector fields $\vec{\psi_1}$ and $\vec{\psi_2}$ in the Landau free energy (6) play the role of two coupled order parameters, and the averaged sublattice magnetizations are obtained trough the equations

$$\vec{m_1}=\frac{S_0}{\lambda_1}\vec{\psi_1}-\frac{S_1}{\lambda_2}\vec{\psi_2};\quad \vec{m_2}=\frac{S_1}{\lambda_1}\vec{\psi_1}+\frac{S_0}{\lambda_2}\vec{\psi_2}. \quad (10)$$

3. Solving the mean-field equations

The initial microscopic Hamiltonian is symmetric with respect to the rotation of all spins through the same angle. The application of Hubbard-Stratonovich transformation for the derivation of Landau free energy, presented in the previous section, preserves the symmetry of the initial Hamiltonian with respect to the auxiliary field variables $\vec{\psi}_{1,2}$, as well, which means that we can find the magnitude and the mutual orientation between the variables $\vec{\psi}_{1,2}$ but not their orientation with respect to crystallographic axes. This may be achieved for a particular magnetic substance by including in the initial microscopic Hamiltonian (1) terms accounting for the magnetic anisotropy. For pure exchange interactions we can introduce the following notations [10]: $\vec{\psi_{1}}_i = |\vec{\psi_1}|\beta_i$ and $\vec{\psi_{2}}_i = |\vec{\psi_2}|\delta_i$, where $|\vec{\psi_1}| = \psi_1$, $|\vec{\psi_2}| = \psi_2$ are the magnitudes of the vector fields, and β_i, δ_i are the respective direction cosines, which fulfil the condition:

$$\sum_{i=1}^{3} \beta_i^2 = 1 \quad \text{and} \quad \sum_{i=1}^{3} \delta_i^2 = 1. \tag{11}$$

Then, the equations of state will be

$$\frac{\partial f}{\partial X_i} = 0, \quad \text{where} \quad X_i = \{\psi_1, \psi_2, \beta_i, \delta_i\}. \tag{12}$$

Solving the above equations with respect to direction cosines β_i, δ_i renders two possible orientations between the vector fields $\vec{\psi_1}, \vec{\psi_2}$ for $K < 0$:

(i) The collinear phase with $\sum_i \beta_i \delta_i = -1$, that is, $\vec{\psi_1}$ and $\vec{\psi_2}$ are antiparallel, and
(ii) The non-collinear phase with $\sum_i \beta_i \delta_i = 0$, that is, $\vec{\psi_1}$ and $\vec{\psi_2}$ are perpendicular.

Hereafter, we will discuss in detail the non-collinear phase 2. The general form of angle obtained between $\vec{\psi_1}$ and $\vec{\psi_2}$ is:

$$\sum_{i}^{3} \beta_1 \delta_i = -\frac{w(\psi_2^2 - \psi_1^2)}{2b\psi_1\psi_2}$$

and is defined only when $\psi_1 \neq 0$ and $\psi_2 \neq 0$. For $K < 0$, the analysis shows that the non-collinear phase exists only when the field variables $\vec{\psi_1}$ and $\vec{\psi_2}$ are of equal magnitudes and mutually perpendicular. Substituting the above expression for the angle of non-collinear phase in (12) we find for the magnitude $\psi = \psi_1 = \psi_2$ of non-collinear phase:

$$\psi^2 = -\frac{t_1 + t_2}{u}. \tag{13}$$

Then, the sublattice magnetization magnitudes calculated using (10) for the non-collinear phase are

$$|\vec{m}_1| = \psi\sqrt{\frac{S_0^2}{\lambda_1^2} + \frac{S_1^2}{\lambda_2^2}}, \quad |\vec{m}_2| = \psi\sqrt{\frac{S_1^2}{\lambda_1^2} + \frac{S_0^2}{\lambda_2^2}}. \tag{14}$$

Note that the sublattice magnetizations are not perpendicular, but form an angle $\gamma = \angle(\vec{m}_1, \vec{m}_2)$ with each other, expressed by

$$\cos(\gamma) = \frac{S_0 S_1(\lambda_1^2 - \lambda_2^2)}{\sqrt{(S_0^2\lambda_2^2 + S_1^2\lambda_1^2)(S_1^2\lambda_2^2 + S_0^2\lambda_1^2)}}.$$

A thorough analysis of the Landau free energy shows that this non-collinear phase has no domain of stability, for $K < 0$, within the exchange approximation used here. We should mention that the free energy (6) is very sensitive to the sign of the inter-sublattices interaction K. When $K > 0$, *i.e.*,

IOP Conf. Series: Journal of Physics: Conf. Series **1186** (2019) 012002 doi:10.1088/1742-6596/1186/1/012002

the interaction between the sublattices is ferromagnetic then there is small domain in which the respective non-collinear phase is stable [12].

For antiparallel $\vec{\psi_1}$ and $\vec{\psi_2}$ it is obvious that the sublattice magnetizations (10) will be also antiparallel. In this case, we may write the resulting equations, that minimize the free energy, for the magnitudes of the field variables ψ_1 and ψ_2 of the collinear phase and $K < 0$ in the following form:

$$t_1\psi_1 + g\psi_1^3 + 3b\psi_1\psi_2^2 - w\psi_2(\psi_2^2 - 3\psi_1^2) = 0, \tag{15a}$$

$$t_2\psi_2 + g\psi_2^3 + 3b\psi_1^2\psi_2 - w\psi_1(3\psi_2^2 - \psi_1^2) = 0, \tag{15b}$$

with the stability conditions given by:

$$t_1 + 3g\psi_1^2 + 3b\psi_2^2 + 3w\psi_1\psi_2 > 0, \tag{16a}$$

$$(t_1 + 3g\psi_1^2 + 3b\psi_2^2 + 6w\psi_1\psi_2)(t_2 + 3g\psi_2^2 + 3b\psi_1^2 - 6w\psi_1\psi_2) - 9[w(\psi_1^2 - \psi_2^2) + 2b\psi_1\psi_2]^2 \geq 0. \tag{16b}$$

We will make some remarks on the dependence of solutions of (15) on the magnitude of the exchange parameters J_1, J_2 and K. When $J_1 < |K|$ and $J_2 < |K|$, the leading interaction is determined by the antiferromagnetic coupling between the two sublattices. This may be dubbed a strong coupling limit for which the eigenvalue

$$\lambda_2 = \frac{1}{2}\left[J_1 + J_2 - \sqrt{(J_1 - J_2)^2 + 4K^2}\right], \tag{17}$$

turns negative. This is equivalent to the inequality $K^2 - J_1J_2 > 0$. The coefficient t_2 in front of ψ_2^2 is positive; see (7), and the field $\vec{\psi}_2$ becomes redundant and the Landau free energy (6) reads

$$\left(\frac{F}{T}\right)_s = f_s = \frac{t_1}{2}\vec{\psi_1}^2 + \frac{g}{4}(\vec{\psi_1}^2)^2. \tag{18}$$

The minimization of the above expression with respect to $\vec{\psi_1}$ yields the solution

$$(\vec{\psi_1})^2 = -\frac{t_1}{g}, \tag{19}$$

which exists and is stable for $t_1 < 0$.

The sublattice magnetizations:

$$\vec{m}_1 = \frac{S_0}{\lambda_1}\vec{\psi_1} \quad \text{and} \quad \vec{m}_2 = \frac{S_1}{\lambda_1}\vec{\psi_1} \tag{20}$$

are antiparallel as $S_1 \sim K/D$ and $K < 0$. The phase described by the above equations corresponds to two antiparallel sublattices with different magnitudes of sublattice magnetizations.

In the weak coupling limit for antiparallel configuration, *i.e.*, when $J_1 > |K|$ and $J_2 > |K|$, or equivalently $J_1J_2 > K^2$, the system of equations (15), together with the stability conditions (16) is hard to solve analytically. To gain insights into the phase behaviour of the considered system we will perform a numerical analysis and the results are presented in the next Section.

4. Results and discussion

The analytical result for sublattice magnetizations (20) in the case of strong antiferromagnetic coupling gives for the magnitude of total magnetization $|\vec{M}| = |\vec{m}_1 + \vec{m}_2|$ the following expression:

$$|\vec{M}| = \frac{S_0}{\lambda_1}\left|1 + \frac{S_1}{S_0}\right| |\psi_1|$$

with $|\psi_1|^2$, from (19). It is found that regardless of the values of the model's parameter, the phase transition is of second order. The total magnetization behaviour with temperature is smooth resembling the one of Weiss ferromagnet with the exception that no saturation is reached at $T = 0$. According to the Néel's classification of ferrimagnets, see [2], the change of magnetization with temperature in the strong coupling limit falls within R-type curve. It is worth noticing that a qualitatively similar curve is obtained theoretically and compared with the respective experimental measurements for $Y_3Fe_5O_{12}$ [4] where two sublattice model with strong antiferromagnetic coupling is considered.

When $J_1 = J_2 = J$ the relation $S_0 = -S_1 = 1/\sqrt{2}$ holds and an antiferromagnetic ordering with $\vec{m}_1 = -\vec{m}_2$ sets in, only if $\vec{\psi}_2 \equiv 0$ and $|\vec{\psi}_1|^2 = -2t_1/u$. The transition temperature to antiferromagnetic ordering is given by $t_c^a = \frac{1}{n}(J + |K|)$.

Further we present the numerical results for the temperature dependence of sublattice magnetizations and the total magnetization of the system in the weak-coupling limit which we define here in the following way: $\mathcal{J}_1 > |\mathcal{K}|$ and $\mathcal{J}_2 > |\mathcal{K}|$. Such a situation is present, for example in some ferrimagnetic compounds like $GdCo_{12}B_6$ [9]. It is experimentally found that the exchange constants within sublattices are ferromagnetic and larger than the antiferromagnetic coupling between the sublattices; moreover there the contribution of the magnetic anisotropy may be disregarded.

Experiments for some R-T compounds where R is a rare earth element and T is a transition one, show that the exchange in the transition metal sublattice is leading in magnitude, while the exchange in the rare-earth ion sublattice can be safely ignored and considered as negligible. The intersublattice interaction is also small for the compounds $DyFe_5Al_7$ [13], $ErFe_{11}TiH$ [14], RCo_2 (R = Tb and Gd and R = Er, Ho, and Dy) [15]. In our notations the relation between the exchange integrals in this case will be $\mathcal{J}_1 > |\mathcal{K}| \gg \mathcal{J}_2$. This does not fall into our assumption of weak coupling and will not be considered here.

In order to solve numerically the equations of state (15) for weak coupling between sublattices we introduce the following dimensionless parameters.

$$t = \frac{T}{\mathcal{J}_1 + \mathcal{J}_2}, \quad \alpha = \frac{\mathcal{J}_1 - \mathcal{J}_2}{\mathcal{J}_1 + \mathcal{J}_2}, \quad \beta = \frac{\mathcal{K}}{\mathcal{J}_1 + \mathcal{J}_2}, \tag{21}$$

with t – the dimensionless temperature. In the above expression we have supposed that $\mathcal{J}_1 > \mathcal{J}_2$ which in view of symmetry in interchanging the sublattices does not limit the consideration; then $\alpha > 0$ and $\beta < 0$ as $\mathcal{K} < 0$.

The weak coupling between the sublattices, namely $\mathcal{J}_1 > |\mathcal{K}|$ and $\mathcal{J}_1 > |\mathcal{K}|$ may be expressed by the parameters from (21) with the relation:

$$\alpha^2 + \beta^2 < 1.$$

The parameter α is a measure for the difference in exchange parameters of the two sublattices and by its definition $0 < \alpha < 1$. The total magnetization of the system is the sum of the two sublattice magnetizations (10):

$$\vec{M} = \vec{m}_1 + \vec{m}_2 = \frac{S_0 + S_1}{\lambda_1}\vec{\psi}_1 + \frac{S_0 - S_1}{\lambda_2}\vec{\psi}_2. \tag{22}$$

Hereafter we will use the following notations for the magnitudes of the sublattice magnetizations and the total magnetization both in the text and in figures:

$$m_1 = |\overrightarrow{m}_1|, \quad m_2 = |\overrightarrow{m}_2| \text{ and } M = |\overrightarrow{M}|$$

The phase transition to the ordered ferrimagnetic state for weak antiferromagnetic coupling occurs at the temperature:

$$t_c = \frac{1}{6}\left(1 + \sqrt{\alpha^2 + \beta^2}\right) \tag{23}$$

which increases when either the difference between the exchange interactions in the sublattices grows, or when the antiferromagnetic coupling becomes larger, or both. The phase transition from the disordered to the ordered phase is of second order.

Notice that within the exchange approximation for the regime of weak coupling defined above with the decrease of temperature a compensation point sets in no matter how small is the difference of the exchange interaction between the sublattices. At the compensation temperature $t_{\text{comp.}}$, the sublattice magnetizations $\overrightarrow{m}_1$ and $\overrightarrow{m}_2$ are equal in magnitude and antiparallel, thus M vanishes. The relation between the field variables magnitudes at $t_{\text{comp.}}$ is given by:

$$\psi_1 = \frac{\lambda_1}{\lambda_2}\frac{(S_0 - S_1)}{(S_0 + S_1)}\psi_2. \tag{24}$$

As the calculations show $\psi_2 < \psi_1$ for all values of α and β, but ψ_1 grows with the decrease of temperature in a monotonic way, while ψ_2 grows more rapidly. The quantity

$$\frac{\lambda_1}{\lambda_2}\frac{(S_0 - S_1)}{(S_0 + S_1)} = \frac{(6t_c)^2}{1 - \alpha^2 - \beta^2}\left(\frac{\sqrt{\alpha^2 + \beta^2} - \beta}{\alpha}\right)$$

is always > 1 as $\beta < 0 \sim K$ and $\alpha > 0$ so at some temperature $t_{\text{comp.}} < t_c$, and respective values of ψ_1, ψ_2 the condition (24) is fulfilled. In figure 1, we depict the change of the net magnetization magnitude M as a function of the reduced temperature t for $\alpha = 0.1$, *i.e.*, $\mathcal{J}_2 = 0.83\mathcal{J}_1$ and different values of β. One can see that the increase of antiferromagnetic coupling between the sublattices

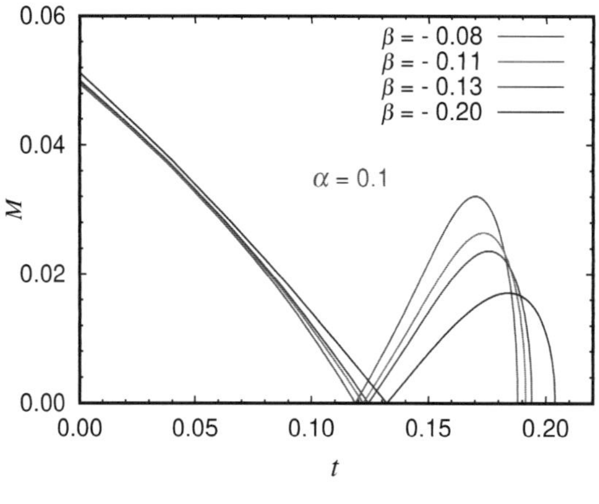

Figure 1. The dependence of net magnetization M on reduced temperature t for fixed α and different values of the antiferromagnetic coupling β.

slightly shifts the compensation temperature to higher values, and M grows more rapidly below the compensation temperature as the temperature decreases. We suppose that within the exchange approximation and in weak coupling limit the key factor for the compensation point to appear is the weakness of antiferromagnetic exchange between the sublattices compared to the ferromagnetic exchange of sublattices 1 and 2, respectively.

Hereafter we turn out attention to the influence of the difference between the magnitudes of exchange interactions within the sublattices, represented by the parameter α on the magnetization M as a function of t and the sublattice magnetizations m_1, and m_2. For $\alpha = 0.08$, *i.e.*, $\mathcal{J}_2 = 0.85\mathcal{J}_1$, the behaviours of M, m_1, and m_2 are shown in figure 2. At t_c the transition is of second order and

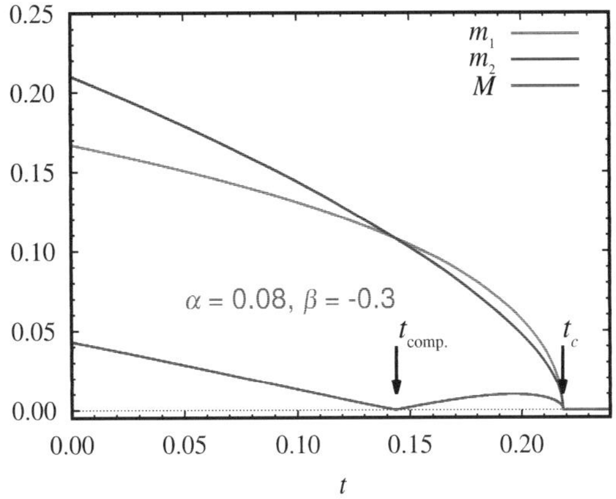

Figure 2. The dependence of the net magnetization M and the sublattice magnetizations m_1 and m_2 on the reduced temperature t for a small difference between sublattice exchange parameters.

when lowering the temperature a compensation point $t_{\text{comp.}}$ shows up closer to t_c than $t = 0$. The sublattice magnetizations vary with temperature in a monotonic way, and in the temperature interval $t_{\text{comp.}} < t < t_c$, the relation between the sublattice magnetizations is $m_1 > m_2$, as expected, since we have $\mathcal{J}_1 > \mathcal{J}_2$. Below $t_{\text{comp.}}$ the magnetization of the weaker sublattice m_2 becomes larger than m_1. For intermediate values of $\alpha = 0.4$ or $\mathcal{J}_2 = 0.43\mathcal{J}_1$, see figure 3, by lowering the temperature below the

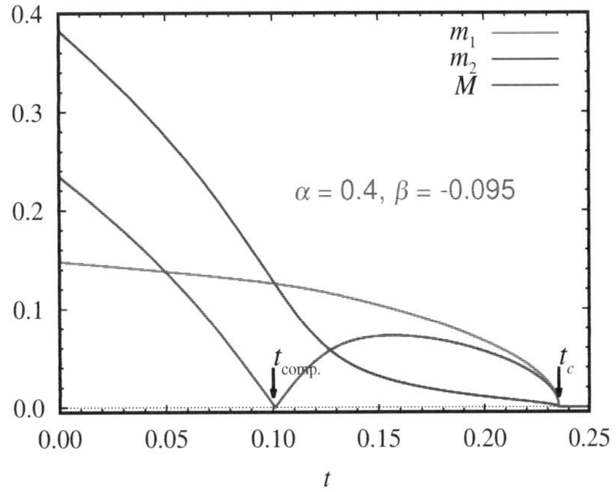

Figure 3. The dependence of the net magnetization M and the sublattice magnetizations m_1 and m_2 on the reduced temperature t for an intermediate difference between the sublattice exchange parameters.

compensation point the total magnetization rapidly grows in non-monotonic way as $t \to 0$, similar to the V-curve according to Néel's classification. The behaviour of the sublattice magnetizations with temperature is quite different; m_1, which is the sublattice magnetization with stronger exchange interaction $\mathcal{J}_1$ grows in smooth way with decrease of temperature, while m_2 for weaker sublattice interaction $\mathcal{J}_2$, below the compensation point grows drastically in a non-monotonic way and in the temperature interval $0 < t < t_{\text{comp.}}$, $m_2 \gg m_1$.

Such behaviour is described in detail in [16] for ferrimagnets with a compensation point for many experimentally investigated substances. The behaviour of M below the compensation point is interpreted by introducing the notion of weak sublattice where depending on the particular substance considered different mechanisms for the explanation of this effect are pointed out. Within our exchange model the effect of weak sublattice is readily seen and mainly depends on the difference $\mathcal{J}_1 - \mathcal{J}_2 \sim \alpha$. When α further grows the compensation point is shifted to lower temperatures and the magnetization m_1 of the stronger sublattice decreases for $t \to 0$. This is illustrated in figure 4 for $\mathcal{J}_2 = 0.19\mathcal{J}_1$.

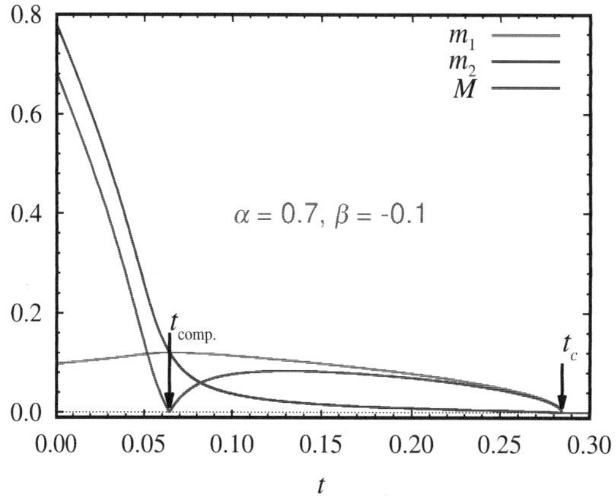

Figure 4. Illustration of the dependence of the net magnetization M and the sublattice magnetizations m_1 and m_2 on the reduced temperature t when the exchange in sublattice 2 is very small compared to that in sublattice 1.

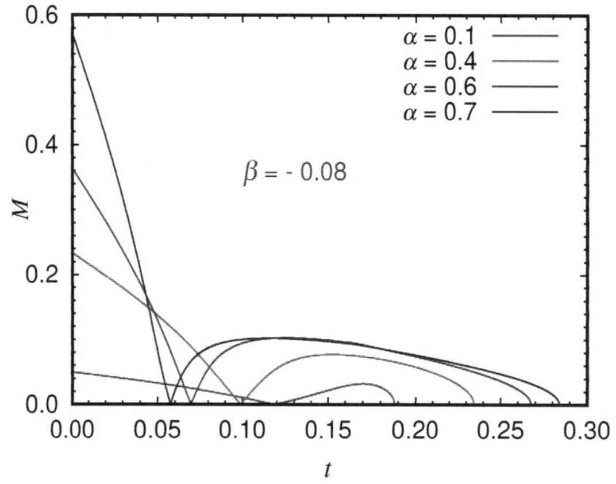

Figure 5. Dependence of the net magnetization M for fixed antiferromagnetic exchange β and distinct values of the difference α between the ferromagnetic exchanges of the two sublattices.

A qualitatively similar behaviour of the total magnetization is observed in experiments with $ErFe_2$ [8] and $GdCo_{12}B_6$ [9], although direct comparison with the experimental curves can hardly be made, as these substances have crystallographic and magnetic structure that differs from the one assumed within our model.

To shed light on the influence of the parameter α on the temperature behaviour of net magnetization we show it in figure 5 for small value of antiferromagnetic coupling $\beta = 0.08$ and different values of the parameter α. As the difference between the magnitudes of sublattice exchange interactions grows the transition temperature is shifted to higher values as expected, see (23) and the compensation temperature is lowered. Another effect is the change of net magnetization behaviour below $t_{\text{comp.}}$ from monotonic to nearly exponential when α increases.

5. Conclusions

We performed a mean-field analysis of two antiferromagnetically coupled Heisenberg models (1) with in-sublattice ferromagnetic interaction on a bcc lattice. The magnetic structure of the collinear ferrimagnet is described by two interpenetrating simple cubic magnetic sublattices shifted with respect to one another along the cube diagonal by $a(1/2,1/2,1/2)$ with a - the lattice constant which interact antiferromagnetically. Our study shows that the behaviour of the net magnetization of the two-sublattice ferrimagnet depends essentially on the difference in the magnetic interactions between the sublattices within the weak coupling limit. The weak antiferromagnetic interaction as defined above between the ferromagnetically ordered sublattices is responsible for the appearance of compensation point when lowering the temperature. The model may be generalized to include the expansion of the effective Hamiltonian up to sixth-order terms in the field vectors on the two sublattices in order to consider the possibility of a first order phase transition to ferrimagnetic state due to the influence of external parameters, such as pressure and change of concentration.

Within our approach the parameters of the Landau free energy can be directly related to the averaged microscopic exchange interactions as described in section 2, and this relation is quite simple in the case of high symmetry crystal structure as the one considered here. It will be of interest also to include the influence of an external magnetic field and to perform calculations for ferrimagnets, for which the relation between exchange interactions fulfills the condition $\mathcal{J}_1 > |\mathcal{K}| \gg \mathcal{J}_2$, that is, when one of the sublattices is very weak.

Acknowledgements

This work was supported by the Bulgarian National Science Fund under contract DN08/18.

References

[1] Skomski R 2008 *Simple models of magnetism* Oxford Graduate Texts (Oxford New York: Oxford University Press)
[2] Néel L 1971 *Science* **174** 985
[3] Andreenko A S, Belov K P, Nikitin S A and Tishin A M 1989 *Sov. Phys. Usp.* **32** 649
[4] von Ranke P J, Alho B P, Plaza E J R, Carvalho A M G, de Sousa V S R and de Oliveira N A 2009 *J. Appl. Phys.* **106** 053914
[5] Kaneyoshi T and Chen J 1991 *Journal of Magnetism and Magnetic Materials* **98** 201
[6] Tucker J W 1999 *J. Magn. Magn. Mater.* **195** 733
[7] Abubrig F 2013 *World J. Condens. Matter Phys.* **03** 111
[8] Chaaba I, Othmani S, Haj-Khlifa S, de Rango P, Fruchart D, Cheikhrouhou-Koubaa W and Cheikhrouhou A 2017 *J. Magn. Magn. Mater.* **439** 269
[9] Isnard O, Skourski Y, Diop L V B, Arnold Z, Andreev A V, Wosnitza J, Iwasa A, Kondo A, Matsuo A and Kindo K 2012 *J. Appl. Phys.* **111** 093916
[10] Shopova D V and Boyadjiev T L 2001 *J. Phys. Stud.* **5** 341
[11] Brout R 1974 *Phys. Rep.* **10** 1
[12] Shopova D V and Boyadjiev T L 2003 *Phys. Lett. A* **311** 438
[13] Gorbunov D, Andreev A and Mushnikov N 2012 *J. Alloys Comp.* **514** 120
[14] Kostyuchenko N V, Zvezdin A K, Tereshina E A, Skourski Y, Doerr M, Drulis H, Pelevin I A and Tereshina I S 2015 *Phys. Rev. B* **92**
[15] Valiev E Z and Teplykh A E 2017 *Phys. Metals Metallogr.* **118** 21
[16] Belov K P 1996 *Phys. Uspekhi* **39** 623

IOP Conf. Series: Journal of Physics: Conf. Series **1186** (2019) 012003 doi:10.1088/1742-6596/1186/1/012003

Surface photovoltage study of GaInNAs layers for photovoltaic applications

V Donchev[1,3], S Georgiev[1], A Aho[2] and M Guina[2]

[1] Faculty of Phsics, Sofia University, 5, blvd. J.Bourchier, BG-1164, Sofia, Bulgaria

[2] Tampere University of Technology, Optoelectronics Research Centre, PO Box 692, FI-33101, Tampere, Finland

E-mail: vtd@phys.uni-sofia.bg

Abstract. The optical and carrier transport properties of GaInNAs layers grown by molecular beam epitaxy on n-GaAs substrates are studied by surface photovoltage (SPV) spectroscopy combined with photoluminescence (PL) measurements data. The SPV spectra clearly show a red shift of the band gap energy with respect to the band gap of GaAs, which is in line with the PL results. The combined analysis of the SPV amplitude and phase spectra has allowed determining the residual doping type in the layer. The minority carrier diffusion length was assessed by means of the method called "constant SPV".

1. Introduction

Dilute nitrides form a novel class of semiconductor alloys, which has emerged from the conventional III-V compounds by incorporating a small amount of nitrogen. The interest in these compounds is due to their potential to alter the energy band gap and the lattice constant, which creates additional flexibility desired in many applications such as multijunction solar cells [1], heterojunction bipolar transistors [2] and high-speed photodetectors [3]. Different experimental techniques have been used to investigate such materials. However surface photovoltage (SPV) spectroscopy has been relatively rarely applied [4–7] although it provides information on the optical absorption spectrum and carrier transport in the sample [8]. To partially fill this gap we present an original study of the optical and electronic properties of GaInNAs layers based on SPV spectroscopy, combined with photoluminescence (PL) results.

2. Experimental

The samples under study were grown by molecular beam epitaxy (MBE). They consisted in an n-GaAs substrate (350 μm), an n-GaAs buffer (200 nm) and a nominally undoped $Ga_{1-y}In_yN_xAs_{1-x}$ dilute nitride layer of 1000 nm. The doping for n-GaAs was 5×10^{17} cm^{-3}. The nominal content of In and N was y ~ 0.06 and x ~ 0.02, respectively.

The SPV measurements were performed at room temperature using the metal-insulator-semiconductor operation mode of the SPV technique [8]. The probe electrode was a semi-transparent SnO_2 film evaporated on the bottom surface of a quartz glass. The excitation was performed using a 250 W halogen lamp along with a SPEX grating monochromator (f=0.25 m, 600 gr/mm) and an optical

[3] To whom any correspondence should be addressed

chopper (94 Hz). The light wavelength was scanned from high to low values. The photon flux was kept constant ($\approx 10^{14}$ $cm^{-2}s^{-1}$) by adjusting the position of a neutral density filter with variable optical density according to the signal of a spectrally flat Golay detector (Oriel IR50).The probe signal with respect to the ground was fed to a high-impedance unity gain buffer and then measured by a Lock-in amplifier (SR830). The reference signal from the optical chopper defined the zero value of the SPV phase. For that purpose we took all precautions to eliminate undesired phase shifts, following the procedures described in [9]. More details about the SPV set-up and measurement procedure can be found in [9,10].

PL spectra were measured at room temperature with Accent RPM 2000 PL mapper by using 532 nm laser excitation.

3. Results and discussions

Figure 1 represents typical SPV amplitude and phase spectra measured with light excitation from the layer side. The sharp increase of the amplitude around 1150 nm corresponds to the onset of the band-to-band transitions in GaInNAs. The band gap assessed from Tauc plot is $E_g = 1.08$ eV. Therefore, the band gap shift ΔE_g with respect to the band gap of GaAs (1.42 eV) is 0.34 eV. Figure 1 represents also the PL spectrum of the same sample. The band gap value estimated from the SPV data is close to the PL peak position (1.117 eV) indicating a good agreement between the two techniques.

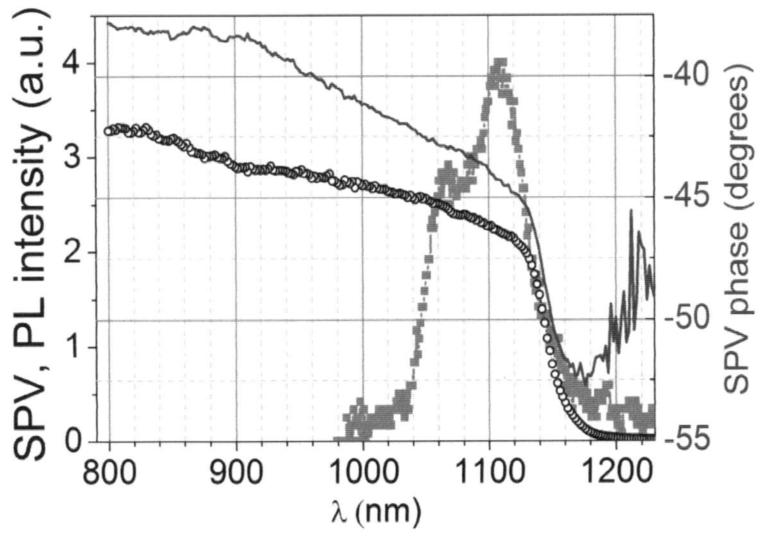

Figure 1. Typical SPV amplitude (circles, left axis) and phase (line, right axis) spectra measured with excitation from the layer side. The PL spectrum of the same sample is presented by squares.

To explain the spectral behaviour of the SPV amplitude we take into account that the SPV phase is in the 4th (2nd) quadrant when electrons move toward the bulk (surface) and/or holes move toward the surface (bulk) [9,10]. This is usually the case where the energy band bending at the surface/interface is upward (downward) in the direction from bulk to surface. In the present case the energy band bending at the interface GaInNAs/n-GaAs is upwards due to the different doping (i-n junction), the space charge region (SCR) being mostly in the lower-doped GaInNAs layer. The SPV generated there has phase in the 4th quadrant in accordance with the expectations. The phase and amplitude in figure 1 change monotonically with decreasing the wavelength, the phase values being in the 4th quadrant. The phase values remain in the 4th quadrant for wavelengths down to 500 nm (not shown here), where the absorption is much closer to the surface. All this leads to the conclusion that the surface SCR gives SPV signal with phase in the 4th quadrant, too and the two signals (from surface and interface) add constructively. Schematically this is shown in figure 2 using the vector model for SPV [10]. According to this model, the SPV vector has a magnitude equal to the amplitude of the SPV signal and an angle with respect to the X-axis equal to the SPV phase. In figure 2 are represented the SPV vectors corresponding to the SPV signals from the GaInNAs/n-GaAs interface (vector A1) and from the GaInNAs surface (vector A2), as well as the overall measured signal (vector A = A1 + A2). The two SPV vectors form an acute angle, therefore their sum is larger than each of them. One observes a continuous increase of the SPV amplitude with decreasing the wavelength, due to the increase of the GaInNAs optical absorption coefficient. This is accompanied with a continuous anti-clockwise change of the SPV phase towards -0 degrees. The fact that the SPV signal from the GaInNAs surface has phase in the 4th quadrant means that the energy band bending at the surface is upwards. This is typical for n-type semiconductors. Thus, we conclude that the nominally undoped GaInNAs layer has n-type residual conductivity.

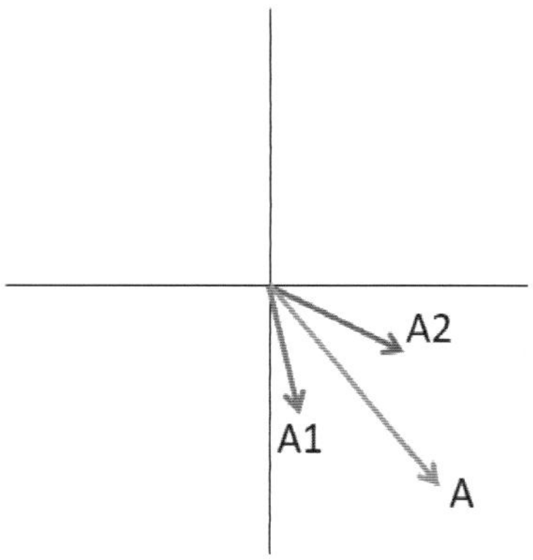

Figure 2. Schematic representation of the SPV vectors corresponding to the SPV signals from the interface (A1), surface (A2) and the overall SPV vector (A=A1+A2) for excitation with wavelength around 1050 nm.

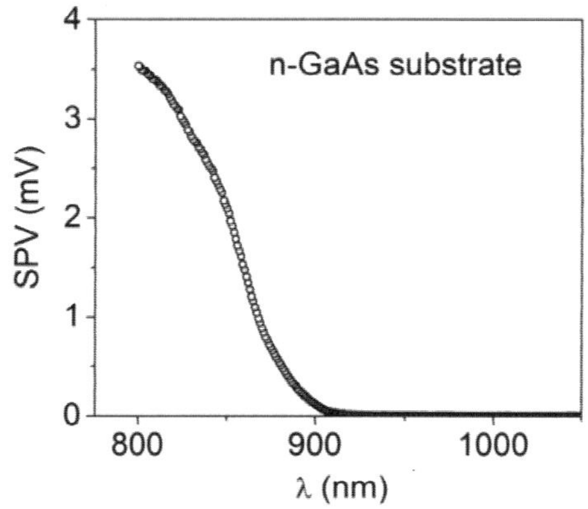

Figure 3. SPV amplitude spectrum of n-GaAs substrate measured at the same experimental conditions as the spectrum of figure 1.

At around 900 nm the SPV amplitude starts increasing faster, while the phase starts increasing slower. This feature is observed in all the spectra measured with different photon fluxes. To explain it the following discussion is proposed. For wavelengths below 900 nm (1.377eV) n-GaAs can already absorb photons, as can be seen from the SPV spectrum of n-GaAs presented in figure 3. Assuming a value around $10^4 cm^{-1}$ for the absorption coefficient of GaInNAs in this energy range, means that the 1000 nm layer still transmits around 30% of the photons to reach the interface, where the band bending in n-GaAs is also upward. So, n-GaAs starts absorbing light and contributing to the SPV signal, too. This generates another SPV signal with vector in the 4th quadrant, thus leading to a stronger overall SPV signal, which is seen as a faster increase of the amplitude in figure 1. To account for the slower anti-clockwise change of the phase in this range, we can use again figure 2, but in this case, vector A2 will corresponds to the signal from GaInNAs and vector A1 – to the signal from n-GaAs. One can assume that the SPV vector related to n-GaAs (A1) is closer to -90 degree than the SPV vector related to GaInNAs (A2), as shown in figure 2. It is easy to understand that increasing A1 will effectively slow down the anti-clockwise rotation of the overall SPV vector (A) and in the same time will increase its magnitude. If this mutual position of the SPV vectors is correct, it means that the SPV process related to GaInNAs is faster than the one related to n-GaAs (vector closer to -0 degrees means faster process

[9,10]). Such statement is in agreement with the shorter recombination time in dilute nitrides as compared to GaAs, reported in the literature [11].

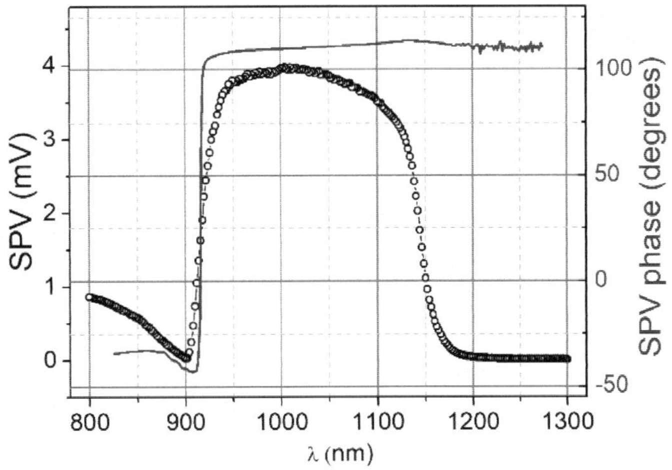

Figure 4. SPV amplitude (symbols, left axis) and phase (line, right axis) spectra of the sample from figure 1 excited from the substrate side.

To check the above conclusions we turned the sample so that the excitation is from the substrate side. Figure 4 represents the SPV spectra measured in this case at the same experimental conditions. The slope in the amplitude around 1150 nm gives the same value of the GaInNAs band gap. The SPV signal in this range mainly comes from the GaInNAs interface SCR. The phase is in the 2nd quadrant because the band bending at the interface is now downward, that is opposite to the previous case (same sample looked from the opposite side). On the other hand the surface of the GaAs substrate has an upward band bending because of the n-type doping. After 950 nm the signal decreases quickly, reaches a minimum at 900 nm and increases again. The phase quickly jumps clockwise toward the 4th quadrant and reaches values around -40 degrees. This behavior is explained by the inclusion of the SPV signal from the n-GaAs substrate with phase in the 4th quadrant on top of the signal from the interface. The interplay of these two SPV signals with nearly opposite phases leads to the minimum in the amplitude spectrum and the phase jump observed around 900 – 920 nm. For wavelengths below 900 nm the SPV process from the n-GaAs surface dominates and the SPV amplitude spectral shape is similar to that of n-GaAs shown in figure 3.

To assess the minority carrier diffusion length in the GaInNAs layer SPV measurements with low photon flux were carried out. For that purpose the method called "constant SPV" or "method A" [8,12] was applied. Although there have been numerous SPV investigations of the minority carrier diffusion length in different materials (see [8] and references therein), this method has been rarely applied to GaInNAs [13]. The method is based on a solution of the continuity equation and two highly reasonable assumptions as follows: i) the SPV is a monotonous function of the excess carrier concentration and ii) the surface recombination velocity is constant if SPV is constant. It consist of keeping the SPV constant for several wavelengths by changing the photon flux Φ. Under such conditions one can show that [8,12]

$$\Phi(\lambda) = cons(\frac{1}{\alpha(\lambda)} + L), \quad (1)$$

where α is the absorption coefficient, L is the diffusion length and λ is the wavelength. By plotting $\Phi(\lambda)$ as a function of $1/\alpha(\lambda)$ a straight line is obtained, which intercepts the abscissa at $-L$. Eq.(1) holds under the following constraints concerning L, α^{-1}, the majority carrier density n_0, the sample thickness d and the SCR width, w [8,12]: i) $\Delta p << n_0$ (low injection level); ii) $\alpha^{-1} << d$ (the illumination of the back SCR can be neglected); iii) $w << L$; $w << \alpha^{-1}$ (the diffusion carrier transport is the major contribution to the SPV signal, and iv) $L/d < 1/4$ (the influence of the back SCR can be neglected).

As a first approximation for the absorption coefficient spectrum of GaInNAs we took that of GaAs at 300 K and translated it in energy by 0.32eV so that the band gap becomes 1.10 eV. To fulfil condition i) we used a filter that decreased the excitation by an order of magnitude and kept SPV=0.51 mV during the measurements. Concerning condition ii) we used wavelengths in the range 900 – 1030 nm for which $\alpha^{-1} < d = 1\mu m$ but stopped at 900 nm to avoid absorption in GaAs. For those values of α^{-1} (0.5 – 0.8 μm) the illumination of the back SCR cannot be completely neglected. However, as explained above, the SPV signals from the surface and from the interface add constructively and therefore the influence of the back SCR should not be crucial for the final result. Concerning condition iii) w of the back SCR was assessed to be < 0.24 μm from the difference of the Fermi levels in GaAs and GaInNAs, and assuming 10^{16} cm^{-3} residual n-type doping of GaInNAs. The width of the surface SCR should be of the same order of magnitude or less. Thus, condition iii) is not fully satisfied, but still w is less than α^{-1} and L. As to condition iv), the ratio L/d was around 0.35. Taking into account the discussion in [8] (p.125) we can conclude that this value is satisfactory for the assessment of the diffusion length.

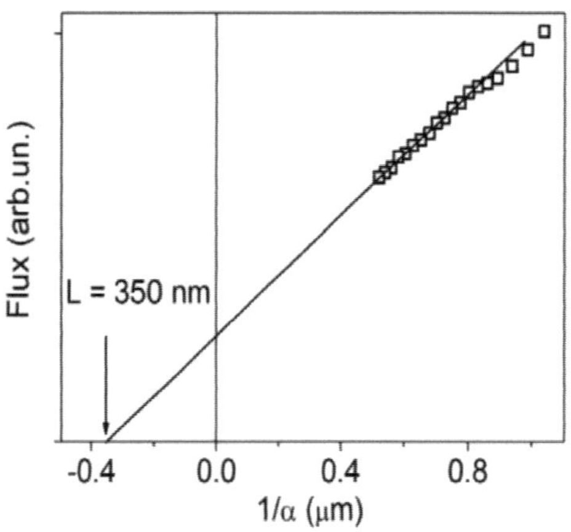

Figure 5. Photon flux versus 1/α curve from which the hole diffusion length is determined.

Figure 5 shows the result ($L = 350$ nm) obtained with the above procedure. The error in the L value calculated using the standard errors of the least square fit coefficients and considering the two ultimate cases for the slope of the straight line is quite small ± 5 nm. However, other uncertainties come from the unknown GaInNAs absorption coefficient, the influence of the back SCR, the close values of w, α^{-1}, and L, and the neglect of the reflectivity spectrum. Therefore, the obtained L value is regarded only as estimate of the hole diffusion length in the GaInNAs layer.

Conclusion

An original SPV spectroscopy study of MBE grown undoped GaInNAs dilute nitride layers is carried out combined with PL results. A reduction of the band gap energy with respect to the band gap of GaAs is evidenced in accordance with the PL results. N-type residual doping in the layers is found from the combined analysis of the SPV amplitude and phase spectra. This way a simple and contactless way is shown for determining the residual doping type of nominally undoped layers grown on substrates with known doping type. The hole diffusion length in GaInNAs is estimated by means of the "constant SPV" method. The study highlights the potential of the SPV spectroscopy for studying dilute nitride materials.

Acknowledgements
This work was supported by the Cost Action MP1406 "Multiscale in modelling and validation for solar photovoltaics (MultiscaleSolar)", the Bulgarian National Science Fund (contract ДКОСТ 01/16) and the European Regional Development Fund within the Operational Programme "Science and Education for Smart Growth 2014 - 2020" under the Project CoE "National center of mechatronics and clean technologies" BG05M2OP001-1.001-0008-C01.

References
[1] Aho A, Polojärvi V, Korpijärvi V-M, Salmi J, Tukiainen A, Laukkanen P and Guina M 2014 Composition dependent growth dynamics in molecular beam epitaxy of GaInNAs solar cells *Sol. Energy Mater. Sol. Cells* **124** 150–8
[2] Asbeck P M, Welty R J, Tu C W, Xin H P and Welser R E 2002 Heterojunction bipolar transistors implemented with GaInNAs materials *Semicond. Sci. Technol.* **17** 898–906
[3] Chen W C, Su Y K, Chuang R W, Yu H C, Chen B Y and Hsu S H 2008 Investigation of InGaAsN MSM photodetectors with transparent ITO Schottky contacts *Semicond. Sci. Technol.* **23** 035027
[4] Bansal B, Kadir A, Bhattacharya A, Arora B M and Bhat R 2006 Alloy disorder effects on the room temperature optical properties of GaInNAs quantum wells *Appl. Phys. Lett.* **89** 032110
[5] Kudrawiec R, Sitarek P, Gladysiewicz M, Misiewicz J, He Y, Jin Y, Vardar G, Mintarov A M, Merz J L, Goldman R S, Yu K-M and Walukiewicz W 2014 Surface photovoltage and modulation spectroscopy of E_- and E_+ transitions in GaNAs layers *Thin Solid Films* **567** 101–4
[6] Donchev V, Asenova I, Milanova M, Alonso-Álvarez D, Kirilov K, Shtinkov N, Ivanov I G, Georgiev S, Valcheva E and Ekins-Daukes N 2017 Optical properties of thick GaInAs(Sb)N layers grown by liquid-phase epitaxy *J. Phys. Conf. Ser.* 794 12013
[7] Donchev V, Milanova M, Asenova I, Shtinkov N, Alonso-Álvarez D, Mellor A, Karmakov Y, Georgiev S and Ekins-Daukes N 2018 Effect of Sb in thick InGaAsSbN layers grown by liquid phase epitaxy *J. Cryst. Growth* **483** 140–146
[8] Kronik L and Shapira Y 1999 Surface photovoltage phenomena: Theory, experiment, and applications *Surf. Sci. Rep.* **37** 1–206
[9] Donchev V, Ivanov T, Germanova K and Kirilov K 2010 Surface photovoltage spectroscopy – an advanced method for characterization of semiconductor nanostructures *Trends Appl. Spectrosc.* **8** 27–66
[10] Ivanov T, Donchev V, Germanova K and Kirilov K 2009 A vector model for analysing the surface photovoltage amplitude and phase spectra applied to complicated nanostructures *J. Phys. D. Appl. Phys.* **42** 135302
[11] Friedman D J, Geisz J F and Ptak A J 2005 Photovoltaic Applications of Dilute Nitrides *Physics and Application of Dilute Nitrides* ed I Buyanova and W M Chen (New York: Taylor & Francis) pp 396–420
[12] Goodman A M 1961 A Method for the Measurement of Short Minority Carrier Diffusion Lengths in Semiconductors *J. Appl. Phys.* **32** 2550–2
[13] Lebedeva N, Repo P, Yli-Koski M, Savin H, Aho A, Tukiainen A, Guina M, Tallian M and Savtchouk S 2011 Experimentally defined electronic properties of InGaAsN: input for solar cell simulations *Proceedings of the 26th European Photovoltaic Solar Energy Conference and Exhibition* pp 621–4

IOP Conf. Series: Journal of Physics: Conf. Series **1186** (2019) 012004 doi:10.1088/1742-6596/1186/1/012004

Inter-granular effects at high magnetic fields of cuprate and iron chalcogenide superconducting materials

K Buchkov [1], M Valkovski [1], D Gajda [2], K Nenkov [3], E Nazarova [1]

[1] Institute of Solid State Physics, Bulgarian Academy of Sciences,
72 Tzarigradsko Chaussee, 1784, Sofia, Bulgaria
[2] Institute of Low Temperatures and Structure Research, Polish Academy of Sciences,
Okólna 2, 50-422 Wroćlaw, Poland
[3] Leibniz Institute for Solid State and Materials Research, (IFW Dresden),
P.O. Box 2700116, D01171 Dresden, Germany
e-mail: buchkov@issp.bas.bg

Abstract. The weak links effects are one of the main challenges for effective power applications of high temperature superconducting materials. Studies of these effects help for their better understanding and subsequent improvement. An overview analysis of the intergranular properties of cuprate ($Y_{0.8}Ca_{0.2}Ba_2Cu_3O_{7-\delta}$) and iron-based chalcogenide ($FeSe_{0.5}Te_{0.5}$) polycrystalline samples was carried out, by means of series of electro-transport experiments at different magnetic fields. The temperature evolution of the Josephson coupling and intrinsic superconductivity effects for the both systems was constructed. The $FeSe_{0.5}Te_{0.5}$ compound shows very stable and superior behavior compared to $Y_{0.8}Ca_{0.2}BCO$ up to the highest magnetic fields (14T) used. We have explored $FeSe_{0.5}Te_{0.5}$ Josephson weak links influence (as a non-linear process) over the resistive transition using different AC current amplitudes and applying the sensitive AC transport third harmonics technique.

1. Introduction

One of the major obstacles before the high power applications of HTSCs (for both cuprates and iron based superconductors (IBS)) is the prominent weak link effects of the grain boundaries which limits the effective critical current density [1]. The layered crystal structure (especially the cuprates) leads to a highly asymmetrical order parameter (*p-d* wave) and strong anisotropy of the superconducting properties. Consequently, the electro-transport capabilities are highly dependent on the grain boundaries misorientation angle [2] and the specific materials morphology (size, shape and packing of the grains). In addition, the grain boundaries regions have a specific defect structure and altered local chemical composition and electronic properties [3]. The grain boundaries as natural defects and their variable composition as well changes the electronic mean free path and affects the Ginzburg-Landau parameter and subsequently the pinning strength at boundaries region [4]. Usually, the grain periphery is also oxygen depleted (for the cuprates) and thus may form a SIS type inter-grain junctions (superconductor-insulator-superconductor) [5].

For the IBS there are more advantageous options for the weak link problem, due to reasons such as the semi-metallic nature of the normal state (compared to the antiferromagnetic Mott insulator state for the cuprates) and the *s*-wave highly symmetric order parameter resulting in a lower anisotropy of the properties [6]. Expectedly, the values of the critical grain misorientation angle and its influence over the current density are effectively superior [7]. In addition the grain boundary junctions for IBS show a metallic behavior resulting in the preferable SNS type connections (superconductor–normal metal–

superconductor) [8]. The overcoming of the high critical current limitation of the weak link in granular type superconductors is one of the most investigated topics. There is a number of developing processing methods: on a first place approaches for morphology texturing control, annealing under hot isostatic pressure (proven effective for IBS and MgB_2) [9,10], chemical modification (substitutions and/or additions) of the grain boundary regions [11,12].

In this study we present a qualitative overview of the weak link behaviour in two HTSC systems in form of polycrystalline samples: the cuprate system – $Y_{0.8}Ca_{0.2}Ba_2Cu_3O_{7-\delta}$ and the IBS – $FeSe_{0.5}Te_{0.5}$. By means of electro-transport experiments we have analysed the resistive transitions to superconducting state in various set of applied high magnetic fields (H_{DC}) and also amplitudes (I_{AC}) of applied AC transport current particularly for $FeSe_{0.5}Te_{0.5}$.

2. Experimental details

The main objects of the present study are two polycrystalline samples: cuprate Y(Ca)BCO and IBS $FeSe_{0.5}Te_{0.5}$. The preparation details, structural and morphological characterization are presented elsewhere: Y(Ca)BCO [13,14] and IBS [15]. The resistive experiments are performed using the AC transport modules of two (9 and 14T) PPMS systems. The samples dimensions: for $Y_{0.8}Ca_{0.2}Ba_2Cu_3O_{7-\delta}$ cross section $A = 0.18$ cm^2 and voltage leads distance $l = 0.34$ cm, for $Y_{0.975}Ca_{0.025}Ba_2Cu_3O_{7-\delta}$ $A = 0.24$ cm^2, $l = 0.44$ cm and for $FeSe_{0.5}Te_{0.5}$, $A = 0.24$ cm^2, $l = 0.3$ cm

One of the samples - $FeSe_{0.5}Te_{0.5}$ was additionally characterized by third harmonics resistive measurements which were challenging from experimental point of view. In order to achieve optimal balance of signal sensitivity and accuracy for the effective third harmonic detection, specific amplitudes and frequencies were selected as well with certain adjustments of the system signal gain settings. For the PPMS the higher harmonics data are acquired in dB referenced to the fundamental component voltage signal. The harmonic contribution is usually less than -50 dB for a clean measurement. For the studied superconducting transition the harmonic signal intensity is within range. While for larger dB range when analysing very small signals, the experimental data could be deformed by false experimental artefacts.

3. Results and discussion

The choice of $Y_{0.8}Ca_{0.2}Ba_2Cu_3O_{7-\delta}$ system for the present study is based on a number of effects resulting from the substitution, with specifically strong influence over the inter-granular properties (depending on the percentage of substitution) [16–18]. In particular, the Ca segregates mainly at the grains periphery which changes locally the electronic structure, charge balance, carrier mean free path, pinning and reduces the anisotropy effects [18]. In addition the Ca substitution alters the morphology (resulting in smaller and spherical shaped grains [14]) and leads to improved eutectics by lowering the melting temperature of the compound and more easy possibility for texturing [19]. The weak link type is transformed from SIS (typical for YBCO) to SNS type, the negative effects of the critical mis-orientation angle are reduced and the overall inter-granular transport capabilities are significantly improved [14,20].

Due to the overdoping the normal state resistance is also reduced so the stability of the superconductor in case of quench is also improved. Unfortunately, there are also several important adverse effects of the substitution (which tend to increase with the level of overdoping/substitution), specifically the critical temperature of $Y_{0.8}Ca_{0.2}Ba_2Cu_3O_{7-\delta}$ is lower compared to YBCO and the inter-grain transition is shifted to much lower temperatures [14].

From practical view the $FeSe_{0.50}Te_{0.50}$ is the most perspective compound of chalcogenide IBS. It is structurally simple (low anisotropy) and chemically stable (in comparison to FeSe and particularly $Re(A)Fe_2Se_2$ systems: Re – rare earth element or A organic radical [21]) and possesses appropriate morphology consisting of compact and partially melted grains (prerequisite for large area of the Josephson contacts). In addition, the structural disorder is in favorable margins as a result of nano-scale phase separation, thus the pinning landscape is very effective due to different types of pinning centers [22]. The critical parameters and particularly the critical current persists up to very high magnetic fields

[23]. This is also as a result from a superior inter-grain current capabilities especially due to the metallic type SNS junctions [8].

The inter-granular properties of polycrystalline $Y_{0.8}Ca_{0.2}Ba_2Cu_3O_{7-\delta}$ sample have been analyzed by series of resistive R vs T experiments at various fixed DC magnetic fields. The results are presented on figure 1 (A) and normally a semi logarithmic scale is used in order to individuate the resistive responses of the intra-grain and weak link contributions (in the margins of experiment noise level ~ 10^{-6} Ω).

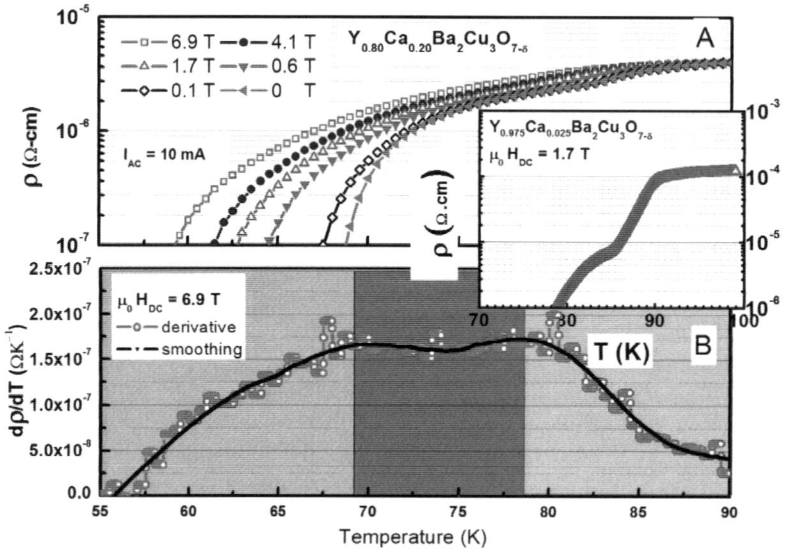

Figure 1. A. Semi logarithmic plot of resistivity vs. temperature for $Y_{0.80}Ca_{0.20}BaCuO$ sample at different magnetic fields. B. Derivative of resistivity vs. temperature for 6.9 T. Inset. Semi logarithmic plot of resistivity vs. temperature for $Y_{0.975}Ca_{0.025}Ba_2Cu_3O_{7-\delta}$ sample at 1.7 T.

For all applied fields the resistive curves have a typical for the intra/inter-grain transitions double step shape. The overall resistive response is dominated by the weak link influence resulting in broadening and shifting of the transitions to lower temperatures with increasing of the applied DC field. The physical phenomenology behind the observed behaviour is well investigated in the literature. It is related with the specific temperature and field evolution of processes in the intrinsic superconductor volume (intra-grain) and the Josephson inter-grain medium [20,24,25]. In order to present the domains of activity of the different transitions, the data are also represented using the R(T) first derivative and marked by the corresponding inflection points (figure 1 B).

At the lowest temperatures (light grey) a stable current percolation path is established through the effectively coupled Josephson links. The rapid increase of the resistance with the temperature is related to penetration of the magnetic flux inside the inter-grain regions (which is controlled by Josephson type vortices). As a consequence the number of non-dissipating (coupled) Josephson contacts (dark grey) decreases till reaching certain limit because the normal state junctions (decoupled) are expected to be independent from the DC field. This is manifested as saturation in the magneto-resistance behaviour. The DC field penetrates also inside the superconducting grains and at the higher temperatures close to T_c, the resistive response is controlled only by the intrinsic vortex dynamics processes (blue area).

For comparison and to emphasize the effect of Ca-substitution on the weak-link behaviour, the R(T) data for sample with a small percentage of substitution (2.5%) are also presented in the inset section of figure 1 for a fixed DC field equal to 1.7T. In this case the transition width (ΔT) is of order of 10K, which is smaller than ΔT for 20% Ca substituted sample, where ΔT~30K. The higher Ca concentration increases the number of grain connections (ρ is reduced more than one order of magnitude), but the length scale of boundary imperfection probably exceeds the coherence length and electron mean free path, which is detrimental for some of the pinning centers and broadens the transition.

On figure 2 A the temperature dependence of the resistivity R(T) for polycrystalline $FeSe_{0.5}Te_{0.5}$ sample (up to $\mu_0 H_{DC}$ = 14T and few are shown for clarity) is presented (semi-logarithmic scale). The resistive curves display slight (about 4K) transition broadening of the (for 14T). The double step transition is observed for all curves but in contrast to the behaviour of $Y_{0.8}Ca_{0.2}Ba_2Cu_3O_{7-\delta}$ in moderate DC fields, the inter- and intra-grain responses are far less distinguishable.

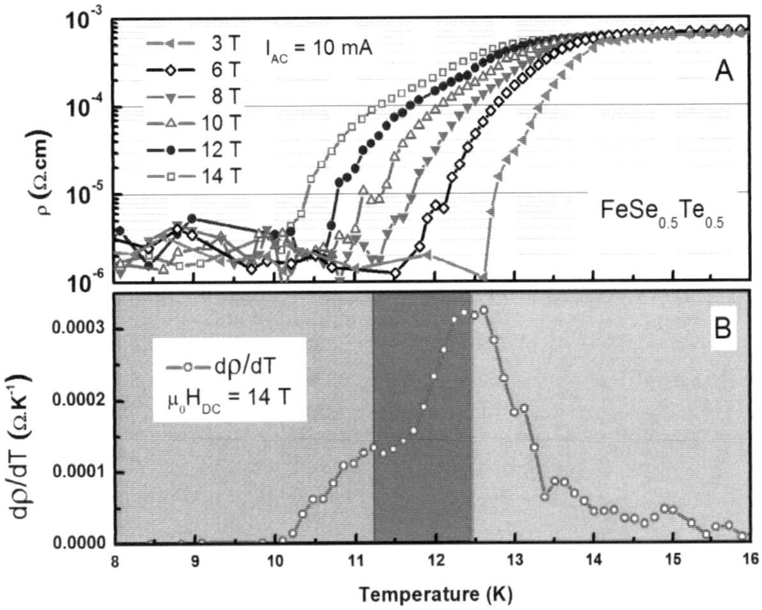

Figure 2. Semi logarithmic plot of resistivity vs. temperature for $FeSe_{0.5}$ $Te_{0.5}$ sample at different magnetic fields. B. Derivative of resistivity vs. temperature for 14 T.

The R(T) derivative main peak splitting is detectable only above 10T and the inter- and intra-grain components separation is most prominent for 14T (presented on figure 2(B). Obviously, for $FeSe_{0.5}Te_{0.5}$ the temperature evolution of the current transport through dominating regimes of coupled, decoupled Josephson links and bulk (grain) volume is in very narrow interval. The lower T_c and smaller thermal fluctuations in comparison with Y(Ca)BCO contribute for this result.

We have investigated the electro-transport behaviour of $FeSe_{0.5}$ $Te_{0.5}$ also from the perspective of AC current amplitude variation (no DC field applied). The results for tree different amplitudes (10, 50 and 100mA) are presented on figure 3A using a semi-logarithmic scale. In this case the double step transition is indistinct and hardly visible. It was impossible to directly individuate the inter-grain domain signal via the differentiation routine. The AC transport response of superconductor is very complex due to the combined effects of various linear and non-linear processes [26,27]. The Josephson Effect is a typical example of a non-linear phenomenon and therefore is expected to cause a higher harmonics generation in the AC signal. Usually, the third harmonics modulus $U_{3\omega}$ is studied because of the highest signal allowing sensitive detection of different non-linear processes. The results are presented on figure 3 B. A stable harmonic signal was achieved only for the highest AC current amplitudes (50 and 100mA).

Particularly a double peak of $U_{3\omega}$ (I_{AC} = 100mA) was detected as a possible fingerprint of the Josephson links activity as the dominating process in the inter-grain domain. The other non-linear processes with influence are the flux pinning and creep relaxation but naturally they control the response in the intrinsic (intra-grain) superconductor volume.

In conclusion we have presented a general overview of the electro-transport properties in the context of weak link influence for two compounds in polycrystalline form: $Y_{0.8}Ca_{0.2}Ba_2Cu_3O_{7-\delta}$ (cuprate) and $FeSe_{0.5}Te_{0.5}$ (IBS). The $Y_{0.8}Ca_{0.2}Ba_2Cu_3O_{7-\delta}$ inter-grain response is strongly affected by the magnetic field with even moderate intensity. The resistive behaviour evolves through the different temperature intervals of the domination of coupled, decoupled Josephson links and bulk (grain) volume. However this system operates in much higher temperatures compared to IBS.

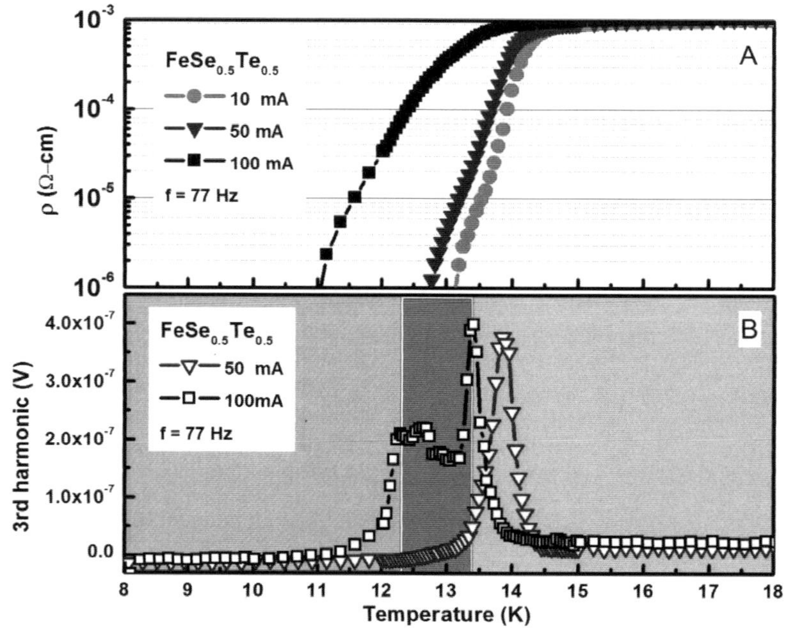

Figure 3. A. Semi logarithmic plot of resistivity vs. temperature for $FeSe_{0.5}Te_{0.5}$ sample at different currents. B. AC transport third harmonic signal vs. temperature for $FeSe_{0.5}Te_{0.5}$ sample at the different AC current amplitudes.

The $FeSe_{0.5}Te_{0.5}$ shows a great stability (in particular for inter-grain) of the resistive transition in very high magnetic fields or current amplitudes (in some extent the intra and inter-grain grain contributions are almost indistinguishable). The Josephson effects influence (as a non-linear phenomenon) over the AC response of $FeSe_{0.5}Te_{0.5}$ was effectively detected by the sensitive third harmonic technique.

Acknowledgments
This work was conducted within the framework of Polish –Bulgarian inter-academical research project.

References

[1] Hilgenkamp H and Mannhart J 2002 Grain boundaries in high-Tc superconductors *Rev. Mod. Phys.* **74** 485–549
[2] Dimos D, Chaudhari P and Mannhart J 1990 Superconducting transport properties of grain boundaries in $YBa_2Cu_3O_7$ bicrystals *Phys. Rev. B* **41** 4038–49
[3] Laval J Y and Swiatnicki W 1994 Atomic structure of grain boundaries in $YBa_2Cu_3O_{7-x}$ *Phys. C Supercond.* **221** 11–9

[4] Suenaga M and Jansen W 1983 Chemical compositions at and near the grain boundaries in bronze-processed superconducting Nb_3Sn *Appl. Phys. Lett.* **43** 791–3

[5] Dravid V P, Zhang H and Wang Y Y 1993 Inhomogeneity of charge carrier concentration along the grain boundary plane in oxide superconductors *Phys. C Supercond.* **213** 353–8

[6] Mazin I I, Singh D J, Johannes M D and Du M H 2008 Unconventional Superconductivity with a Sign Reversal in the Order Parameter of $LaFeAsO_{1-x}F_x$ *Phys. Rev. Lett.* **101** 057003

[7] Katase T, Ishimaru Y, Tsukamoto A, Hiramatsu H, Kamiya T, Tanabe K and Hosono H 2011 Advantageous grain boundaries in iron pnictide superconductors. *Nat. Commun.* **2** 409

[8] Sarnelli E, Adamo M, Nappi C, Braccini V, Kawale S, Bellingeri E and Ferdeghini C 2014 Properties of high-angle Fe(Se,Te) bicrystal grain boundary junctions *Appl. Phys. Lett.* **104** 162601

[9] Gajda G, Morawski A, Rogacki K, Cetner T, Zaleski A J, Buchkov K, Nazarova E, Balchev N, Hossain M S A, Diduszko R, Gruszka K, Przysłupski P, Fajfrowski Ł and Gajda D 2016 Ag-doped $FeSe_{0.94}$ polycrystalline samples obtained through hot isostatic pressing with improved grain connectivity *Supercond. Sci. Technol.* **29** 095002

[10] Gajda D, Morawski A, Zaleski A, Kurnatowska M, Cetner T, Gajda G, Presz A, Rindfleisch M and Tomsic M 2015 The influence of HIP on the homogeneity, J_c, B_{irr}, T_c and F_p in MgB_2 wires *Supercond. Sci. Technol.* **28** 015002

[11] Glowacki B A 1998 Texture development of HTS powder-in-tube conductors *Supercond. Sci. Technol.* **11** 989–94

[12] Klie R F, Buban J P, Varela M, Franceschetti A, Jooss C, Zhu Y, Browning N D, Pantelides S T and Pennycook S J 2005 Enhanced current transport at grain boundaries in high-Tc superconductors *Nature* **435** 475–8

[13] Buchkov K, Nenkov K, Zaleski A, Nazarova E and Polichetti M 2012 Fundamental and 3rd harmonic AC magnetic susceptibility of over-doped polycrystalline $Y_{1-x}Ca_xBa_2Cu_3O_{7-\delta}$ (x=0.025 and x=0.20) samples *Phys. C Supercond.* **473** 48–56

[14] Nazarova E, Nenkov K, Zaleski A, Buchkov K and Zahariev A 2012 Investigations of the overdoped state in polycrystalline $R_{1-x}Ca_xBa_2Cu_3O_{7-\Delta}$ samples (R=Y, Eu, Gd, Er) *Superconductivity: Theory, Materials and Applications* (Nova Science Publishers, Inc.) pp 327–61

[15] Sedky A, Nazarova E, Nenkov K and Buchkov K 2017 A Comparative Study Between Electro and Magneto Excess Conductivities in FeTeSe Superconductors *J. Supercond. Nov. Magn.* **30** 2751–62

[16] Cheng C H and Zhao Y 2007 Repair of grain boundary by preferential-doping in $YBa_2Cu_3O_{7-y}$ *Phys. C Supercond. its Appl.* **463–465** 174–7

[17] Berenov A, Farvacque C, Qi X, MacManus-Driscoll J ., MacPhail D and Foltyn S 2002 Ca doping of YBCO grain boundaries *Phys. C Supercond.* **372–376** 1059–62

[18] Laval J Y and Orlova T S 2002 Microstructure and superconducting properties of sintered DyBaCuO ceramics doped by Ca *Supercond. Sci. Technol.* **15** 314

[19] Buchkov K, Nazarova E, Gurova K, Zahariev A, Sechenski H and Nenkov K 2012 Reduction of YBCO melting temperature by simultaneous Ca substitution and Ag addition *Optoelectron. Adv. Mater. Rapid Commun.* **6** 1061–3

[20] Schmehl A, Goetz B, Schulz R R, Schneider C W, Bielefeldt H, Hilgenkamp H and Mannhart J 1999 Doping-induced enhancement of the critical currents of grain boundaries in $YBa_2Cu_3O_{7-\delta}$ *Europhys. Lett.* **47** 110–5

[21] Krzton-Maziopa A, Pomjakushina E V, Pomjakushin V Y, von Rohr F, Schilling A and Conder K 2012 Synthesis of a new alkali metal-organic solvent intercalated iron selenide superconductor with Tc ≈ 45 K. *J. Phys. Condens. Matter* **24** 382202

[22] Rößler S, Cherian D, Harikrishnan S, Bhat H L, Elizabeth S, Mydosh J A, Tjeng L H, Steglich F and Wirth S 2010 Disorder-driven electronic localization and phase separation in superconducting Fe1 + y Te0.5Se0.5 single crystals *Phys. Rev. B* **82** 144523

IOP Conf. Series: Journal of Physics: Conf. Series **1186** (2019) 012004 doi:10.1088/1742-6596/1186/1/012004

[23] Lei H, Wang K, Hu R, Ryu H, Abeykoon M, Bozin E S and Petrovic C 2012 Iron chalcogenide superconductors at high magnetic fields *Sci. Technol. Adv. Mater.* **13** 054305

[24] Ekin J W, Braginski A I, Panson A J, Janocko M A, Capone D W, Zaluzec N J, Flandermeyer B, de Lima O F, Hong M, Kwo J and Liou S H 1987 Evidence for weak link and anisotropy limitations on the transport critical current in bulk polycrystalline $YBa_2Cu_3O_x$ *J. Appl. Phys.* **62** 4821–8

[25] Vanderbemden P, Bradley A D, Doyle R A, Lo W, Astill D M, Cardwell D A and Campbell A M 1998 Superconducting properties of natural and artificial grain boundaries in bulk melt-textured YBCO *Phys. C Supercond.* **302** 257–70

[26] Cheenne N, Mishonov T and Indekeu J O 2003 Observation of a sharp lambda peak in the third harmonic voltage response of high-T_c superconductor thin films *Eur. Phys. J. B - Condens. Matter* **32** 437–44

[27] Ossandón J G, Giordano J L, Esquinazi P, Kempa K, Schaufuss U and Sergeenkov S 2006 Non-linear response of ac conductivity in narrow YBCO film strips at the superconducting transition *J. Phys. Conf. Ser.* **43** 655–8

IOP Conf. Series: Journal of Physics: Conf. Series **1186** (2019) 012005 doi:10.1088/1742-6596/1186/1/012005

Impact of substrate temperature on surface and grain boundary reflection in thin chromium nanofilms

L Udachan[1*], L Shiva[2], N Ayachit[2], A Banagar[3], S Kolkundi[3], S Bhairamadagi[4]

[1*]Channabasaveshwar Arts, Science and Commerce College, Bhalki , Dist: Bidar, Karnataka, India.
[2]Department of Physics, Rani Channamma University, Belagavi-591 156, Karnataka, India.
[3]Government First Grade College, Shahapur-585 223, Dist : Yadgir, Karnataka, India.
[4]Government Degree College, Kalaburagi- 585 105, Karnataka, India.

E-mail: ulingappa@gmail.com

Abstract. The impact of substrate temperature on surface and grain boundary reflection is studied as a function of thickness of chromium nanofilms in the range, 10-80 nm. Substrate temperature happens to be one of the important and crucial deposition parameters, which regulates many physical properties of evaporated thin films. Hence, we have deposited the chromium nanofilms at four different substrate temperatures (Ts = 22, 100, 150 & 1800C) in order to study and analyze its effect on surface and grain boundary reflections and other electrical properties. In this work a qualitative analysis of both surface and grain boundary reflections of thin chromium nanofilms with independent variation of thickness is reported. The grain boundary reflection coefficients (R) have been calculated by estimating the resistivity data of the chromium nanofilms using Mayadas-Shatzkes and Fuchs-Sondheimer models. The substrate temperature dependent resistivity of chromium films exhibit an unusual behavior.

1. Introduction

Thin chromium films have found many applications for example it was the first metal to be investigated as a thin film resistor material, and is used in photo-masks, integrated circuits, optical beam splitters, semi- reflection coating and the magnetic recording disks. Other applications exist such as thin film capacitor plates and antimagnetic shielding materials [1].

In the present article, the impact of substrate temperature (Ts) on the electrical resistivity, surface and grain boundary reflections of chromium nanofilms is explored in the thickness range 10-80 nm. The thin nanofilms of chromium have been grown by thermal evaporation in vacuum onto glass substrates as it is capable of yielding good quality thin films. The majority of physical properties of thin films utilized for practical applications depend, to a great extent, on thickness as well deposition parameters. The deposition conditions include substrate temperature [2-5], substrate materials [6-8], deposition pressure [9], deposition rate [10], application of DC field [11, 12] during the formation of the film, etc. play a vital role in the nucleation, growth and other characteristics of nano-films. Tailoring of deposition parameters is the main topical issue for their application in electronics. The goal of the present work is the investigation of the impact of substrate temperature on the electrical properties including surface and grain boundary reflection of chromium nano-metric films deposited by resistive heating method in vacuum, onto glass substrates, because of their smooth surface and dielectric nature. The reason for selecting this thickness range, 10-80 nm, is that we have first estimated the conduction electron mean free path in bulk chromium [13] from its resistivity data, which is around 15.2 nm. The microstructure of chromium films have been investigated from AFM & SEM images. Initial EDS analysis usually involves the generation of an X-ray spectrum from the entire scan area of the material under investigation and the material is identified as chromium.

2. Theoretical section

The electrical resistivity of thin films, based on the Fuchs-Sondheimer(FS) theory [14,15] is given by

$$\rho = \rho_0 \left[1 + \frac{3}{8} \frac{(1-p)}{\lambda} \right], \qquad \lambda > 0.1 \quad \text{.........................(1)}$$

where ρ = resistivity of thin film,

ρ_0 = the resistivity of the infinitely thick film,

λ = the ratio of the film thickness(t), to the conduction electron mean free path (l),

p = the specularity parameter.

The second term on the Right Hand Side (RHS) of Eqn. (1) is the contribution to the electrical resistivity from size effect. Taking into account of grain boundary scattering, which is predominant in very thin films, Mayadas - Shatzkes modified the above equation as [16]

$$\rho = \rho_0 \left[1 + \frac{3}{8} \frac{(1-p)}{\lambda} + \frac{3}{2} \alpha' + \ldots \right] \quad \text{...(2)}$$

where $\alpha' = l(R) / d(1-R)$ = the scattering power of grain boundaries which depends upon the electron mean free path(l) and average grain size(d) and R being the grain boundary reflection coefficient. The third term on the RHS of Eqn. (2) is contribution to the electrical resistivity from grain boundary scattering.

3. Experimental Section

Chromium nano-thin films have been successfully deposited onto soda lime glass. The experimental method of preparation are given elsewhere [17,18]. The electrical resistivity measurements have been performed, in-situ, carefully, by the standard four point probe technique[19].

4. Results and Discussions

Figure 1 shows the graph of electrical resistivity (ρ) versus nano film thickness (t) for chromium films grown on glass substrates at different substrate temperature (T_S = 22, 100, 150 & 180^{0}C). It is evident from the same figure that the electrical resistivity is quite large for lower thickness films and decreases for higher thicknesses, and finally attains a constant values of about 51, 95, 140 and 72 x 10^{-8} Ωm , respectively after a thickness of 40 nm.

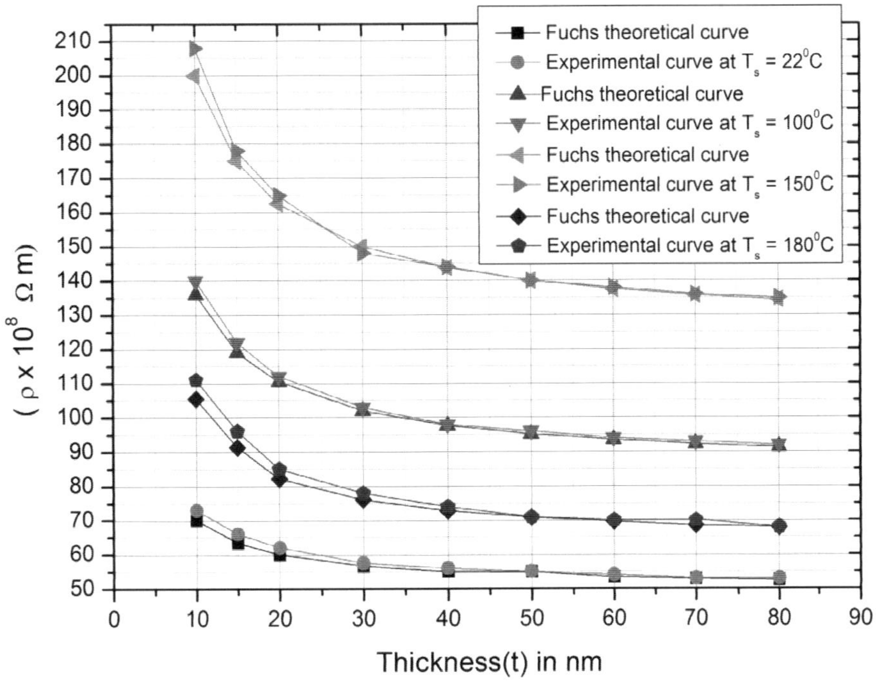

Figure 1. Plot of Electrical Resistivity(ρ) against chromium nano-film thickness(t) grown at different T_S. (= 22, 100, 150 & 180^0 C), based upon F-S theory and our experimental results.

Figure 2 shows (ρ x t) vs t graphs, which have been plotted for chromium nano-films deposited simultaneously, onto glass substrates at different substrate temperatures (22, 100, 150 & $180)^0$ C.

The slopes of these graphs, according to Eqn.(2), give ρ_0, the infinitely thick film resistivity, as 49.6, 85.9, 126, and 63.1 x 10^{-8} Ω m, respectively for the films deposited at T_S = 22, 100 , 150 and 180^0 C, and the intercepts on the (ρ x t) axis give $3\,\rho_0\, l\,(1-p)\,/\,8$, as 260 and 492, 712, and 424 x 10^{-17} Ωm^2 . We tried to fit our experimental data on the basis of FS theory by assigning different values for p ; the best fit is obtained for p = 0.1, 0.2, 0.15 and 0.25 respectively for the films grown at different substrate temperatures and these values are given in Table 1. The resistivity ρ_0 (12.9, 85.9 & 63.1 x 10^{-8} Ω m) of an infinitely thick chromium nano-film, grown at different T_S, estimated from the Fig. 2 are quite high in comparison with the bulk resistivity of chromium [20].

IOP Conf. Series: Journal of Physics: Conf. Series **1186** (2019) 012005 doi:10.1088/1742-6596/1186/1/012005

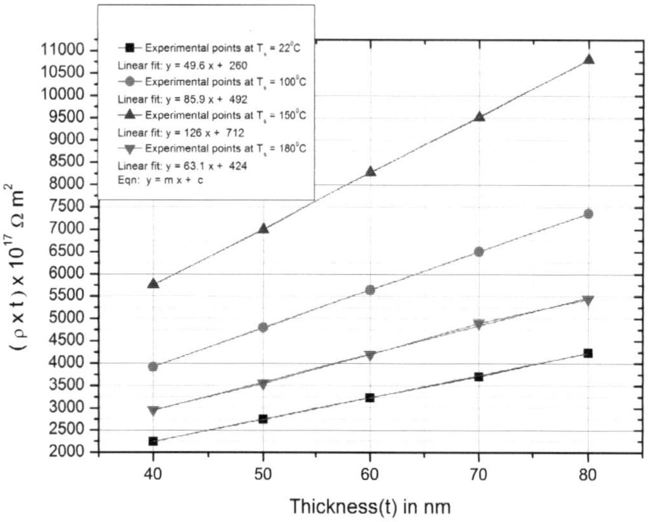

Figure 2. Plot of (ρ x t) versus (t) for chromium nanofilms grown at different T_S. (= 22, 100, 150 & 180^{0}C) for the thicknesses > 30 nm.

Table 1. (for Chromium nanofilms)

Sl. No.	Parameter	Substrate Temperature(Ts)			
		22^0 C	100^0 C	150^0 C	180^0 C
	(t) in nm	R_1 x 10^3	R_2 x 10^3	R_3 x 10^3	R_4 x 10^3
1	10	25	15.3	7.4	23
2	15	33	18	12.1	30
3	20	34	12	16	24
4	30	15.6	12	20	24
5	(R- average) R_{av}	26.9	14.32	9.3	25.3

	Parameter	Substrate Temperature(Ts)			
		22^0 C	100^0 C	150^0 C	180^0 C
6	ρ_0	49.6 x 10^{-8} Ω m	85.9 x 10^{-8} Ω m	126 x 10^{-8} Ω m	63.1 x 10^{-8} Ω m
7	$l\,(1\text{-}p)$	14 nm	16 nm	15.1 nm	18 nm
8	l	15.3 nm	20 nm	17.8 nm	24 nm
9	p	0.1	0.2	0.5	0.25

The curves based on FS theory are shown in Fig. 1 by the lower curves, for the chromium films deposited at (T_S.= 22, 100, 150 & 180^{0}C)., while the upper ones represent the experimental curves.

We obtained good agreement between FS theory and experimental data for higher thicknesses > 30 nm for all the films grown at different substrate temperatures. Making use of the values ρ_0 and p, values of l have been calculated for the films grown at different substrate temperatures which are given in Table I.

Even though there is perfect agreement between F-S theoretical and our experimental curves for higher thickness films, there is, however, some deviation from the theoretical curves for lower thickness films. This is attributed to the fact that the F-S theory takes into account the variation due to size effect and disregards the grain boundary scattering, which is predominant in lower thickness films. Similar deviation at lower thicknesses between FS theory and experimental data has also been reported for palladium [21], samarium [22], manganese [23], yttrium [24], ytterbium [25] and tin [26] films.

We tried to fit our experimental data with MS theory. The theoretical curves based on MS theory are denoted by the upper curves in Figure 1, for chromium films grown at different T_S. It is clear from the Figure 1 that there is good agreement between the MS theory and the experimental data for thinner films where grain boundary scattering is predominant. To calculate the grain boundary reflection coefficient(R), we assumed the grain diameter to be equal to the thickness of the film, i.e., up to 30 nm for glass at different T_S, below which the experimental data deviate from the FS theoretical curves. However, for higher thicknesses (> 30 nm) the grain diameter becomes very large in comparison with the mean free path of the conduction electron and hence contribution to the film resistivity from the grain boundaries become negligible.

The difference between the resistivity from the F-S theoretical curve and the experimental data gives ($3\alpha'\rho_0/2$). Using the values of (ρ_0) and (l) estimated from the Fuchs theory, we obtained the value of α' and from this the constant R_{av} (R average) are found to be 26.9, 14.32, 9.3 and 25.3 x 10^{-3} ,respectively, for the films grown at different substrate temperatures, 22, 100, 150 and 180^0 C, which are mentioned in the Table 1. These are used in the equation (2) to obtain MS theoretical curve.

Figure 3 shows the variation of grain boundary reflection coefficient(R) with substrate temperature(T_S) for chromium nanofilms.

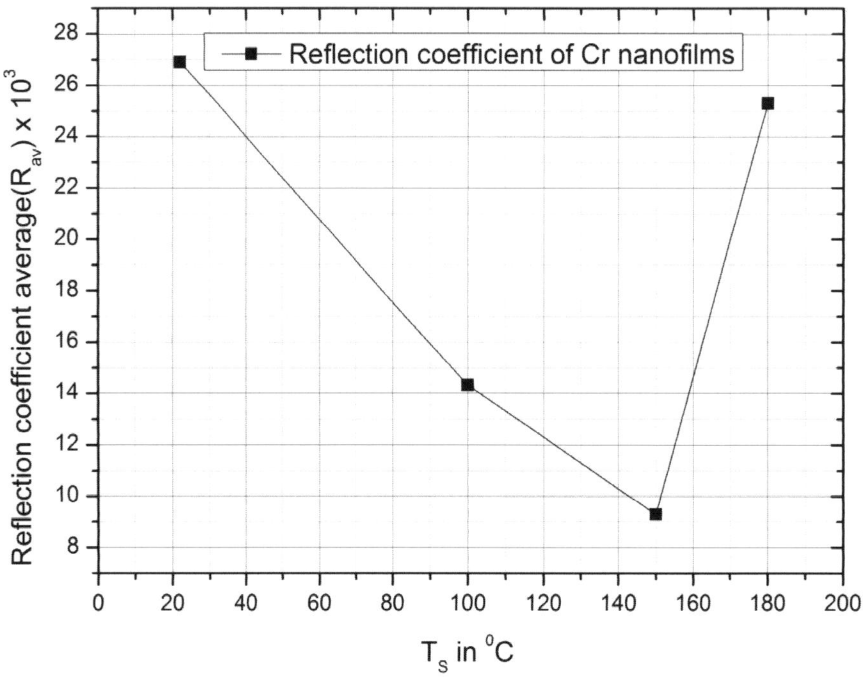

Figure 3. Reflection coefficient (R) versus substrate temperature (Ts) for thin chromium nano-films.

The growth and structure of evaporated films are very much affected by the nature of the substrates, smoothness/roughness and the binding force between them (substrates), evaporated atoms and substrate temperature. An increase in the binding force between the substrate and evaporated atoms usually decreases the surface mobility of evaporated atoms and hence increases the population of critical nuclei. This in turn enhances the film adhesion. Therefore, the film resistivity may be reduced considerably.

The theoretical curves based on the M-S theory are shown in Fig. 1. It is clear from this figure, that there is good agreement between M-S theory and our experimental data for chromium films coated at different Ts, for thinner films where grain boundary scattering is predominant. The average grain boundary reflection coefficient , R_{av} , as mentioned in the Table I for Cr coated at substrate temperatures 22, 100 & 150^{0}C appears to decrease with the increase in Ts up to 150^{0}C and then after wards it increases at Ts=180^{0}C.The conduction electron mean free path(l), infinitely thick film resistivity(ρ_0) and specularity parameter*(p)* also follows the same path or order in respect of Ts. However, the grain boundary reflection coefficient varies with Ts, in the reverse order, that is with an increase in Ts, the grain boundary reflection coefficient first decreases up to Ts = 150^{0}C and then after words R increases as shown in Table I and Fig. 3. Similar results have been reported for Ni-Si silicide films [27]. It is reported that (ρ_0) and (l) are different and dependent on Ts for Antimony [28].film. The variation in the reflection coefficient with Ts may be due to the enrichment of impurities in the grain boundaries[29].

5. Conclusion

It is found that the Fuchs-Sondheimer theory considers only the size effect avoiding the grain boundary scattering. The grain boundary scattering mechanism is predominant in thin nanofilms below the thickness of the electron mean free path which has been taken into account by Mayadas - Shatzkes. Our experimental curves fit well with Mayadas – Shatzkes theoretical curves.

We have calculated the values of ρ_0, ρ , p and l for chromium films grown at different Ts. By tuning the grain size during the film formation, we discriminate whether the dominant collision mechanism controlling the resistivity of nanofilms is electron-surface or electron -grain boundary scattering.

With the use of the electron conduction model proposed by Mayadas-Shatzkes, we have calculated the fraction (p) of electrons which are specularly reflected at the surface of the film and the grain boundary reflection coefficient(R), at different Ts. We have also deduced the electron mean free path from these results.

We found that the factors correlated to the electrical resistivity of the nano- films also govern the grain size and grain boundary reflection.

The grain boundary theory of Mayadas-Shatzkes reproduces our experimental observations faithfully even at different Ts. The grain boundary reflection plays a significant role in electron transport mechanism in chromium nanofilms [30].

Acknowledgements
One of the authors, Udachan L. A[a] is very much thankful to Dr. B. G.Hegde, Associate Professor, Rani Channamma University, Belagavi, Karnataka, India for constant support and encouragement.

References

[1] Rauf A, Ahmed K, Nasim F, Khan A N and Gul A, 2016 IOP Conf. Series: Materials Science and Engineering 146 012013 doi:10.1088/1757-899X/146/1/012013.
[2] Angadi M A & Udachan L A, 1981 *Journal of Materials science* **16** 1412.
[3] Shadia J Ikhmayies, Naseem M Abu El-Haija and Riyad N Ahmad-Bitar, 2010 The Royal Swedish Academy of Sciences *Physica Scripta* **81** 015703.
[4] D. Jhansi Ranir, A. Guru Sampath Kumar & T. Subba Rao, 2017 *Journal of Coatings Technology and Research* 14 971-80.
[5] Salah Abdul-JabbarJassim, Abubaker A. Rashid AliZumaila, Gassan Abdella Ali, Al Waly, 2013 *Results in physis* **3** 173-80.
[6] Feroz A. Mir, Ikram M & Ravi Kumar, 2011 *Solid State Sciences* **13** 1994-1999.
[7] Cao M, Li L, Zhang B L, Huang J, Tang K. Cao H, Sun Y, Shen Y, 2012 *Journal of Alloys and Compounds* **530** 81-84.
[8] Justyna Krzak-Ros, Jaroslaw Filipiak , Celina Pezowicz, Agnieszka Baszczuk, Miroslaw Miller, 2009 *Acta of Bioengineering and Biomechanics* **11** 21.
[9] L U Shiva, Ayachit N.H, Udachan L. A[a] , Banagar A V, Kolkundi S S, Bhairamadagi S S 2018 Base pressure, annealing temperature and ageing effects on the electrical resistivity of chromium and tin nanofilms submitted to Metallofizika I Noveishie Tekhnologii Journal.
[10] Angadi M A & Udachan L A, 1981 *Thin Solid Films* **78** 299-302.
[11] Angadi M A & Udachan L A, 1981 *Journal of Physics D: Applied Physics* **14** L81-83.
[12] J. W. Lim and M. Isshiki, 2006 *Journal of Applied Physics* **99** 094909 doi.org/10.1063/1.2194247.
[13] Charles Kittel 2010 *Introduction to Solid State Physics* (Wiley India Pvt. Ltd., New Delhi) pp 158.
[14] Fuchs K 1938 The conductivity of thin metallic films according to the electron theory of metals *Proc. Cambr. Philo. Soc.* **34** p 100.

[15] F. H. Sondheimer, 1950 *Phys. Rev.* **80** 401.
[16] Mayadas A. F. & Shatzkes M., 1970 Electrical-resistivity model for polycrystalline films: the case of arbitrary reflection at external surfaces *Physical Review B* **1** 1382-89.
[17] Udachan L A, Shivaprasad S M, Ashrit P V and Angadi M A, 1980 *Physica Status Solidi(a)* **60** K 191.
[18] L U Shiva, Ayachit N H, Udachan L A , Sunagar S M, 2017 *International Journal of Research in Engineering & Technology*, **6** 75-80 doi.org/10.15623/ijret.2017.0613016.
[19] Thakur M K, Singh O P, Thakur Karuna, Gaur K S, Muhanthan N, Singh S , Singh D , Vijayan N, Singh H K and Singh V N, 2016 *Advanced Materials letters* **7** 525.
[20] *Hand Book of Physics and Chemistry* 1974 (Chemical Rubber Company Press, Cleveland, Ohio) 54th edition p F155.
[21] Shivaprasad S.M, Udachan L. A & Angadi M.A, 1980 *Physics Letters* **78A** p 187-188.
[22] Kumar J & Srivastava O N, 1972 **13** S29-S33.
[23] Shivaprasad S M, & Angadi M A & Udachan L A, 1980 *Thin Solid Films* **71** L1-L4.
[24] Ashrit P V, Shivaprasad S M and Angadi M A, 1980 *Thin Solid Films* **72** L5.
[25] Ashrit P V and Angadi M A, 1981 *Physica Status Solidi(a)* K77-K82.
[26] Angadi M A and Udachan L A, 1980 *Thin Solid Films* **78** 299.
[27] Artunc G U, 2014 *RTA Microelectronic Engineering* **113** 86-92 https://doi.org/10.1016/j.mee.2013.07.019.
[28] Deschacht D, Boyer A, 1985 *Journal of Material Science* **20** 807-811.
[29] Wu W, Brongersma S H, Hove M V and Maex K, 2004 *Applied Physics Letters* **84** 2838 doi.org/1063/1.1703844.
[30] Bandopadhyay S K & Pal A K 1979 *Journal of Materials Science* **14** 1321-1325.

A study on nucleation, growth and grain boundary reflection in thin tin nanofilms

L Shiva[1*], N Ayachit[1], L Udachan[2], A Banagar[3], S Kolkundi[3], S Bhairamadagi[4]

[1]Department of Physics, Rani Channamma University, Belagavi-591 156, Karnataka, India.
[2]Channabasaveshwar Post Graduate Studies & Research Centre, Bhalki-585 328, Dist: Bidar, Karnataka, India.
[3]Government First Grade College, Shahapur-585 223, Dist : Yadgir, Karnataka, India.
[4]Government Degree College , Kalburagi-585 105, Karnataka, India.

E-mail: shivaudachan8@gmail.com

Abstract. The present work incorporates a study on nucleation, and growth due to the changes in the nano-structural properties as a function of thickness and grain boundary reflection in tin thin nanofilms Thickness happens to be one of the important and crucial nanofilm preparative parameters which dictates majority of properties like, nano-structure, electrical, optical and morphological texture of evaporated tin nanofilms. Hence, we have selected tin nanofilms in the thickness range 20-160 nm, in order to study the thickness dependence of their nano structural and electrical properties, in the temperature range 77-450 K, in detail. All the films were grown in a conventional vacuum coating unit under a pressure of ~10-6 Torr at room temperature of 22 0C, onto cleaned glass substrates. The tin film nano-structure was analyzed by Transmission Electron Microscope(TEM). The nano structural properties can be tuned to the desired level and application by the proper choice or proper combination of deposition parameters. And it is found to depend upon the nature of the substrate material, environment and the binding force between the substrate material and evaporated tin atoms.

1. Introduction

The present work is the extension of the earlier work reported on thin tin films[1,2]. Thin film science has received tremendous attention in the recent years for their applications in diverse fields such as, microelectronic industries, military weapons systems, space science, solar energy utilization, cell phones, smart devices, computer note books and gas sensors etc.[3-8]. Thickness plays an important role in thin films as it is a significant parameter, which affects the optical, electrical, nano-structural properties considerably. Thin film properties mainly depend on the preparative conditions, film nano-structures, presence of defects and impurities.

Tin has low resistivity, low melting point, high temperature coefficient of resistance and negative thermo-electric power in the bulk form. Tin develops a milky white surface because of its un-specular reflection. Therefore tin is rendered unsuitable for mirror coatings. However, it is used as one of the components in solder material and its oxides and compounds are widely used in the fabrication of solar cells and liquid crystal displays[9,10] and other semiconductor devices. In view of the above mentioned interesting properties of tin metal in the bulk form, a study on nucleation, growth and electrical properties in thin tin nanofilm form has been undertaken herewith. There are several steps involved in the growth process from nucleation to final film formation stage.. By the knowledge of the nanostructure, it is possible to select a material of specific thickness for specific application in order to suit it for conduction or insulation mechanisms. By varying the deposition parameters the electrical, optical and nano-structural properties of films can be regulated.

2. Theoretical section

The electrical resistivity of thin films, based on the Fuchs-Sondheimer(FS) theory [11,12] is given by

$$\rho = \rho_0 \left[1 + \frac{3}{8} \frac{(1-p)}{\lambda} \right], \qquad \lambda > 0.1 \qquad (1)$$

where ρ = resistivity of thin film ,

ρ_0 = the resistivity of the infinitely thick film ,

λ = the ratio of the film thickness(t) , to the conduction electron mean free path (l) ,

p = the specularity parameter.

The second term on the Right Hand Side(RHS) of Eq.(1) is contribution to the electrical resistivity from size effect. Taking into account of the grain boundary scattering, which is predominant in very thin films, Mayadas - Shatzkes (1970) modified the above equation as [13]

$$\rho = \rho_0 \left[1 + \frac{3}{8} \frac{(1-p)}{\lambda} + \frac{3}{2} \alpha' + \dots \right] \qquad (2)$$

where $\alpha' = l\,(R) \,/\, d\,(1 - R)$ = the scattering power of grain boundaries which depends upon the electron mean free path(l) and average grain size(d) and R being the grain boundary reflection coefficient. The third term on the RHS of Eqn. (2) is contribution to the electrical resistivity from grain boundary scattering.

3. Experimental Section

The vacuum coating unit used in the present investigation is the 'Hind High Vacuum Coating Unit, Model 12A4D', to grow thin films of tin in this work. The film thickness was controlled by means of an in-built Quartz Crystal Digital Thickness Monitor (Model DTM-101). Tin was evaporated from a molybdenum boat, and the distance between the boat and the glass substrate was around ~22 cm. Tin of purity 99.99% was thermally evaporated by resistive heating technique on to cleaned glass slides. The glass slides were cleaned initially in chromic acid, ultrasonically and at last by the ionic bombardment method. Thin tin films in the thickness range 20-160 nm have been grown on to thus cleaned glass substrates at the rate of 0.5 nm/s under a pressure of ~10^{-6} Torr at room temperature of 22^0C. After growing the films, they were taken out of the vacuum chamber for nano-structural analysis and temperature dependent resistivity measurements in the range 77-450 K. The sample preparation technique in order to view image for TEM analysis is given below :

3.1 For TEM

Nano-films were grown in the conventional vacuum chamber on freshly cleaned glass substrates for their structural studies. The films were removed from the glass substrates by immersing them in distilled water. Afterwards, these films were mounted on carbon coated copper grids(having 200 meshes/cm^2). All the films were examined with Hitachi Transmission Electron Microscope for bright field image.

3.2 For grain boundary reflection measurement

We have initially measured the resistance of tin nanofilms by four probe technique {12], at temperatures 77, 300 & 450 K as a function of thickness,. For observing the effect of temperature on ρ_0, l and p ; the plots ρ vs t have been drawn at temperature 77, 300 and 450 K , in the thickness range 40-160 nm for tin nanofilms, which is shown in Fig. 1, since tin exhibits a positive TCR, Fig. 2 shows the plot of (ρ x t) vs t at 77 , 300 and 450 K for tin nano films. These are found to be straight lines yielding values for slopes as ρ_0, and intercepts give values for $3\,\rho_0\,l(1\text{-}p)/8$, which are given in Table I. The grain boundary reflection coefficient(R) was determined from the resistivity data and its average values at temperatures 77, 300 & 450 K are listed in the same table I.

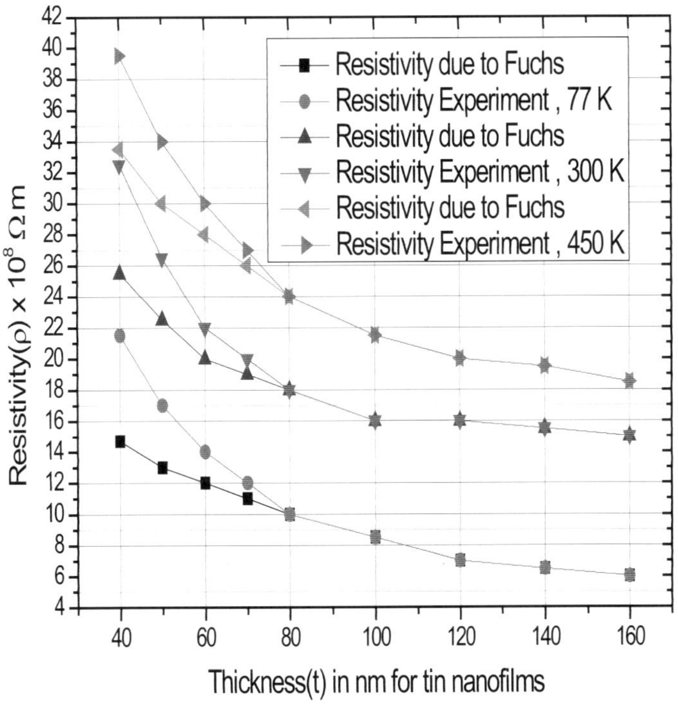

Figure 1. Resistivity , ρ (x 10^8 Ω m) Versus Thickness (t) for tin nano films based upon Fuchs – Sondeimer theory(lower curve) and Experimental(upper) curve based on Mayadas - Shatzkes theory.

IOP Conf. Series: Journal of Physics: Conf. Series **1186** (2019) 012006 doi:10.1088/1742-6596/1186/1/012006

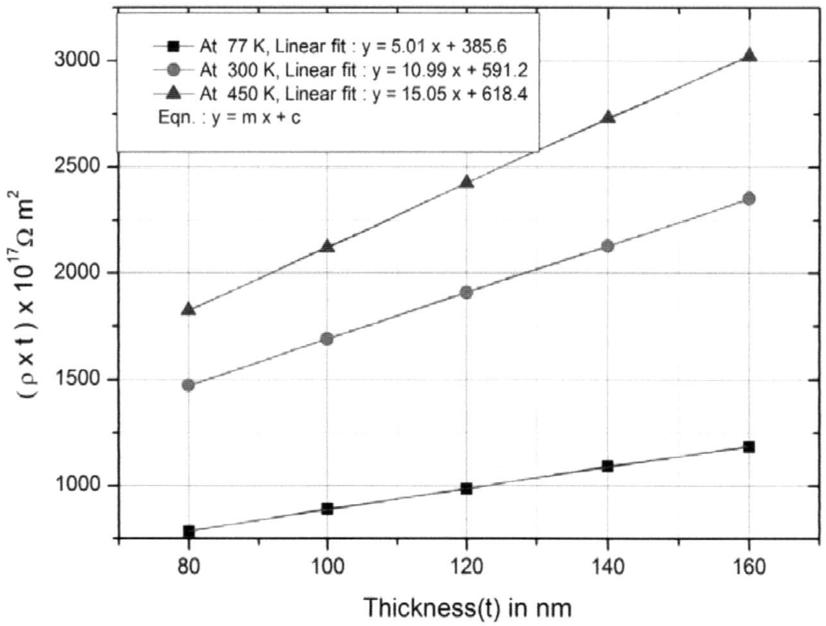

Figure 2. ρ x t , (x 10^{17} $\Omega\, m^2$) versus Thickness (t) for tin nano films (t > 80 nm).

The symbols used in the graph are the actual measured data points.

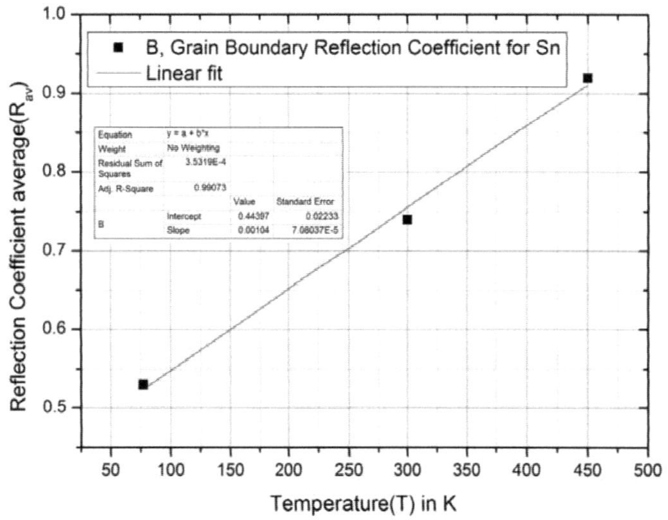

Figure 3. Grain Boundary Reflection coefficient average (R_{av}) Versus thickness (t) for tin nano films with linear fit.

Table 1. (For tin nanofilms)

Temperature, K	Infinitely thick tin film resistivity (ρ_0) x 10^8 Ω m	Conduction Eletron mean free path (l) in nm	Specularity Parameter (p)	Average grain boundary reflection coefficient (Rav)
77	5.01	196	0	0.53
300	11	143.4	0	0.74
450	15	111	0	0.92

4. Results and Discussion

4.1 The morphology and nanostructure were analyzed using TEM. The TEM images of Sn films reveal that the observed morphological and nano-structural modification induce changes in electrical properties. It is found that there was no electrical continuity of the tin films below the 40 nm thickness. The initial stages of the growth of the film is formation of nucleation centres/islands structures as shown in Fig. 4 (a). No physical connections between the nuclei/islands exist and hence no electrical continuity is achieved. The small islands are dense populated. As the thickness increases, the smaller islands merge with bigger islands and well crystallized with various nano crystalline morphology as depicted in Fig. 4 (b) [15]. Both grains and sub-grains are clearly visible in the TEM nanographs. Similar type of results were obtained by T. M. Rajkumar et.al. for Cd chalcogenide films [14]. As the thickness increases the island structures merge with forming a semi continuous or quazi-continuous structure, with bigger voids, which has been shown in the Fig. 4 (c). Further if we increase the thickness, again the quazi-continuous structures merge with neighboring ones and the voids between the structures will be reduced as shown in the Fig. 4 (d). With the further increase in the thickness almost the entire structure becomes continuous and the film became opaque, to the incident electron beam in the TEM and thus film helps in the conduction of current.

IOP Conf. Series: Journal of Physics: Conf. Series **1186** (2019) 012006 doi:10.1088/1742-6596/1186/1/012006

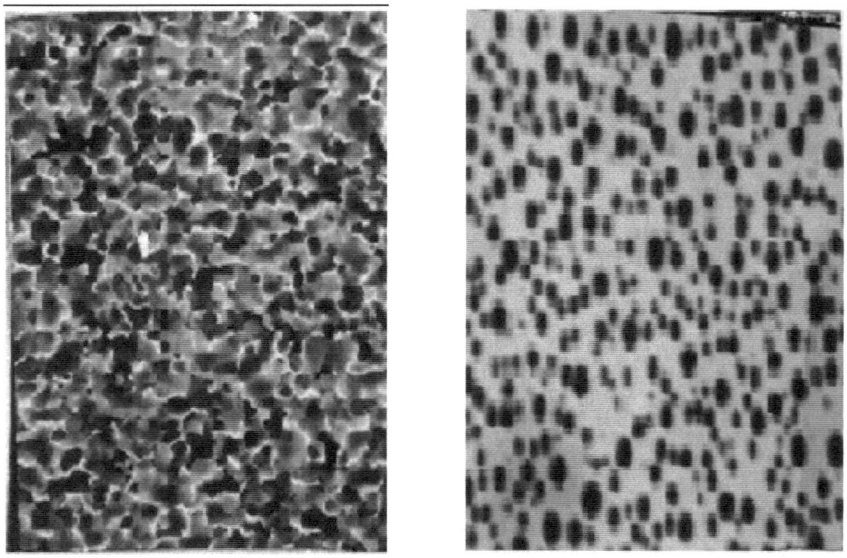

(a) Sn – 20 nm thickness x 57,000 (b) Sn –30 nm thickness x 91,200

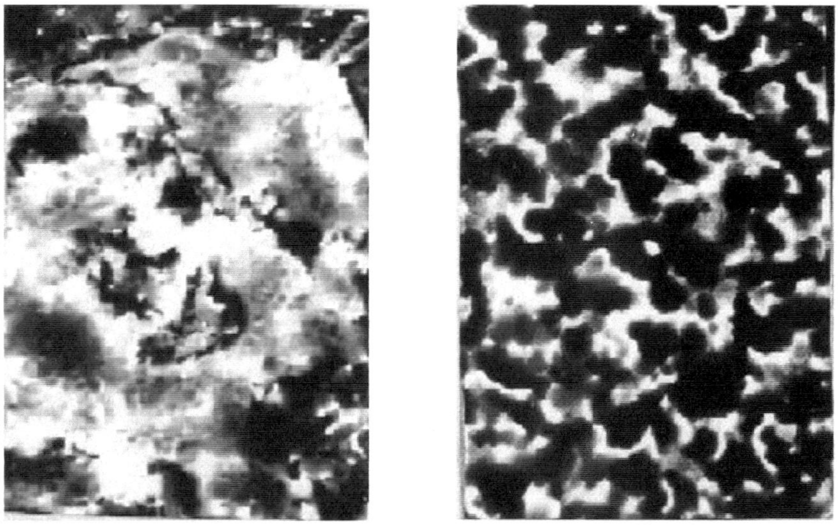

(c) Sn – 40 nm thickness x 9,16,600 (d) Sn – 40 nm thickness x 1,18,000

Figure 4. TEM images of thin tin nanofilms for different thicknesses and magnifications.

4.2 *Grain Boundary Reflection Coefficient Analysis*

In case of tin films, the thickness dependence of electrical resistivity graph approaches the saturation value of $15 \times 10^{-8}\,\Omega$ m after about a thickness of 100 nm, for the resistivity measurement done at 300 K, 8 & 19 $10^{-8}\,\Omega$ m respectively, for the measurements taken 77 & 450 K, as shown in the Fig. 1. However, we obtained the infinitely thick film resistivity ($\rho_0 = 11 \times 10^8\ \Omega$ m) which is same as that of the bulk tin value ($\rho_B = 11 \times 10^8\ \Omega$ m) [15] from the plot of (ρ x t) vs t (Fig. 2). Tin being a metal of low melting point in the bulk form, has a low resistivity even in the thin film form. Because , low melting point metals produce large crystal, preferentially oriented to the substrate[18]. This has been confirmed by the TEM study on tin films. We have estimated the average island (crystal) size as 31.1 nm for 30.0 nm thickness tin nanofilm from Fig. 1(b).

We did not see the electrical continuity of tin films below 30 nm thickness which has also been reported by Pal and Sen (1977) [19] , Kong et al. (1977) [20] and Stolecki et al. (1979) [21]. Although there is a fairly good agreement between the Fuchs theory and our experimental points of resistivity data, for higher film thicknesses, there is a wide difference between the Fuchs theoretical curve and our experimental data for lower thickness films (< 80 nm). This is mainly due to the grain boundary scattering and is predominant in lower thickness films. The Fuchs theory considers only size effect but disregards the scattering of electrons from grain boundaries.

So, Mayadas-Shatzkes (M-S) theory [13] takes into account the grain boundary scattering in addition to the Fuchs size effect. The M-S theory is in good agreement with our experimental ρ - data throughout the thickness range. Similar type of deviation between the Fuchs and experimental curves for the resistivity data has been reported for other substances like Yb [22], Ni[23], Pd [24], and Sm [25] films. By knowing the difference between the Fuchs and M-S theoretical ρ - curves, we have estimated the value of R , the grain boundary reflection coefficients, for the ρ-t data, measurements done at temperatures 77, 300 and 450 K .The values of R calculated closely agrees with that reported by Pal and Sen (1977) [21].

For the tin films. M – S theory holds good to our experimental data for p = 0 , for all the ρ versus t data at temperature 22, 300 and 450 K, indicating total diffuse scattering of conduction electrons. From this value of p , we have calculated the electron mean free paths (l) which have been given in Table I. It is clear from the table I that as the temperature of Sn nano film is decreased, the value of conduction electron mean free path increases due to decrease in the amplitude of thermal vibrations of the Sn-lattice. The average grain boundary reflection coefficient decreases with the decrease in temperature, again due to decrease in the amplitude of lattice vibrations, as shown in Fig. 3.

5. Conclusion

We have depicted the thickness dependence of nano-structural properties of thin tin films starting from nucleation, growth and final film formation mechanisms. The structure of the nanostructure is found to depend upon the thickness and preparative conditions. The observed morphological and nano-structural modifications induce changes in the electrical properties of tin films. By judicial adjustments of deposition parameters, one can grow films of specific nanostructure/thickness for specific applications as required for optical and electrical devices. The F-S theory considers only size effect and neglects grain boundary scattering. The M-S theory considers the grain boundary scattering also. Our experimental results for tin nano films closely fits with M-S theory for the entire range of film thickness and temperature. The electron mean free path and grain boundary reflection coefficient are found to be temperature dependent in tin nano films.

Acknowledgements

One of the authors Udachan L. A[2] is very much thankful to Mr.B.G.Hegde , Rani Channamma University, Belagavi, Karnataka, India for constant support and encouragement.

References

[1] Angadi M A & Udachan L A, 1983 *Journal of Materials Science Letters* **2** 379.
[2] L U Shiva, Ayachit N H, Udachan L A, A V Banagar, S S Kolkundi, S S Bhairamadagi, 2018 *Bulgarian Journal of Physics* **45** 35.
[3] Shigesato Y & Paine D C, 1994 *Thin Solid Films* **238** 44.
[4] Majumdar S, Chakraborty S, Devi P S & Sen A, 2008 *Materials Letter* **62** 1249.
[5] E. Elangovan & K. Ramamurthi, 2004 *Thin Solid Films* **476** p 231.
[6] Rao M C, Ravindranadh K & Srikanth K S, 2016 **8** p 677.
[7] Ali S M, Muhammud J,. Hussain S T, Bakar S A, 2013 *Materials in electronics* **24** 2432.
[8] Nagirnayak S V, Lutz V A, Donstova T A and Astrelin I M, 2016 *Nanoscale Research Letters* **11** 343
[9] Tiwari D, Chaudhuri T K, Shripathi T, Deshpande U, Sathe V G, 2014 *Materials in electronics* **25** 3687.
[10] Nadaf L I, Venkatesh K S 2016 *Journal of Applied Chemistry* **9** 2278
[11] Fuchs K, 1938 *Proc. Cambr. Philo. Soc* **34** 100
[12] F. H. Sondheimer, 1950 *Physics Review* 80 401
[13] Mayadas A F & Shatzkes M, 1970 *Physical Review B* **1** p 1382
[14] Thakur M K , Singh O P, Thakur K, Gaur K S, Muhanthan N, Singh S, Singh D, Vijayan N, Singh H K and Singh V.N, 2016 *Advanced Materials letters* **7** p 525
[15] Shreekanthan_K N, Bangera_K V, Shivakumar G K & Mahesha M G, 2006 *Indian Journal of Pure & Applied Physics* **44** 705
[16] Rajakumar T M, Uma V & Chandramani R, 2013 *Indian Journal of Pure & Applied Physics* **51** 131
[17] *Hand Book of Physics & Chemistry* 1973-74 (CRC press 54th Edition) p F-155
[18] Levinstein H, *Journal of Applied Physics* **20** 306
[19] Pal A K and Sen P 1977 *Journal of Materials Science* **12** 1472
[20] Kong W M, Polan J S & Rose K, 1977 *Journal of Applied Physics* **48** p 5170
[21] Stolecki B, Borodziuk-Kulpa A and Wesolowska, 1979 *Thin Solid Films* **56** 299
[22] Ashrit P V and Angadi M A, *Physicsa Status Solidi(a)* **63** K77
[23] Angadi M A & Udachan L A, 1981 Thin Solid Films **79** 149
[24] Shivaprasad S M, Udachan L A, Angadi M A, 1980 *Physics Letters* **78A** 187
[25] Ashrit P V, Angadi M A, 1980 *Journal of Less Common Metals* **72** 317

IOP Conf. Series: Journal of Physics: Conf. Series **1186** (2019) 012006 doi:10.1088/1742-6596/1186/1/012006

Supplementary data for Tin

Fig. 1 Resistivity, ρ (x 10^8 Ω m) Versus Thickness (t) for tin nano films based upon Fuchs – Sondeimer and Mayadas – Shatzkes theories.

Thickness (t) in nm	Resistivity due to Fuchs (ρ_f) x 10^8 Ω m	Experimental resistivity value at 77 K (ρ_e) x 10^8 Ω m	Resistivity ρ due to Fuchs (ρ_f) x 10^8 Ω m	Experimental resistivity value at 300 K (ρ_e) x 10^8 Ω m	Resistivity ρ due to Fuchs (ρ_f) x 10^8 Ω m	Experimental resistivity value at 450 K (ρ_e) x 10^8 Ω m
40	14.5	21.5	25.8	30	30.6	39.5
50	12.6	17	22.8	26.5	27.5	34
60	11.4	14	20.8	22.5	25.4	30
70	10.5	12	19.5	20.5	23.9	26
80	9.8	9.8	18.4	18.4	22.8	22.8
100	8.9	8.9	16.9	16.9	21.2	21.2
120	8.2	8.2	15.9	15.9	20.2	20.2
140	7.8	7.8	15.2	15.2	19.5	19.5
160	7.4	7.4	14.7	14.7	18.9	18.9

IOP Conf. Series: Journal of Physics: Conf. Series **1186** (2019) 012006 doi:10.1088/1742-6596/1186/1/012006

Fig. 2 ρ x t , (x 10^{17} Ω m^2) versus Thickness (t) for tin nano films (t > 80 nm).

Thickness(t) in nm	ρ x t , (x 10^{17} Ω m^2) at 77 K	ρ x t , (x 10^{17} Ω m^2) at 300 K	ρ x t , (x 10^{17} Ω m^2) at 450 K
80	784	1472	1824
100	890	1690	2120
120	984	1908	2424
140	1092	2128	2730
160	1184	2352	3024

IOP Conf. Series: Journal of Physics: Conf. Series **1186** (2019) 012006 doi:10.1088/1742-6596/1186/1/012006

Fig. 3 Grain Boundary Reflection Coefficient average (R_{av}) Versus thickness (t) for tin nano films with linear fit.

Temperature in K	R_{av}
77	0.53
300	0.74
450	0.92

IOP Conf. Series: Journal of Physics: Conf. Series **1186** (2019) 012007 doi:10.1088/1742-6596/1186/1/012007

Langmuir-Blodgett Films from Fluorescently Labelled Phospholipids Deposited on Surface Acoustic Wave Devices

George R Ivanov[1] and Ivan Avramov[2]

[1]Department of Physics and University Laboratory "Nanoscience and Nanotechnology", University of Architecture, Civil Engineering and Geodesy, blvd. Hr. Smirnenski, 1, 1164, Sofia, Bulgaria and NanoBioSense Ltd.
[2]Georgi Nadjakov Institute of Solid State Physics, Bulgarian Academy of Sciences, Tzarigradsko Chussee blvd, 72, Sofia, Bulgaria

E-mail: george@at-equipment.com

Abstract. Over the years several new effects in fluorescently labelled Langmuir monolayers or Langmuir-Blodgett (LB) films, deposited on solid support plates, have been reported. Mostly layers from Dipalmytoyl Phosphatidyl Ethanolamine head labelled with Nitrbenzoxadiazole (DPPE-NBD) have been studied. This molecule behaves exactly as the DPPE molecule which is part of biological membranes. So it is expected that DPPE-NBD molecules can serve as an appropriate matrix for implementation of selectively reacting proteins or enzymes while preserving their function. This combination can act as an active layer in biosensor applications for operation in either gas or liquid environments and this molecule could further increase the system sensitivity in such applications. Addition of the chromophore NBD head group also stabilizes the film and allows for multilayer deposition which is not possible for phospholipids. In this study, LB monolayers were deposited on glass substrates at different surface pressures and with or without Cadmium ions in the water subphase. They were studied using fluorescence spectroscopy and Atomic Force Microscopy (AFM). In view of possible biosensor applications, LB layer by layer deposition was performed on Rayleigh Surface Acoustic Wave (SAW) resonant devices working at 440 MHz. These can effectively be used to transduce the biosensor signal from the active layer by providing a mass proportional frequency output. Layers from DPPE-NBD that might also be suitable for possible SAW based biosensor applications were deposited at a surface pressure of 10 mN/m. AFM spectroscopy reveals the phase coexistence of liquid and solid phases as well as higher bilayer nanosized cylinders. Fluorescence intensity was strongly self-quenched at this pressure and fluorescence lifetime spectroscopy shows a complex behaviour with a 3-rd order exponential decay. A frequency downshift of 10 KHz per monolayer was measured with the coated SAW device which results in a deposited mass of about 1 ng/monolayer.

1. Introduction

One of the best methods for the creation of supramolecular architectures in a well-controlled manner is the method of Langmuir and Blodgett [1]. The authors of this method received the Nobel Prize in the thirties of the last century but the interest in it has been growing significantly starting from the eighties. This can be seen from the exponential growth of publications and from international conferences named after the method. This interest can be attributed in part to the fact that optoelectronics and molecular electronics have become frontier areas of material science. In both areas there are limitations to inorganic materials and therefore, ordered organic materials become increasingly important to a variety of practical applications. Another major area where LB films have promising applications is the area of chemical and biological sensing and this is our focus based on 3 new effects studied in LB monolayers of fluorescently labelled phospholipids [2, 3, 4].

Some of the most precise chemical sensors nowadays utilize bulk acoustic wave (BAW) resonators and surface acoustic wave (SAW) devices as highly sensitive microbalances measuring extremely small quantities of chemical analytes attached to their active surface area. If that area is coated by a bioactive layer that can trap certain biological substances, a selective biosensor is obtained. In the case of a BAW resonator as a gravimetric sensor device the relationship between its frequency shift and the attached mass is given by the Sauerbrey equation. The sensor's frequency shift is proportional to the square of the frequency but this equation assumes that the attached mass is rigid and is able to follow the substrate deformations during oscillation. However, when measuring big organic molecules in liquids this is not the case, moreover, the selected resonant frequency of the BAW resonators rarely exceeds 5-10 MHz for optimum operation in liquids. These factors limit the usability of BAW sensors for biosensing applications. SAW devices typically operate in the 30 MHz to 1 GHz range and provide much higher sensitivity compared to their BAW counterparts but, with exception of the leaky SAW mode, their operation is more appropriate for gas phase environments.

In this study we tested Rayleigh surface acoustic wave (RSAW) two-port resonators (TPR) for their ability to tolerate LB films as possible selective biosensing layers. There are several reasons why we decided to use these devices for such applications:

1. Compared to leaky wave delay lines that are widely used as biosensors, RSAW TPR have much lower insertion loss and higher Q factor resulting in substantially lower noise and detection limits in the sensor system;
2. They occupy much less substrate area and cost less accordingly;
3. They are fabricated on thermally compensated quartz cuts and feature superior thermal stability – an issue of great practical importance in real-life sensor systems.

2. Materials and Methods

DipalmitoylPhosphatidylEthanolamine head labelled with NitroBenzoxaDiazole (DPPE-NBD) was purchased from Avanti Polar Lipids (USA) in chloroform solution and was with 99.5 % claimed purity. A 5th generation of a Langmuir-Blodgett system designed by Advanced Technologies Ltd. (Bulgaria) [5] was used for this experiment. The system has several world innovations but what is important for the experiment described here is that the noise of the surface pressure sensor is less than 0.03 mN/m and constant surface pressure control accuracy is in the ± 0.1 mN/m range. Thus a precise measure on the surface area change during the LB deposition is possible and hence an estimate of the deposited mass of the substance. LB film monolayers were deposited initially on microscope cover glass slides which were tested to be smooth enough for Atomic Force Microscopy (AFM) studies. Optical spectroscopy measurements were performed in the Institute of Electronics – BAS on Horiba Scientific Fluorolog-3 spectrophotometer. AFM measurements were performed in the Faculty and Chemistry and Pharmacy in Sofia University on a Bruker Nanoscope V instrument. The test RSAW device used here, operate at 440 MHz and use a gold electrode structure to avoid possible corrosion when coming in contact with various liquids. Design details and uncoated performance are described in [6] in detail.

3. Results and Discussion

The isotherm of the DPPE-NBD molecule just prior to deposition of the first layer on the SAW device is shown on figure 1. The liquid-expanded and liquid-condensed phase coexistence region starts at around 12 mN/m at a temperature of 25° C. The surface pressure of this region strongly depends on temperature. The Langmuir film isotherm from DPPE-NBD was studied in detail using additional surface potential measurements and Brewster angle microscopy [7]. In the phase coexistence region circular domains of the solid phase coexist with the liquid phase. With increasing surface pressure, the solid phase domains grow in diameter and eventually make a contact with each other. At this point the surface pressure starts to increase sharply.

LB monolayers were deposited on an up stroke on glass slides at a slow withdrawal speed of 0.02 mm/s and measured by fluorescence spectroscopy (figure 2) and AFM (figure 3). The deposition was carried out at the following conditions: sample 1 (NBD_PE_1) from pure water subphase and deposition surface pressure of 5 mN/m - in the liquid phase; sample 2 (NBD_PE_2) – pure water and pressure of 11.5 mN/m; sample 3 (NBD_PE_3) – pure water and pressure of 35 mN/m; sample 4 (NBD_PE_4) with a presence of CdCl in the water subphase with a concentration of 1 μg/l (5 times below the allowable concentration in the drinking water) and pressure of 8 mN/m; sample 5 (NBD_PE_5) – same subphase as the previous sample and pressure of 35 mN/m. The strong self-quenching of the fluorescence can be seen on figure 2 when the molecules come in close proximity to one another in the solid phase (sample 2). The fluorescence is restored at higher pressures because the monolayer contains very high 30 to 100 nm in height cylinders [4]. So large part of the molecules escaped in the vertical dimension and they seem to be in the liquid phase with strong fluorescence. It can be seen in the case of sample 4 that Cadmium ions dequench the self-quenched fluorescence which is another unexpected new phenomena for this molecules.

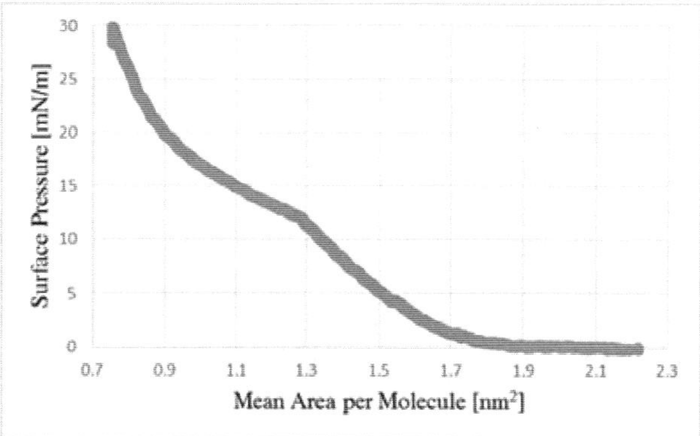

Figure 1. Surface pressure isotherm of a Langmuir film from DPPE-NBD at 25° C prior to the deposition on the SAW resonator. The chloroform was allowed to evaporate for 25 minutes before the compression was started. In order to avoid kinetic effects the compression velocity was very low and equal to 0.04 mm/s.

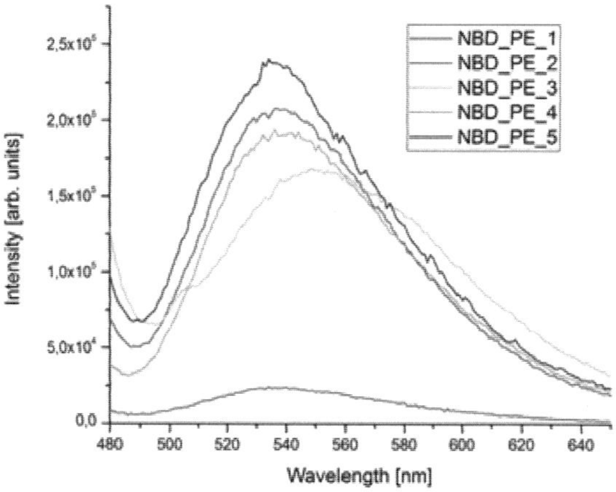

Figure 2. Fluorescence spectroscopy of LB monolayers from DPPE-NBD deposited at different surface pressures and samples 4 and 5 with Cd^{+2} ions in the water subphase.

The phase coexistence region of the solid and the liquid phase can be clearly seen on the AFM picture and cross-section of sample 2. Also, bilayer height cylinders can be clearly seen. This is another one of the new effects for this molecule that we observed. It should be noted that the Equilibrium Spreading Pressure for this molecule is at 19.6 mN/m [4] so this monolayer deposited at 11.5 mN/m should be in thermodynamic equilibrium. From the point of view of sensor applications this bilayer cylinders with diameters in the 50 to 200 nm range lead to a much better developed surface which can significantly increase the sensitivity of a future sensor. Further, our previous AFM experiments show that the bilayer cylinders remain stable for at least 50 days [4].

As a next step, we made subsequent depositions of LB layers on the surface of the RSAW resonant device. Surface pressure in the middle of the “horizontal” plateau of the liquid – solid phase coexistence region was used. The same monolayer substance was used for all subsequent depositions. The first monolayer was deposited on the upstroke. It was found to have strong adhesion to the device surface and could not be removed later on from the device even by using the glue technique. Subsequent layers were removed from the device with this technique. A very slow deposition speed of 0.02 mm/s was used with a 1 minute relaxation time before reversing the direction of substrate movement.

Results for the surface pressure as a function of time diagram and mean area per molecule as a function of time diagram are shown on figure 4 for the deposition of layer 2 and 3. Initially the Langmuir film monolayer is compressed to 1 mN/m to compensate for the monolayer relaxation during the measuring process of the 1st LB monolayer. Then, the constant surface pressure maintenance software is switched on and it can be seen that above 1000 seconds, the surface pressure is very stable. Also, the monolayer is very stable and there is almost no change in the area-per-molecule before the deposition starts around 1200 seconds. Then, one can see the change in area per molecule when deposition of the both – the 2nd and 3rd LB monolayers was performed. From here, one can calculate the amount of substance deposited on the entire substrate. The immersed in the water subphase plate on which the resonator was attached had a surface area of around 145 mm^2. Assuming an active resonator area of 0.1 mm^2 it can be calculated that around 1 ng of DPPE-NBD was deposited with these 2 layers. A more precise evaluation on the active resonator surface is needed in order to compare results from this calculation with results from the resonant frequency change.

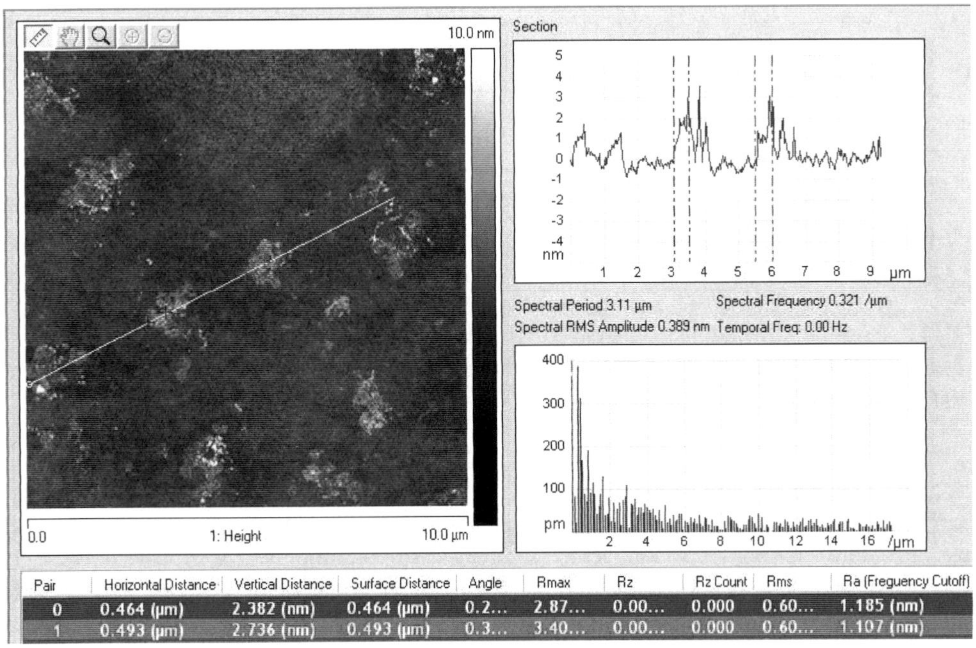

Figure 3. AFM of an LB monolayer from DPPE-NBD – sample 2 described in the text.

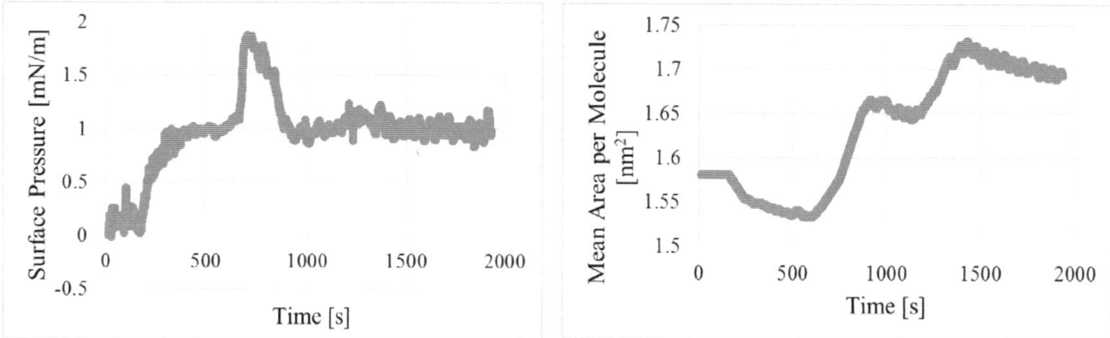

a) **b)**

Figure 4. The deposition of layers 2 and 3 from DPPE-NBD on the RSAW resonator: a) surface pressure as a function of time; b) mean are per molecule as a function of time.

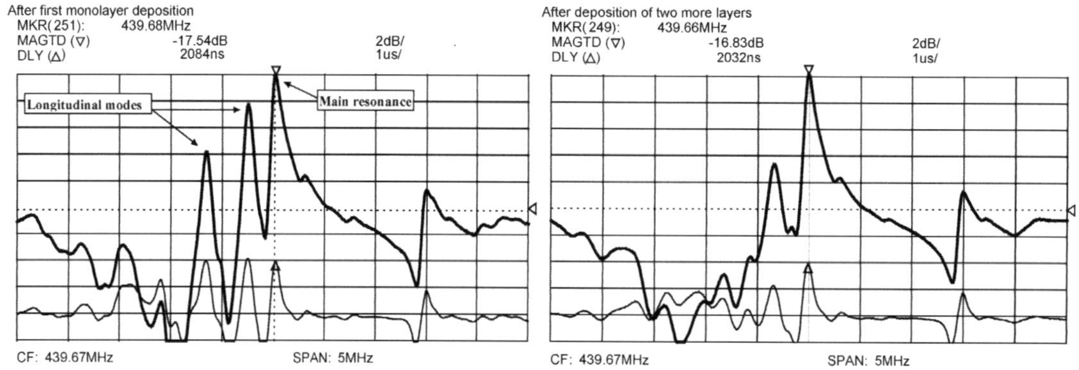

a) **b)**

Figure 5. Frequency (upper curves) and group delay (lower curves) responses of the two-port RSAW resonator with gold electrode structure after a): deposition of a single LB monolayer and b) deposition of two more monolayers on top of the first one.

Table 1. Electrical characteristics of the RSAW devices after the LB layer depositions.

Depositions	Resonant frequency f_o (MHz)	Insertion loss *IL* (dB)	LM suppression (dB)	Group delay τ_g (ns)	Loaded Q_L
First LB layer	439,68	17,54	2,2	2084	2880
Two more LB monolayers	439,66	16,83	6,3	2032	2807

In figure 5 we compare the electrical characteristics of one RSAW TPR after the first monolayer deposition (figure 5 a)) and then after two more layer depositions on top of the first one (figure 5 b)). The frequency and group delay data were measured with a vector network analyzer and the characteristics indicative of the resonant behavior after coating are summarized in table 1. The frequency downshift between the two depositions is 20 kHz which, according to the mass sensitivity of 20 kHz/ng of these devices, derived in [8], results in about 2 ng deposited mass. Keeping in mind that two LB layers were deposited in the case of figure 2 b), we can conclude that the absolute weight of the phospholipids monolayer used is about 1 ng. We also observe an improvement of the longitudinal mode (LM) suppression in the acoustic devices between figure 5 a) and b) meaning that the LB film is a highly nonlinear elastic material affecting the wave propagation properties. A slight improvement in insertion loss and a negligible degradation in loaded Q ($Q_L=\pi f_o \tau_g$) indicate that monolayers of phospholipid LB films are very well tolerated by RSAW TPR. This makes them suitable for biosensing applications.

4. Conclusion

LB films of DPPE-NBD were deposited on glass slides and studied with fluorescence spectroscopy and AFM. The present study confirms previously observed new effects of fluorescence self quenching in the solid phase and dequenching if there are Cadmium ions in the water. Also, bilayer height cylinders were observed. Successful deposition of LB layers on RSAW device was performed. For possible biosensor applications only one LB monolayer is needed. Present experiment shows that it is firmly attached to the surface of the SAW device, thus robust and stable over time sensors are to be expected. The change of frequency can easily be measured by connecting the SAW device to an oscillator loop and measuring its frequency shift with a frequency counter. In this study, we measured a 10 kHz per monolayer frequency down shift which is an excellent sensitivity for possible biosensor applications keeping in mind that such sensor oscillators provide a short term stability of a few parts in 10^{-10}/s. Also, an unexpected improvement in insertion loss was observed which means that the thin LB nanofilm techniques is very well tolerated by SAW resonant devices.

Acknowledgments

This work was in part supported from contract BN-204/17 with Centre for Research and Design from the University of Architecture, Civil Engineering and Geodesy.

References

[1] Petty M C 2007 *Molecular Electronics. From Principles to Practice* (New York: Wiley–Interscience)

[2] Ivanov G R 1992 First observation of fluorescence self-quenching in Langmuir films *Chem. Phys. Lett.* **193** 323

[3] Lukes P J, Ivanov G R, Petty M C, Yarwood J, Greenhall M H and Lvov Y 1994 Deposition and structural properties of Langmuir-Blodgett films from a fluorescently labeled phospholipid *Langmuir* **10** 1877

[4] Ivanov G R, Georgiev G and Lalchev Z 2011 "Fluorescently Labeled Phospholipids - New Class of Materials for Chemical Sensors for Environmental Monitoring ", in "*Relevant Perspectives in Global Environmental Change*", ed Julius Ibukun Agboola (London: InTech International) chapter 6 pp 89 - 112

[5] Ivanov G R and Polevska Z 2017 Computerized system for investigation of organic monolayrs at the air-water interface and for deposition on nano thin layers folowing the Langmuir and Blodgett method, Proc. 27th Int. Scientific Symposium Metrology and Metrology Assurance, (Bulgaria) 260

[6] Avramov I D, Voigt A and Rapp M 2005 Rayleigh SAW resonators using gold electrode structure for gas sensor applications in chemically reactive environments, *IEE Electronics Letters* **41** 450 - 52

[7] Ivanov G R and Polevska Z 2017 First observation of 3D aggregates in a single-component Langmuir film below the equilibrium spreading pressure *MATEC Web of Conferences* **9** 01003

[8] Avramov I D and Stahl U 2017 On the Mass Sensitivity of Rayleigh Surface Acoustic Wave (RSAW) Resonators *40th Int. Spring Seminar on Electronics Technology (ISSE)* ISSN 2161-2536, DOI 10.1109/ISSE2017.80001982

On Measuring the Characteristic Length of the Cooperative Molecular Dynamics in the Glass-Forming Liquids

Hristo Solunov

Department of Physics and Technology, University of Plovdiv,4000 Plovdiv, Bulgaria
E-mail: solunov@uni-plovdiv.bg

Abstract. An analysis of the methods for measuring the characteristic length of the cooperative molecular dynamics in glass-forming liquids is submitted. The equations and methods for measuring the size of the cooperatively rearranging regions in post Adam-Gibbs theories are compered, and it is proved that they give one and the same number, but with different interpretation for the relaxed molecular units. The interpretation that the rearranging units defined by Wunderlich are smallest structural units, named beads, is supported by the predicted "universal" value for the cooperatively rearranging range at the glass-transition temperature, as well as from a precise modeling for the relaxations by the generalized entropy theory. It has been found that the characteristic length estimated by random first order transition theory, as well as by the "four-point" dynamic susceptibility method, measured as the number of the bead's diameters coincide with the length of the cooperatively rearranging region in Adam-Gibbs theory extended at the molecular level.

1.Introduction

The structure relaxation time of supercooled liquids exhibits fast non-Arrhenius temperature dependence which is yet assumed as an unexplained feature of glass-formation [1-6]. Despite its unsolved origin, this temperature dependence is expected to be related to a cooperative effect. A theory of such cooperatively rearranging has been put forward by Adam and Gibbs (AG) [7]. AG defined a size of cooperatively rearranging region (CRR) as the number of molecules, rearranging into another configuration by surmounting simultaneously the individual potential barriers independent of its environment. AG assumed that a critical lower limited size of such regions, with two configurations, and entropy $s_c^* = kln2$, where k is Boltzmann's constant governs the relaxation properties of the liquids. By this assumption AG obtained a celebrated relation between the dynamics, presented by relaxation time τ(T) and thermodynamics, presented by configuration entropy $\Delta S = S_l - S_c$ where S_l is entropy of the liquid and the S_c is entropy of crystal or glass

$$\tau(T) = A_G exp\left(\frac{C}{T\Delta S}\right) \quad (1)$$

where T is the temperature and C and A_G are constants. This macroscopical equation has been supported by many experimental results, making AG theory one of the widely accepted and mentioned in all review articles [1-6,8-11,13-15].

AG have not written explicitly an equation for the size of the cooperatively rearranging region (CRR) and have not estimated the number of the molecules in such a range. Therefore, the molecular level of the theory has not been clarified and, due to this fact, several post AG theories [2-4,10,13-15] are devoted to developing the molecular level of the theory.

An equation for measuring the size of the CRR has been suggested [10,13,14] and another method for measuring this size has been applied [8,9]. The molecules have been assumed as bases dynamical unit in some studies [8,9,10,],while in other studies the beads .[2,3,13,14]. The beads in the simplest case coincide with the structural units; the chemical groups the ions and the atoms[16,17], but in the general they are different [18].
Several theoretical models explain the nature of the cooperative molecular dynamics in supercooled liquids by the term "characteristic length" [1,2,3,5,15].
The aim of present study is to compere the suggested methods for measuring the size of the CRR and the cooperative lengths from post AG theories.

2. Measuring the size of the cooperatively rearranging region

2.1 Measuring the size of the cooperatively region by molecular entropy theory

In an extension by the author [13,14], the molecular level of the AG theory, which will be cited with the abbreviation as Molecular Entropy (ME) theory for the most probable size of the CRR j_s, was obtained

$$j_s(T) = \frac{T}{T-T_0} \tag{2}$$

where T_0 is the temperature from the empirical Vogel-Fulcher Tamman (VFT) formula,

$$\tau(T) = Aexp\left(\frac{B}{T-T_0}\right) \tag{3}$$

where A and B and T_0 are empirical constant.

In ME theory two additional expressions equivalents to equation (2) have been obtained $j_s(T) = \Delta C_p(T)/\Delta S(T)$ in the case of a linear temperature dependence of the configuration heat capacity $\Delta C_p(T)$ [14] and in the case of a hyperbolic dependence $j_s(T) = T_g \Delta C_p(T_g)/\ T_K \Delta S(T)$ where T_g is glass-transition temperature and T_K is Kausmman temperature, which in AG theory is accepted equal to VFT temperature T_0 [19].

In analyses of the heat capacity jump at glass-transition temperature $\Delta C_p(T_g)$ Wunderlich [16] observed that if molecules were regarded as constructed by structure units named beads, then the value of $\Delta C_p(T_g)$=11.3 *J/molbeadK* appears as ''universal''. Wunderlich defined the smallest structural units as -CH_2-, -CH_3, -OH, or -O- as beads. Accepting the beads conception Bestul and, Chang [17] defined the ''universal'' rule for the configuration entropy $\Delta S(T_g) = 2.9$ J/*molbeadK*. Replacing both values in $j_s(T_g)$ in the case of the linear temperature dependence of $\Delta C_p(T)$ one obtains $j_s(T_g) = 3.9$ *beads* and in the hyperbolic dependence, accepting $T_g/T_k = 1.24$, as an average value from investigated liquids [7], for $j_s(T_g)$ one obtains 4.8 *beads*. Hence, the ''universal'' value for the CRR at T_g is obtained around 4-5 *beads*. The measured values [10,13,14] are around this ''universal'' value and supported the hypotheses that the beads are rearranging units [14].

2.2.Measuring the size of the cooperatively rearranging region by the maximal value of the configuration entropy.

Another method for measuring the size of the CRR has been suggested by Takahara et al [8], using AG definition for the entropy s_c^* of the critical low limited size of the CRR z*

$$s_c^* = \frac{\Delta S}{N_A} z^* \tag{4}$$

where N_A is Avogadro number. From equation (4) it follows that $N_A s_c^*$ is a constant since AG accepted $s_c^* = kln2$. Hence if this constant can be estimated by using the measured $\Delta S(T)$ the $z^*(T) = N_A s_c^*/\Delta S(T)$. Takahara et al observed that $\Delta S(T)$ as a function of T comes to a saturated, maximal value at temperatures far above the fusion temperature T_f and accepted the extrapolated value, when $T \rightarrow \infty$ to be $N_A s_c^*$. For 3-brompentane this constant was estimated between 80-100 $J/molK$, and the upper value is listed in Table 1. In a later study [9] it has been assumed that $\Delta S(T) = N_A s_c^* - K/T^2$, where K is a constant and by fitting the last 50 K of $\Delta S(T)$ the $N_A s_c^*$ was estimated. Within the uncertainty cited by them the upper values of $N_A s_c^*$ for o-terphenyl and 1-butene are also given in Table 1. In ME theory equation (4) is replaced by a rule of the individual constant number of configurations (RICNC) [18]

$$N_A s_{cj} = j_s(T)\Delta S(T) = nRlnw_{js} \quad (5)$$

where n is the number of the beads in the molecule R is the gas constant and w_{js} is an individual number of configurations in the most probable CRR of every liquid. The values of j_s, calculated from equation (2) at the glass-transition temperature T_g and at the fusion temperature T_f multiplied by corresponding configuration entropies [11] are listed also in Table 1.

It will be mentioned that Richert and Angell [11], comparing VTF equation with the macroscopic AG equation (1) found that the $\Delta S(T)$ will obey

$$\Delta S = S_\infty\left(1 - \frac{T_K}{T}\right). \quad (6)$$

As $T_K = T_0$ by using equations (2) and (6) one obtains

$$j_s(T)\Delta S(T) = S_\infty. \quad (7)$$

The values of the S_∞ [11] are also given in the table. Hence independent of the mining introduced in $constant = j_s(T)\Delta S(T)$ it is obvious that if this constant is estimated to be one and the same for given liquid the measured values of $j_s(T)$ will coincide.

Table 1. The maximum of configuration entropy estimated in the References listed in the table and the rule of individual constant number of configurations

Liquids	S_∞^a (J/molK)	$j_s(T_g)\Delta S(T_g)$ (J/molK)	$j_s(T_f)\Delta S(T_f)$ (J/molK)	$N_A s_c^{*b}$ (J/molK)
o-terphenyl	137	137.2	137.3	131
Salol	138 4	138.4	138.4	---
2MTHF[c]	97.5	97.4	97.6	---
n-propanol	72.2	70.0	70.4	---
3-brompentane	99.0	100.8	100.8	100
1-butene	---	68.1	96.5	70

[a] S_∞ from Ref. [11],
[b] $N_A s_c^*$ from References [8,9]
[c] 2-Methylthetrahydrophuran

As it can be seen from the table the values of the constants are in good agreement. Hence, measured $j_s(T)$ by equation (2) and the ones estimated by Takahara's method are practically equivalent, if the constant K is correctly estimated. However, the interpretation is different as Takahara et al following AG equations interpreted $j_s(T)$ as the number of the molecules, and in Ref. [9] the s_c^* has been also interpreted per molecule. In AG s_c^* is the configurational entropy of the CRR with smallest number of configurations equal to two, while in the [8,9] it should be the configurational entropy of the CRR with the largest number of configurations. Both statements can coincide only in the case when the minimal and the maximal values are one and the same or in the case of validity of the $w_{js} = constant$, as it is assumed in ME theory [13].

In Generalized Entropy Theory (GET) a similar method for the estimation of the size of the CRR as the one of Takahara et al is used [4]. In GET a configuration entropy density is defined as $s_c = \Delta S/N_l$,where ΔS is the configuration entropy of the system with N_l number of lattice sites. The size of the CRR is measured by $z^* = s_c^*/s_c(T)$, where s_c^* is the high-temperature limit of $s_c(T)$. In modeling the properties of polymers by GET [4] the estimated size of the CRR at T_g is 4.46 for a polymer with flexible backbone and flexible side groups and 5.53 for the ones with flexible backbone and rigid side branches. These values are in the range of the ones measured by equation (2) [14]. The size of the CRR in this theory [4] should be measured in the ''united atom groups'', usually known as functional chemical groups, which in Ref. [4] coincide with the empirical beads [16,17]. Hence, the model of liquids and the values and the interpretation of z^* in GET are practically the same as in ME theory[13,14].

2.3. Confirmation of the formula for the measuring the size of the cooperatively rearranging region by the measuring the activation energies.

The molecular equation for the most probable relaxation time in ME theory is

$$\tau(T) = \tau_0 exp\left(\frac{j_s(T)u}{kT}\right) \tag{8}$$

where τ_0 is the vibration time, k is Boltzmann constant and u is the average value from the individual potential barriers of the rearranging beads during the relaxation time. The equation (10) is an extension of the original AG molecular equation since the critical small size of the CRR, z^* is replaced by $j_s(T)$ and the meaning is changed. Two configurations, accepted by AG, agree with statement that the relaxation units overcome their individual potential barriers essentially simultaneously. In this way, the relaxation time has the meaning of a waiting time a fluctuation of enthalpy to rearrange in bloc the molecules included in the CRR. In ME theory meaning of $\tau(T)$ is the living time of the CRR [14]. In ME theory comparing equation (8) with VTF equation (3) for the $j_s(T)$, the equation (2) has been obtained and

$$u = kB. \tag{9}$$

As the equations for $j_s(T)$ and u are obtained simultaneously, if the estimated value of u coincides with the experimentally measured one then an additional support for the correctness of equation (2) will be obtained. The estimated values for u are in the range of the measured values for the rotations around $R_1C\text{-}CR_2$ bonds, where R_1 and R_2 are different chemical groups [20]. An additional evidence for the correctness of the measured values of the u by ME theory are estimated by GET, where the proper values for the activation energy in polymers were obtained to be $\Delta\mu/k_B = 2000 - 2600\ K$ (16.6-21.6) *kJ/mol* [4]. The values in this range have been measured for polymers by equation (9) [14].

The correctness of the equation (2) indirectly seems to be confirmed from the expression for the free activation energy of RFOT theory [2], given by

$$F^{++} = k_B T_k D\left(\frac{T}{T-T_K}\right) \tag{10}$$

where $k_B = k$ and D is the constant in VFT formula presented as $\tau(T) = \tau_0 exp\, DT_K/(T - T_K)$.
Having in mind $DT_K = B = u/k$ the equation (10) became

$$F^{++} = u\left(\frac{T}{T-T_K}\right) \tag{11}$$

IOP Conf. Series: Journal of Physics: Conf. Series **1186** (2019) 012008 doi:10.1088/1742-6596/1186/1/012008

and by the equation (2) we obtain

$$F^{++} = uj_s(T) \tag{12}$$

or accepting $j_s(T)$ given by equation (2) the activation energy of RFOT theory appears to be equal to that of ME theory and thus $\tau(T)$ will be the same.

The separate measuring of $j_s(T)$ and u in ME theory makes it possible the activation energies per bead u, the cooperative activation energy $j_s(T)u$ and the apparent activation energy $j_s^2(T)u$ to be correctly interpreted. Then it follows that the glass transition temperatures T_g and the crossover temperatures of Mode Coupling theory T_c, to be identified at the energetic scale [19,20,21]. In analogy to the energies, the relaxation volumes have been interpreted and for first time the relaxation volume of a bead has been measured [21]. The correct interpretation of $j_s(T)$ gives the possibility of measuring the number of configurations in the CRR [18] and the processes at T_g, T_c as well as the fusion temperatures to be guessed in the entropic mode [22]. The methods for measuring the number of the beads in the molecules also have been found [22].

2.4 The size of the cooperatively rearranging region and the characteristic length of the cooperative molecular dynamics.

In several theories the size of the CRR is not measured but a "characteristic length" is measured [1-3, 12,15] which is an indirect measure of this size. In the case of AG theory, the length of the size of CRR form a cooperative or characteristic length. The "universal" cooperative length in ME theory in this term will be 4-5 bead diameters and this predication is supported by the measured values in the range 3-9 bead diameters [20].

In RFOT theory the characteristic length at the T_g have been estimated $\xi/r_0 = 4.5$ where ξ is the length of the critical configuration droplet size and r_o is the mean lattice space [2]. In this theory a hard sphere model of the liquids is used, and as the beads are regarded as relaxing units, thus the interpretation of r_o should be the diameter of a bead. So, in this estimation of RFOT theory the characteristic length corresponds to the length of "universal" number of beads in the size of the CRR of ME theory.

In the "four-point" susceptibility method a "dynamic heterogeneity length" [5,12] while in Donts' method a "cooperativity length" [15] are measured. Both these methods are based on the fluctuation-dissipation theorem. In both methods the dynamic heterogeneity is related to the distribution of the relaxation times. The "dynamic heterogeneity length" measured by the "four-point" susceptibility method is in the order of 4.6-5.2 bead diameters [12] and it is in the range of that, measured by ME theory [14,20]. These methods are not directly related with the AG theory, but by using the configurational heath capacity in the interpretation of the experimental results [12, 23, 24] a common point with AG theory is observed. In both these methods through the fluctuation dissipation theorem a "correlation volume" is introduced. The number of molecules $N_{corr,4}$ [24, 26] or N_α [25] measured in the "correlation volume" in the order of 75 and more molecules is debatable as such "correlation volumes" are rejected by experiments [23]. Both Dont's method and the "four-point" susceptibility method measured only one parameter a "characteristic length" and not some energy of activation. A lack of correlation between fragility m and $N_{corr,4}$ has been found by experiments [24],while a close correlation between fragility and the size of the CRRS is observed [14,20].

3. Conclusions

The mathematical expression for measuring and interpretation of the size of cooperatively rearranging region presented only in the Molecular Entropy theory (extended at molecular level Adam-Gibbs theory), is supported by the subsequent methods for the measuring or the estimation this region as well as by measuring characteristic length by post-Adam-Gibbs theories. Indirectly this formula is supported by measuring the activation energies of the relaxation processes and the suggested kinetic and thermodynamic methods for measuring the "fragility" of the glass-forming liquids. The large number of physical properties of the glass-forming liquids explained on energetical, volumetric and entopic scales on the basis of this formula seem to push the most probable size of the cooperatively rearranging region at a favorable position for long-term searching of a cooperative scale governing the glass-formation.

References

[1] Berthier L, Biroli G, Bouchaud P, Cipellety L and van Saarloos W 2011 *Dynamical Heterogeneities in Glasses, Colloids and Granular Materials* (Oxford University Press)
[2] Xia X and Wolynes P G 2000 *PNAS* **97** 2990
[3] Lubchenko V and Wolynes P G 2007 *Ann. Rev. Phys. Chem.* **58** 235
[4] Dudowich J, Freed K F and Douglas J F 2008 *Adv. Chem. Phys.* **137** 125
[5] Berthier L, and Biroli G 2011 *Rev. Mod. Phys.* **83** 587
[6] Zanotto E D and Mauro J C 2017 *J. Non-Crystal. Solids* **471** 490
[7] Adam G and Gibbs J H 1965 *J. Chem. Phys.* **43** 139
[8] Takahara S, Yamammuro O and Matsuo T 1955 *J. Phys. Chem.* **99** 9589A
[9] Yamammuro O et al 1998 *J. Phys. Chem.* B **102** 1605
[10] Johari G P 2000 *J. Chem. Phys.* **112** 8958
[11] Richert R and Angell C A 1998 *J. Chem. Phys.* **108** 9016
[12] Berthier et al 2005 *Science* **310** 1797
[13] Solunov Ch(H) A 1995 *Balkan Phys. Lett.* **3** 26
[14] Solunov Ch(H) A 1999 *Eur. Polymer J.* **35** 1543
[15] Huth H, Beiner M and Donth E 2000 *Phys. Rev.* B **61** 15092
[16] Wunderlich B 1960 *J. Phys. Chem.* **64** 1052
[17] Bestul A B and Chang S S 1964 *J. Chem. Phys.* **40** 2731
[18] Solunov H 2017 *Comp. rend. Acad. bulg. Sci.* **70** 935
[19] Solunov Ch(H) A 2004 *Nonlinear dielectric Phenomena in Complex Liquids*, (*NATO Science series. II Mathematics, Physics and Chemistry).* Eds Rzoska S J & Zhelezny V P (Kluwer Academic Publisher, Boston, 157 275)
[20] Solunov H 2006 *J. Non-Cristal. Solids* **352** 4871
[21] Solunov Ch. (H.) A 2002 *J. Phys.: Condense. Matter* **14** 7297
[22] Solunov H 2017 *J. Phys.; Conf. Ser.* **794** 012030
[23] Casalini R, Zhu L, Baer E and Roland C M 2016 *Polymer* **88** 133
[24] Roland C M, Santangelo P G, Robertson C G and Ngai K L 2003 *J. Chem. Phys.* **118** 10351
[25] Saiter A, et al 2013 *Phys. Rev.* E **88** 042605
[26] Capaccioli S, Ruocco G and Zamponi F 2008 *J. Phys. Chem.* B **112** 10653

Dose-dependent genotoxicity of ammonia-modified graphene oxide particles in lung cancer cells

K Hristova-Panusheva[1], M Keremidarska-Markova[1], T Andreeva[1], G Speranza[2], D Wang[3], M Georgieva[4], G Miloshev[4] and N Krasteva[1]

[1]Institute of Biophysics and Biomedical Engineering, Bulgarian Academy of Sciences, Acad. Georgi Bonchev Str., Bl. 21, Sofia 1113, Bulgaria
[2]University of Trento, Via alla Cascata, 56/C, 38123 Povo, Trento, Italy
[3]Medical School in Southeast University, 87 Dingjiaqiao road Gulou District Nanjing 210009, China
[4]Institute of Molecular Biology "Acad. R. Tsanev", Bulgarian Academy of Sciences, Acad. Georgi Bonchev Str., Bl. 21, Sofia 1113, Bulgaria
e-mail: natalia.krasteva@yahoo.com

Abstract. Graphene oxide (GO), the water soluble form of 2D graphene, has received much attention because of its attractive properties for a wide range of applications and products. Surface modification with different functional groups can improve GO biocompatibility for further biomedical applications. In the present study we have evaluated genotoxicity of pristine and ammonia-modified graphene oxide (GO-NH_2) nanoparticles (NPs) in a human lung epithelial cell line, A549, exposed for 24 h to different concentrations of NPs (0.1, 1, 10, 20 and 50 μg/ml). Quantification of reactive oxygen species (ROS) indicated that exposure to higher concentrations of both types of NPs resulted in enhanced ROS generation. The observed comet tail migration in the method of Single Cell Gel Electrophoresis in the cells treated with 20 and 50 μg/ml GO and GO-NH_2 indicated presence of damages in DNA. Cell cycle analysis showed that after treatment of A549 cells with increasing concentrations of NPs for 24h the percentage of cells in G0/G1 phase of the cell cycle decreased while the percentage of cells in G2/M increased. The presented results suggest that ammonia-modified GO NPs applied at concentrations higher than 20 μg/ ml induced stronger toxicity effect in A549 cells compared to pristine GO and that the use of low concentrations of GO and GO-NH_2 NPs is important to avoide adverse biological effects.

1. Introduction

Graphene-based nanomaterials possess remarkable physiochemical properties suitable for diverse applications in electronics, telecommunications, energy and healthcare [1-3]. The human and environmental exposure to nanomaterials is increasing due to progress in the synthesis, characterization and large-scale production of graphene as well as the subsequent development of graphene-based biomedical and consumer products [1, 4, 5]. Recently, there is an increasing awareness of nanoparticles pollution as an environmental problem [6]. A large number of *in vitro* and *in vivo* toxicological studies have evaluated the interactions of graphene-based nanomaterials with various living systems such as microbes, mammalian cells and animal models [1, 7]. A significant number of studies have examined the short- and long-term toxicity of nanomaterials at the cellular level.

Graphene oxide (GO) is one of the most important graphene derivatives known for its exceptional electrical, chemical and mechanical properties [8]. Due to its attractive physicochemical properties for biomedical applications toxicity of GO is subject of multiple *in vitro* and *in vivo* research [9, 10]. Some papers have demonstrated that GO has low cell toxicity but results achieved so far are inconsistent. [11, 12]. Ideally, therapeutic nanoparticles should be nontoxic and invisible to the immune system. Hence, much efforts have been devoted to improve biocompatibility of GO-based nanomaterials. It is proposed that surface modification could be helpful to decrease the toxicity of GO as a component of nanomedicine because GO contains many oxygen containing groups (hydroxyl, carbonyl groups, carboxylic and epoxy group) making it easy for functionalization [13].

In this study we compare the toxicity of pristine GO and ammonia-modified GO at molecular level in a human alveolar cell line, A549 after exposure to the increasing concentrations of NPs. We used lung cells as a cell model for our toxicological evaluation because many biodistribution studies have shown that GO accumulates predominantly in the lungs [14, 15]. The aim was to understand if ammonia modification of GO would improve its toxicity and if the concentration of NPs in the range of 0.1-50 μg/ml is toxic in a dose-dependent manner. Only few studies report the toxicity response of pristine and functionalized graphene at molecular level. This sort of information is still largely lacking in the literature, though the health effects of NPs have attracted considerable attention.

2. Materials and Methods

2.1. Characterization of GO and GO-NH2 particles

The GO samples were characterized by X-ray photoelectron spectroscopy (XPS, Axis DLD Ultra, Kratos, Manchester, UK), Atomic Force Microscopy (AFM, Bruker Inc.) and Nanosizer (Zetatrac instrument, S3500; Microtrac, Largo, FL) in order to determine chemical composition, particle size and thickness, particle size distribution and zeta potential. For cellular experiments stock solution of 1.0 mg/ml of GO (Graphenea, Spain) and GO-NH_2 (Sigma-Aldrich, Germany) particles in distilated water was prepared. Just prior to the cell exposure the stock solutions were sonicated for 1 h then NPs were added to the cells in different concentrations.

2.2. Cell culture

A549 cells were purchased from American Tissue Culture Collection (ATCC) and maintained in DMEM medium supplemented with 10% foetal calf serum and antibiotic/antimicotic solution at 37°C in a humidified atmosphere of 5% CO_2. For genotocixity evaluation experiments, the confluent cells were harvested and seeded in a concentration of $1x10^5$ cells/well in 6-well plates then cells were incubated 24h before to be exposed to the increasing concentrations of NPs.

2.3. Cell morphology

Overall cell morphology was observed under phase contrast microscope (Axiovert 25, Carl Zeiss, Germany) and phase contrast pictures were done at the 24 h after exposure of the cells to the NPs using a digital camera.

2.4. Neutral Comet Assay (nCA)

Briefly, $1x10^3$ cells were mixed with 0.7% (f.c.) of low-gelling agarose and were layered as microgels on microscopic slides. Slides were then lysed in 146 mM NaCl, 30 mM EDTA, pH7, 10 mM Tris-HCl, pH7 and 0.1% N-lauroylsarcosine at 10 °C for 20 min and we electrophoresed for 20 min at 0.46 V/cm. Results were visualized under a fluorescent microscope after staining of gels with SYBR green. Results are quantified by Comet Assay specialized software CometScore.

2.5. Cell cycle analysis

FACS was performed on A549 cells fixed with 76% of cold ethanol immediately after incubation and left at -20 °C for a night. Cells were pelleted by centrifugation, washed in PBS buffer and treated with 100 μg/ml RNAse A for 30 min at 37 °C followed by staining with 50 μg/ml of propidium iodide for 30 min in the dark. 50 000 cells were counted through flow cytometry, detecting red fluorescence at excitation wavelength of 488 nm. The light scattering was detected as well. The results were quantified by FlowJo V10.

2.6. Determination of reactive oxygen spices (ROS)
Intracellular ROS was measured based on the intracellular peroxide-dependent oxidation of 2,7-dichlorofluorescein diacetate (DCFA-DA, Sigma-Aldrich) to form the fluorescent compound 2,7-dichlorofluorescein. The cells were seeded onto 24-well plates at a density of 3 x10^4 cells per well and cultured for 24h. After washing twice with phosphate-buffered saline (PBS) fresh media containing different concentration of GO and GO-NH_2 particles were added and incubated for another 24h. Then the cells were washed intensively with PBS and fresh serum-free medium containing 20 μM DCFA-DA was added to each well. After 30 min incubation at 37°, the samples were rinsed with PBS and 1 ml of PBS was added to the wells. Then the cells were scraped and the fluorescence intensity was determined using a spectrofluorometer with excitation at 485 nm and emission at 530 nm.

3. Results

3.1. Physico-chemical characterization of GO and GO-NH_2 NPs
As can be seen from figure 1 pristine and ammonia-modified GO differ in their size and chemical composition. XPS analysis (figure 1 C) revealed the same major characteristic peaks of C1s spectra, including C–C (284.7 eV), C–O (286.2 eV), C=O (287.8 eV) and O–C=O (289.1 eV) in both GO NPs but in GO-NH_2 the peak intensities were much weaker than those in GO, pointing a reduction in the oxygen content as a result of ammonia modification. The existence of nitrogen element in GO-NH_2 was confirmed by the appearance of N1s peak. All this indicates both, the de-oxygenation of GO and the incorporation of nitrogen in the ammonia-modified GO. AFM images of sonicated GO-NH_2 particles showed that GO-NH_2 flakes had an irregular shape and lateral dimensions of few microns (figure 1 A). The thickness of the single GO layer is around 0.9 nm and most of GO sheets exist as single or few layers according to the datasheet. DLS measurements demonstrated that the size of GO sheets was larger than that of GO-NH_2 sheets. The pristine GO particles consist of two fractions with quite different sizes: 250 ± 68 nm (9.7% of particles) and 1.5 ± 0.7 μm while the average particle size of GO-NH_2 NPs was 560 ±300 nm (figure 1 B). Zeta potentials of GO and GO-NH_2 particles measured in aqueous solutions showed negative values for GO (−24,5 ± 0.4 mV) and positive for GO-NH_2 NPs (38.5± 2.8 mV) (data not shown).

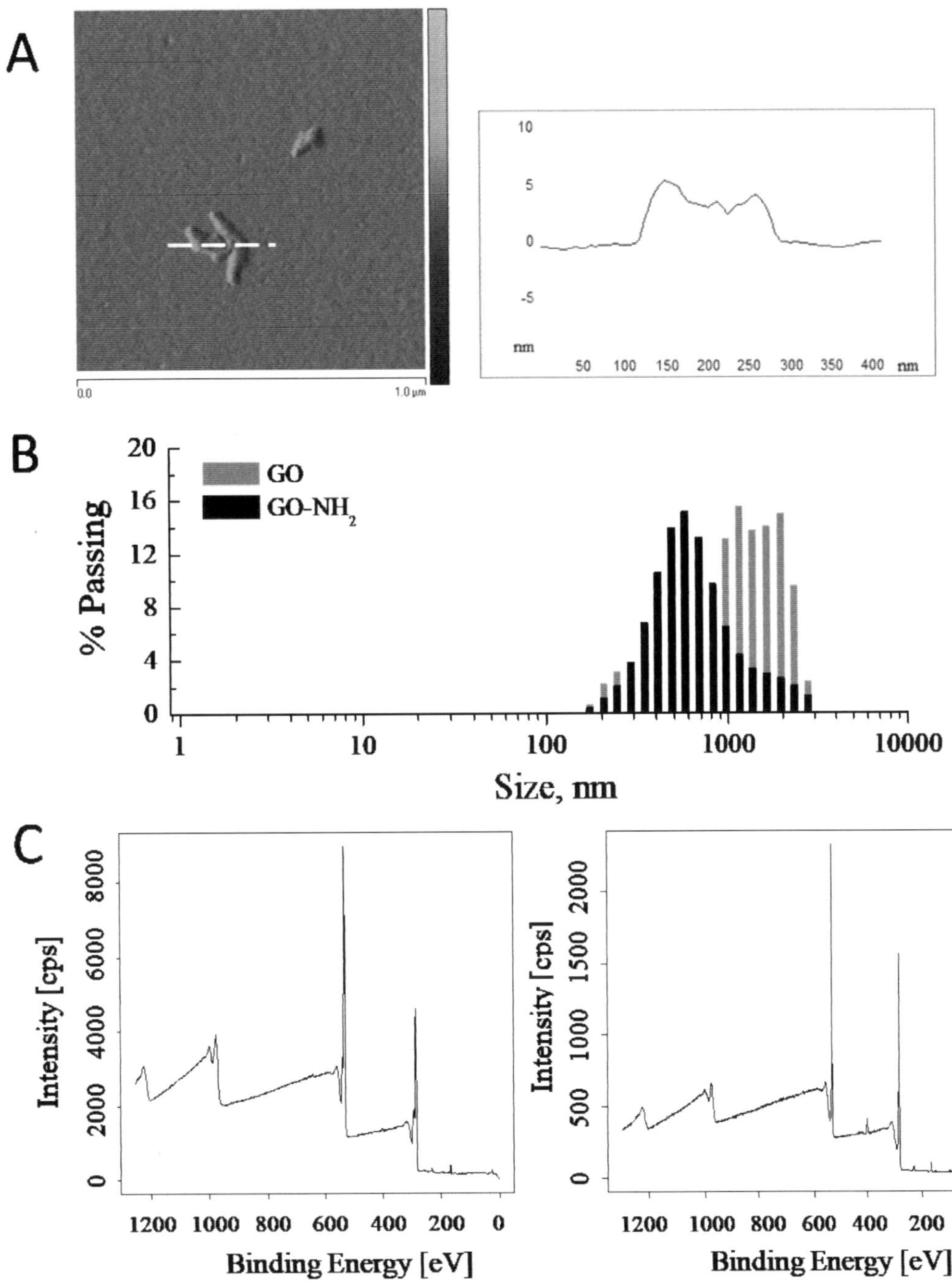

Figure 1. Physiochemical properties of pristine GO and ammonia-modified GO (A) AFM analysis of ammonia-modified GO after sonication (B) Size distribution of GO and ammonia-modified GO after sonication (C) XPS spectrum of GO and ammonia-modified GO.

3.2. Cell morphology

The morphology is one important indicator of the status of cells. The cell morphological changes after GO exposure were recorded to demonstrate the effect of both types of GO NPs on A549 cells directly (figure 2). There was no obvious difference between morphology of cells treated with GO and GO-

NH_2 and the control cells. Most cells were tightly adhered to the substrate and had normal spindle shape. However, in the cells exposed to GO-NH_2 NPs the number of the attached cells seems to be smaller (Fig.2, lower panel). Also, at the highest concentrations of GO (20 and 50 µg/ml) and GO-NH_2 treated cells (50 µg/ml) less attached cells were observed suggesting a detachment of cells as a result of NPs exposure.

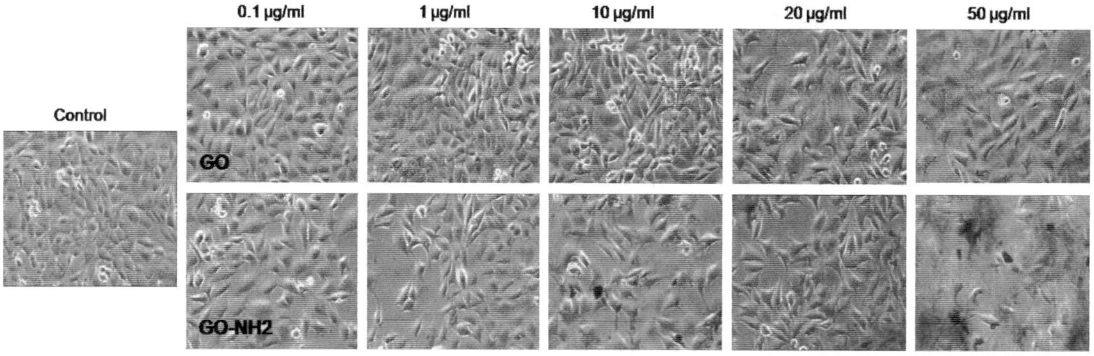

Figure 2. Phase-contrast micrographs of A549 cells, incubated for 24 hours with pristine GO particles (upper panel) and ammonia-modified GO particles (lower panel); magnification 20x

3.3. ROS levels

Measuring the release of ROS produced by cells in response to particle treatment is a well-established method to evaluate toxicity. As shown in figure 3 both, GO and GO-NH_2 NPs applied at concentration of 20 and 50 µg/ml enhanced the formation of intracellular ROS in A549 cells. The mean fluorescence intensity of cells exposed to 50 µg/ml GO-NH_2 was 3.0 times higher than control and for cells exposed to pristine GO was 1.8 times higher pointing that ammonia-modified GO causes more serious oxidative stress at the highest concentrations. Exposure to lower doses of GO-NH_2 particles, up to10 µg/ml did not induce generation of ROS. In cells treated with lower doses of GO (0.1-10 µg/ml) an increase in ROS production was observed but the values were not statistically significant ($p>0.05$).

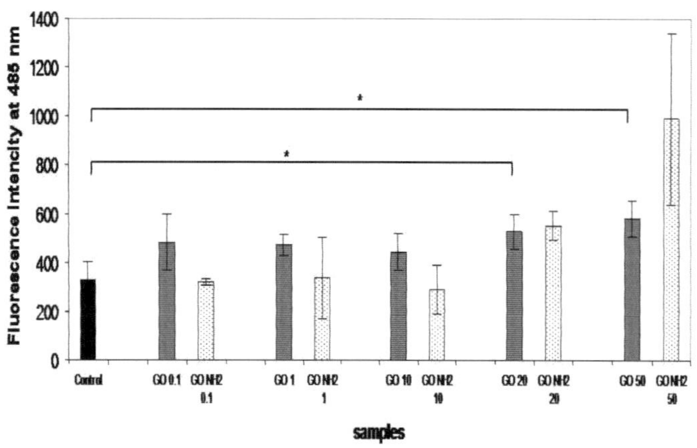

Figure 3. ROS generation in A549 cells induced by GO and GO-NH_2 particles with different concentrations

3.4. Genotoxicity evaluation

Genotoxic potential of NPs was assessed by measuring DNA double strand DNA breaks first using the Single Cell Gel Electrophoresis, also called Comet assay at neutral pH. The results from the the neutral CA (nCA) demonstrated that cells treated for 24 hours with different concentrations of GO

NPs showed presence of comets resembling cells undergoing apoptosis (figure 4). This means that GO and GO-NH_2 particles induced DNA damages and prove to have a genotoxic potential. The genotoxic potential of the NPs however did not increase proportionally to the concentrations. Although, the greatest genotoxic potential have been demonstrated in the cells treated with the highest particles concentrations of 50 µg/ml. Compared to the control, GO and GO-NH_2 particles demonstrated a genotoxic effect on A549 cells as among both types of GO particles ammonia-modified GO were more genotoxic.

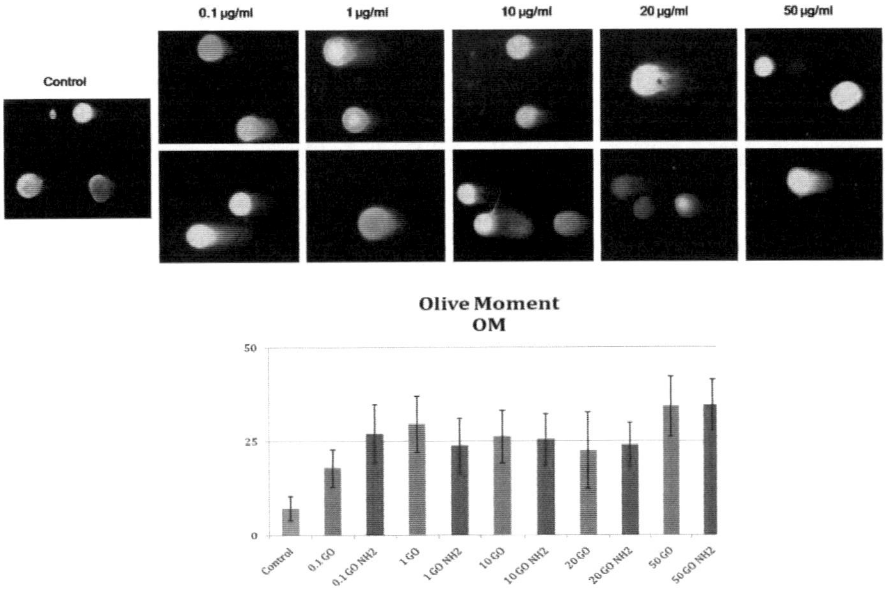

Figure 4. Neutral Comet Assay of A549 cells, treated for 24 hours with GO nanoparticles with different concentrations. A) Comet images of the A549 cells stained with SYBR green and observed under fluorescent microscope; upper panel – cells treated with pristine GO; lower panel – cells treated with ammonia-modified GO particles; magnification 250x B) Comet Assay data quantitation by specialized CA software CometScore – a graph representing the parameter Olive Moment (OM). This parameter essentially represents the product of the percentage of total DNA in the tail and the distance between the centres of the mass of head and tail regions [Olive moment = (tail mean-head mean) x % of DNA in the tail].

Further, we investigated whether GO particles can alter the cell cycle of human lung cells. FACS analysis of A549 cells exposed to GO NPs for 24h and the followed data analysis by FlowJo software showed that after treatment the percentage of viable cells decreased compared to the control cells in both, GO and GO-NH_2 NPs (figure 5A). The distribution of cells throughout different phases of the cell cycles showed an interesting correlation. At lower concentrations of the NPs there was a small percentage of cells blocked in S and G2/M phases of the cell cycle when treated with GO-NH_2 (compare GO 0.1 vs 0.1 µg/ml GO-NH_2 and GO 1 vs 1 µg/ml GO-NH_2). At higher GO concentrations 10-50 µg/ml this correlation was inverse with the GO-NH_2 (Figure 4B). At these high concentrations of GO and GO-NH_2 the percentage of cells in the whole population was smaller than in the control pointing to a cytotoxic and genotoxic effect.

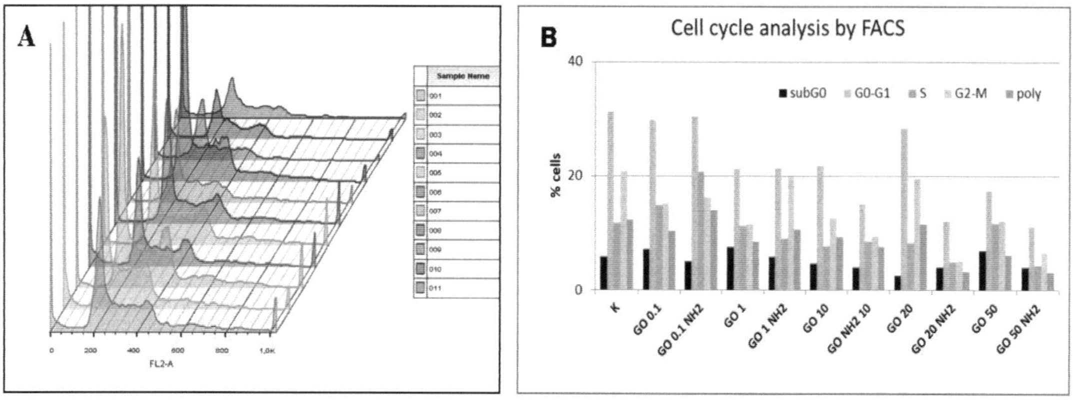

Figure 5. FACS Cell cycle analysis of A549 cells treated with GO nanoparticles for 24 hours. A) Distribution of cells according to FACS data quantitation, assessed by FlowJo V10. B) Histograms of FACS analysis of each treated probe including the control cells (grey histogram).

4. Discusion

The aim of this study was to determine if ammonia-modified GO and pristine GO NPs are toxic at molecular level and if their effect on A549 cells is dose-dependent. The results collectively indicate that aminated GO exert slightly stronger genotoxic effect compared to pristine GO but only at higher concentrations (20 and 50μg/ml). When designing our experiments we supposed that ammonia modified GO would have an improved toxicity profile compared to pristine GO because NH_2 groups are known for their positive effect on cell adhesion, growth and functions of normal cells when have been used to modify biomaterials for cell culture substratum [16]. Besides, ammonia modification reduces oxygen-containing groups on the surface of the GO particles which groups are considered as a factor increasing nanomaterial toxicity. However, the reports on toxicity effect of reduced or modified with different functional groups GO are very controversial [17, 18, 12]. To date only limited studies have focused on the toxicity of aminated GO in cancer cells while the cytotoxicity and especially genotoxicity of aminated GO in normal cells still has not been investigated. Ming Xu et al. for example, have demonstrated a considerably reduced recognition of GO-NH_2 by macrophage sarcoma cells [13].

Based on physico-chemical characterization and data sheet of pristine GO we found that both particles, pristine and ammonia-modified differed not only in nitrogen content but also in size and zeta potential. GO are larger in size and negatively charged unlike GO-NH_2 particles which are smaller and positively charged. It is well known that particles size and surface charge are important factors for the toxicity effect of NPs because both regulate the uptake of nanomaterials from the cells. Positive zeta potential of NPs has been shown to increase the internalization of polymeric nanoparticles in the cells [19] suggesting that GO-NH_2 have been taken up more efficiently by cells than negatively charged pristine GO NPs. However, internalization of both NPs, GO and GO-NH_2 can be hampered from their size (560 nm for GO-NH_2 and 1,5μm for GO) probably due to the size limit of endosomal vesicles that restricts the size of NPs to be endocytosed. Moreover, GO and especially GO-NH_2 aggregate when being added into the culture medium and the aggregation may take the consequences too. Thus the size, charge and the aggregation of GO NPs could inhibit their endocytosis. The GO and GO-NH_2 NPs only partially can be taken up and induces toxicity in A549 cells. Considering that both GO NPs hardly enter into the A549 cells, GO more possibly interact with the cells on the cellular surface or via other pathway indirectly.

One possibility for indirect interaction is that GO and GO-NH_2 absorb the nutrients in culture medium and then the depletion of nutrients induces the oxidative stress and toxicity to A549 cells. Such toxicity mechanism has been reported in a study of CNTs [20]. Guo et al. reported that the depletion of nutrients by the absorption of amino acid and other biological molecules onto CNTs led to severe toxicity to HepG2 cells. We found however increased ROS levels only in cells treated with the highest concentrations of GO and GO-NH_2 particles suggesting that this interaction is less likely. Moreover, Ming Xu et al. have measured the amount of proteins from the culture medium adsorbed on pristine GO and GO-NH_2 and have found that the pristine GO adsorbed higher amount of proteins than GO-NH_2 suggesting a higher cytotoxicity for GO which is inconsistent with our results.

Another possibility is that GO influences the cell adhesion ability of A549 cells. This mechanism is most probable because we found a decreased attachment of A549 cells exposed to GO-NH_2 particles as well as at the highest concentrations of both tested NPs. Although GO hardly enters A549 cells we found that the apoptosis of GO- and GO-NH_2 -treated A549 cells is much greater than that of the control cells. GO and GO-NH_2 induced statistically significant ROS generation only at high concentration. Oxidative stress is a well recognized toxicological mechanism of various nanoparticles [21,12]. At high dose, GO induces ROS generation, and this may contribute to the viability decrease of both GO particles at high concentration. Also, GO and GO-NH_2 induced DNA strand breaks at any of the studied doses but more pronounced at the highest doses. To make such a mechanism clear, more efforts are required.

5. Conclusions

In summary, the present study demonstrates that both tested NPs, pristine GO and GO-NH_2 exert a genotoxic effect at highest concentrations of 20 and 50 μg/ml pointing a dose-dependent genotoxicity effect in lung cancer A549 cells. Among both GO NPs ammonia-modifed GO exert a more adverse effect on cell attachment and induced more intracellular ROS generation.

Acknowledgements

This work was supported by the National Research Foundation, bilateral project Bulgaria-China [Grant number DNTS/ 01/6/2016].

References

[1] Lalwani G, D'Agati M, Khan A M and Sitharaman B 2016, *Adv Drug Deliv Rev.* **105**, p.109
[2] Skoda M, Dudek I, Jarosz A, and Szukiewicz D 2014 *J Nanomater* p.1
[3] Ferrari A C, Bonaccorso F, Fal'ko V, Novoselov K N, Roche S et al. *Nanoscale*, **7**, p.4598
[4] Marcano D, Kosynkin D, Berlin J, Sinitskii A, Sun Z, Slesarev A, Alemany L, Lu W and Tour J 2010 *ACS Nano* **4,** p.4806
[5] Jasim D, Lozano N and Kostarelos K 2016 *2D Mater.* **3**, 014006
[6] Fojtů M, Teo W Z and Pumera M 2017 *Environ Sci-Nano* **8**, p.1617
[7] Ou L, Song B, Liang H, Liu J, Feng X, Deng B, Sun T and Shao L 2016 *PartFibre Toxicol* **13** p.57
[8] Wu S, An S and Hulme J 2015 *Int J Nanomed* **10** p.9
[9] Wan C and Chen B 2011 *Biomed Mater.* **6,** 055010
[10] Garcia-Alegria E, Iluit M, Stefanska M, Silva C, Heeg S, Kimber S, Kouskoff V, Lacaud G, Vijayaraghavan A and Batta K 2016 *Sci Rep.*, **6** 25917
[11] Chang Y, Yang S-T, Liu J-H, Dong E, Wang Y, Cao A, Liu Y and Wang H 2011*Toxicol Lett.* **200**, p. 201
[12] Bengtson S, Kling K, Madsen A-M, Noergaard A W, Jacobsen N R, Clausen P A, Alonso B, Pesquera A, Zurutuza A, Ramos R, Okuno H, Dijon kanWallin H and Vogel U 2016 *Environ Mol Mutagen.* **57** p. 469
[13] Xu M, Zhu J, Wang F, Xiong Y, Wu Y, Wang Q, Weng J, Zhang Z, Chen W and Liu S 2016 *ACS Nano* **10** p. 3267
[14] Zhang X, Yin J, Peng C, Hu W, Zhu Z, Li W, Fan C and Huang Q 2011 *Carbon* **49** p.986
[15] Yang K, Zhang S, Zhang G, Sun X, Lee S T and Liu Z 2010 *Nano Lett.* **10** p.3318
[16] Li H L, Zhang H, Huang H et al 2013 *Mol. Cells.* **35** p. 436
[17] Tabish T A, Pranjol M Z I, Hayat H, Rahat A A M, Abdullah T M, Whatmore J L and Zhang S. *Nanotech.* 2017 **28** 504001
[18] Gurunathan S, Han J W, Park J H and Kim J H. 2014 *Inter J Nanomed.* **9** p.1783

IOP Conf. Series: Journal of Physics: Conf. Series **1186** (2019) 012009 doi:10.1088/1742-6596/1186/1/012009

[19] Slowing I, Trewyn B and Lin V 2006 *J Am Chem Soc*, **128**, p. 14792
[20] Guo J X, Zhang X, Li Q N and Li W X 2007 *Nucl Med Biol* **34** p.579
[21] Lewinski N, Colvin V and Drezek R 2008 *Small* **4** p.26

IOP Conf. Series: Journal of Physics: Conf. Series **1186** (2019) 012010 doi:10.1088/1742-6596/1186/1/012010

Model-independent X-ray reflectometry analysis of phospholipid thin films on liquid substrates

Yu Volkov[1,2 *], A Tikhonov[2,3], V Asadchikov[1], B Roshchin[1], V Honkimäki[4] and M Blanco[4]

[1]FSRC "Crystallography and Photonics" RAS, Moscow, Russia
[2]Institute for Solid State Physics RAS, Chernogolovka, Russia
[3]Institute for Physical Problems RAS, Moscow, Russia
[4]European Synchrotron Research Facility, Grenoble, France

E-mail: * neko.crys@gmail.com

Abstract. We present a brief review of application of a model-independent analysis of X-ray reflectometry data to our recent investigations of thin films of phospholipids DSPC and DMPS on a surfaces of water and colloidal silica hydrosol. Comparison of the results against analytical models of phospholipid membranes known in literature allowed us to extract more extensive information on the samples in question. The effect of spontaneous ordering of lipid films on a surface of colloidal silica is discussed. We also demonstrate the possibilities to influence the structure of prepared membranes by enriching the substrate with alkali ions.

1. Introduction

Planar phospholipid systems are of interest both for condensed matter physics and biophysics, as they are considered as the simplest models of a cell membranes [1, 2]. Preparation of macroscopic samples of phospholipid films in practice is limited by the effect of spontaneous curvature of lipid bilayer in an aqueous medium; for this reason, most structural studies of lipid bilayers and multilayers were performed for vesicles and lamellar aggregates [3, 4]. However, the method of obtaining a macroscopically flat lipid multilayers on a substrate of silica hydrosol was proposed in [5].

The colloidal solution of SiO_2 nanoparticles in water stabilized by a small amount of NaOH (also known as silica hydrosol) forms a strongly polarized air/liquid boundary [6]. The gradient of electric potential at the surface exists due to the difference between the potentials of the electric image forces for strongly negative charged nanoparticles and Na^+ cations. In this work, we present a review of our investigations on the effect of spontaneous ordering within a membranes of standard phospholipids 1,2-distearoyl-sn-glycero-3-phosphocholine (DSPC) and 1,2-dimyristoyl-sn-glycero-3-phospho-L-serine (DMPS) deposited on various aqueous substrates.

The structure of lipid films and near-surface layers has been obtained by the X-ray reflectometry method, with the use of a model-independent approach to the reconstruction of distribution of electron concentration [7, 8]. Its significant advantage over the more widely used model-based approaches (see, for example, [9]) is that it does not require any *a priori* assumptions on a shape of the structure under study. The reconstruction method consists in an extrapolation of asymptotic behavior of the angular dependence of reflectivity factor $R(q_z)$ over the region of high q_z, which allows one to significantly reduce the number of possible solutions. In general case, only two unique distributions of dielectric permittivity $\varepsilon(z)$ (and, respectively, electron concentration $\rho(z) = \pi(1-\varepsilon(z))/(r_o\lambda^2)$ [10]) exist that satisfy both the experimental reflectivity $R_{exp}(q_z)$ measured in a limited range of q_z and the estimated asymptotic $R(q_z \to \infty)$ at the same time. The problem of uniqueness for the obtained solution is discussed in-depth in [7].

2. Preparation of samples and experimental setups

Samples were prepared and studied at room temperature $T = 295$K in a hermetically sealed cell with transparent windows according to the method described in [5]. Phospholipid solutions in chloroform (~ 50 mmol/L) were deposited by the use of a calibrated syringe on a surface of liquid substrate placed in a polytetrafluoroethylene dish with diameter of 100 mm. In each case the amount of lipid in a droplet has been estimated for the formation of about ten lipid monolayers after its spreading among the surface; any excess of the surfactant is accumulated in macroscopic three-dimensional aggregates being in equilibrium with the adsorbed film.

Most of the experimental measurements of X-ray specular reflectivity R were performed at synchrotron station ID31 at ESRF [11]. Monochromatic beam with photons wavelength $\lambda = 0.1747 \pm 0.0003$ Å (energy $E \approx 71$ keV) has been focused to size $10 \times 250\,\mu$m with integral intensity $\sim 10^{10}$ photons/s. Reflected intensity signal has been collected by 2-D CCD detector MaxiPix [12], matrix size 256×256 pixels with resolution $55\,\mu$m/px, then integrated over the exposure area. However, part of the measurements has been performed at FSRC "Crystallography and Photonics" RAS using a multifunctional butterfly-type diffractometer with laboratory X-ray tube [13]. Monochromatic beam with photons $\lambda = 1.5405 \pm 0.0001$ Å ($E \approx 8$ keV), focused to 0.1×10 mm, provided integral intensity 3×10^5 photons/s. Signal has been collected by 0-D scintillation detector Radicon SCSD-4.

Comparison between the data sets obtained with different sources is shown by figure 1 (circles, squares and triangles were measured at ID31, diamonds were measured at laboratory setup). It demonstrates that at least for the case of samples under discussion the laboratory diffractometer has been able to provide experimental results comparable to the synchrotron beam at the cost of substantial increase in measurement times, while on the other hand considerably lowering radiation damage to the samples.

3. Results and discussion

Structure of the transient layer at the air/liquid interface of colloidal silica has been studied previously in [6, 14], however the X-ray reflectivity data in these was interpreted according to the 4-layer analytical model. Example of experimental reflectivity curve $R(q_z)$ for the hydrosol containing silica nanoparticles of diameter $D = 12$ nm (30 wt% SiO_2, 0.5 wt% NaOH, pH$\sim$10.5) is shown in figure 1 (circles). Model-independent $\rho(z)$ distribution reconstructed from that data, shown in figure 2a (black line), is in a very good agreement with analytical model taken from [6] (color lines); it is possible to discern the sub-layers of suspended alkali ions, depleted region, and loose monolayer of nanoparticles. Calculation of the surface concentration of Na^+ from the model-independent profile yields $\sim 4 \times 10^{18}$ m^{-2}, which corresponds to the values estimated independently in [14]. Behavior of SiO_2 nanoparticles in the loose monolayer, including the effect of condensation of nanoparticles at air/liquid interface in the presence of phospholipids, was discussed further in our work [15].

Kinetics of the spontaneous ordering within phospholipid film deposited on a surface of silica hydrosol was investigated in our publications [16] and [17]. Reflectivity $R(q_z)$ from the membrane of DSPC (or $C_{44}H_{88}NO_8P$) on the hydrosol with 12 nm silica nanoparticles is shown in figure 1 (squares); presence of the periodic peaks on the curve indicates a regular structure. Reconstructed model-independent distribution (figure 2b, black line) contains five-layer stack

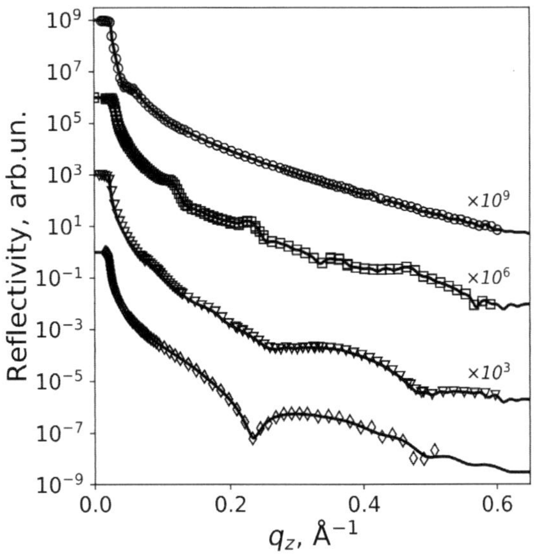

Figure 1. Reflectivity curves measured for the surface of 12-nm sol (circles), DSPC multilayer on 12-nm sol (squares), DSPC bilayer on 7-nm sol enriched with NaOH (triangles), DMPS monolayer on water (diamonds). The solid lines are simulated from the reconstruction.

with period of (68±1) Å, which corresponds to the doubled length of DSPC molecule (2×34.1 Å), however their electron concentration exceeds theoretical value for the bilayer by 15–20%.

Decomposition of the analytical model is also shown in figure 2b in colors; parameters for bilayer has been chosen according to [6], bilayers in the stack implied to be separated with a small amount of water. Assuming the lipid molecules in bilayers are in crystal phase with area per hydrocarbon chain $A = 20.8$ Å^2, the excess number of electrons per lipid molecule is 86±2, which corresponds to nine Na^+ ions. In [17] we proposed a hypothesis that the accumulation of Na^+ is caused by the electroporation effect, as the electric field intensity at the surface of hydrosol exceeds the critical electroporation value for lipid membranes by few orders of magnitude. Other aspects of the spontaneous ordering of lipid multilayer according to that hypothesis is also studied and discussed in [17].

Dependence of the structure and properties of lipid membranes on the properties of a hydrosol substrate was briefly considered in [16]. Reflectivity $R(q_z)$ from the thin membrane of DSPC on the hydrosol containing SiO_2 particles of diameter $D = 7$ nm and enriched with additional NaOH (16 wt% SiO_2, 1.3 wt% NaOH, pH$\sim$12) is presented in figure 1 (triangles); the respective $\rho(z)$ distribution (figure 2c, black) shows the collapse of the whole volume of lipid into a bilayer of thickness $\approx$ 70 Å.

Comparison with a theoretical model (figure 2c, colors), giving the thicknesses, electron densities and area-per-unit for hydrocarbon chains and polar groups the same as above, yields an excess number of electrons for the polar groups is 10±1, which equals to 1 Na^+ ion per lipid molecule. However, the relative orientation of lipid molecules in the bilayer is inverted to the one in the multilayer stack. Another layer of Na^+ ions is present on lipid/substrate interface, and a layer of SiO_2 particles are found adjacent to that interface within the substrate, their concentration being considerably lower than for the case of a clean surface of hydrosol.

It should be noted that even for the simpler samples, such as classical Langmuir films, simultaneous application of model-independent approach along the model-based method can provide more extensive description of the lipid membrane. In [18] and [19] we demonstrated the

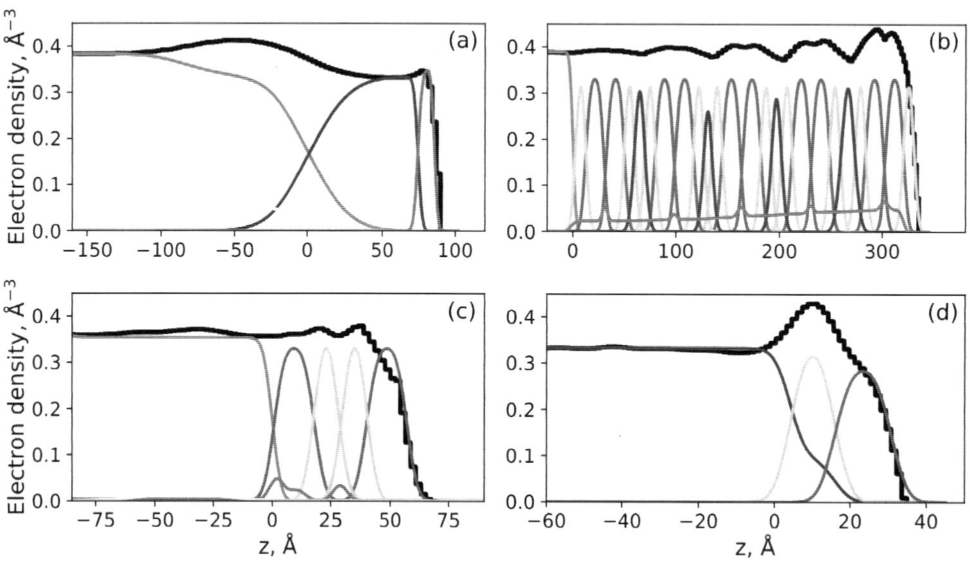

Figure 2. *(Colors online)* Electron density distributions $\rho(z)$ for (a) 12-nm silica sol, (b) DSPC multilayer on 12-nm sol, (c) DSPC bilayer on 7-nm sol and (d) DMPS monolayer. Model-independent reconstruction is denoted by black lines. Components of analytical models by-color: lipid polar groups (light green), hydrocarbon chains (dark green), Na^+ ions (red), H_2O (blue), SiO_2 particles (yellow), and bulk hydrosol (magenta).

possibility of such analysis applied to the two-dimensional liquid/gel phase transition in Langmuir monolayer. Figure 1 (diamonds) shows the reflectivity curve for the DMPS (or $C_{34}H_{65}NO_{10}PNa$) membrane prepared on a substrate of a KCl solution ($\sim$ 10 mmol/L) in deionized water (pH= 7), with the estimated area per lipid molecule $A \approx 45$ Å^2. Figure 2d compares the result of model-independent reconstruction (black) against the decomposition of two-layer theoretical model (colors). In the regions corresponding to the hydrocarbon chains both the model and "free-form" solutions are in perfect agreement, which confirms the initial estimation of A value. However, in the region of serine polar groups model-independent electron concentration exceeds the theoretical one by $\approx$27%, presumably by the effect of hydration of lipid polar groups on water. Integration of the excessive electron concentration yields the hydration level of 5 molecules of H_2O per lipid molecule. In [19] this combined approach allowed us to experimentally observe the changes in hydration level and orientation of hydrocarbon chains at the different stages of phase transition within DMPS monolayer.

4. Conclusions

To summarize, in a series of papers [15, 16, 17, 18, 19] we demonstrated the perspectives of model-independent approach to the analysis of X-ray reflectivity data in application to the studies of macroscopically flat phospholipid membranes on a liquid substrates. It is shown that lipid films on a colloidal silica sols exhibit a spontaneous ordering effect. Investigations on the kinetics of such spontaneous ordering reveal the dependence of the structure and configuration of lipid membrane on the electric properties of air/substrate interface and concentration of alkali ions within the substrate. Possibilities for the quantitative study of a more complex processes, namely two-dimensional phase transitions in lipid membranes, with combined X-ray reflectivity analysis, have been justified for the usage of laboratory sources.

Acknowledgements
Experiments at the station ID31 have been performed under the research projects SC-4246 and SC-4461 of European Synchrotron Research Facility, Grenoble, France. Theoretical part of the work is supported by the Russian Scientific Foundation (project #18-12-00108).

References

[1] Small D M 1986 *The Physical Chemistry of Lipids* (New York: Plenum Press)
[2] Kucerka N, Mieh M-P and Katsaras J 2011 *Biochim. Biophys. Acta* **1808**, 2761
[3] Petrache H, Gouliaev N, Tristram-Nagle S, Zhang R, Suter R M and Nagle J F 1998 *Phys. Rev.* E **57**, 7014
[4] Forster G, Schwieger C, Faber F, Weber T and Blume A 2007 *Eur. Biophys. J.* **36**, 425
[5] Tikhonov A M 2010 *JETP Lett.* **92**, 356
[6] Tikhonov A M 2007 *J. Phys. Chem.* C **111**, 930
[7] Kozhevnikov I V 2003 *Nucl. Instrum. Meth. Phys. Res.* A **508**, 519
[8] Kozhevnikov I V, Peverini L and Ziegler E 2012 *Phys. Rev.* B **85**, 125439
[9] Tikhonov A M and Schlossman M L 2007 *J. Phys.: Condens. Matter* **19**, 375101
[10] Henke B L, Gullikson E M and Davis J C 1993 *Atom. Data Nucl. Data Tables* **54**, 181
[11] Honkimäki V, Reichert H, Okasinski J and Dosch H 2006 *J. Synchrotron Rad.* **13**, 426
[12] Ponchut C, Rigal J M, Clément J, Papillon E, Homs A and Petitdemange S 2011 *J. Instrumentation* **6**, C01069
[13] Asadchikov V E *et al.* 2005 *Instrum. Exper. Tech.* **48**, 364
[14] Tikhonov A M 2009 *J. Chem. Phys.* **130**, 024512
[15] Asadchikov V E, Volkov V V, Volkov Yu O, Dembo K A, Kozhevnikov I V, Roshchin B S, Frolov D A and Tikhonov A M 2011 *JETP Lett.* **94**, 585
[16] Tikhonov A M, Asadchikov V E and Volkov Yu O 2015 *JETP Lett.* **102**, 478
[17] Tikhonov A M, Asadchikov V E, Volkov Yu O, Roshchin B S, Monakhov I S and Smirnov I S 2016 *JETP Lett.* **104**, 873
[18] Asadchikov V E, Tikhonov A M, Volkov Yu O, Ermakov Yu A, Rudakova E B, D'yachkova I G and Nuzhdin A D 2017 *JETP Lett.* **106**, 534
[19] Tikhonov A M, Asadchikov V E, Volkov Yu O, Roshchin B S and Ermakov Yu A 2017 *JETP* **125**, 1051

IOP Conf. Series: Journal of Physics: Conf. Series **1186** (2019) 012011 doi:10.1088/1742-6596/1186/1/012011

Calix[4]amine Langmuir-Blodgett thin film sensing properties against volatile organic compounds

S Şen[1], R Çapan[2], Z Özbek[3], M E Özel[4], G A Stanciu[5], F Davis[6]

[1] Department of Physics, Faculty of Arts and Science, Çanakkale Onsekiz Mart University, Çanakkale, Türkiye
[2] Department of Physics, Faculty of Arts and Science, Balikesir University, Balikesir, Türkiye
[3] Department of Bioengineering, Faculty of Engineering, Çanakkale Onsekiz Mart University Çanakkale, Türkiye
[4] Vocational School, Işık University, İstanbul, Türkiye
[5] Department of Physics, Center for Microscopy, Microanalysis and Information Processing, University "Politehnica" of Bucharest, Bucharest, Romania
[6] Department of Engineeringand Design, University of Chichester, Upper Bognor Road, Bognor Regis, West Sussex, UK

E mail:sozmaya@comu.edu.tr

Abstract. Thin films of CBAMINE were deposited at air-water interface by the method of Langmuir-Blodgett (LB) technique onto a suitable substrate. Atomic force microscopy technique was used to characterize its thin film properties. The results indicate that a uniform LB film monolayer from the water surface to a glass or quartz crystal substrates deposited with a transfer ratio of over 96 %. Gas sensing properties and thickness of the LB thin films of CBAMINE were investigated using Surface plasmon resonance (SPR) technique. Its vapour sensing properties were investigated for different volatile organic compounds. Reversible changes in the optical behaviour were observed and thin films of this material are highly selective for chloroform vapour with fast response and recovery times.

1. Introduction

Carbon-containing volatile organic compounds (VOCs) are quickly evaporated into the atmosphere once emitted. While coming from certain solids or liquids, VOCs are released as gases into the atmosphere. VOCs are a source of both indoor and outdoor pollution. As an indoor pollution, VOCs may be named to as volatile organic chemicals. Indoors, VOCs evaporate under normal indoor atmospheric conditions according to temperature and pressure. VOCs may also be very dangerous as the outdoor pollutants

As with other pollutants, the extent and nature of the health effect will depend on many factors including level of exposure and length of time exposed [1]. Due to this reason, gas sensor technology has advanced quite rapidly during past few decades [2-4].

In this study, the SPR method [5] was preferred for VOCs sensing examination of the medium. It is one of the most sensitive detection methods for changes of thicknesses and refractive index in ultra-thin films and is a convenient optical transduction technique widely utilised in molecular recognition. In our study, the CBAMINE material, a calixarene derivative, was used as the receptor [6]. The chemical structure of this molecule is shown in figure 1. Molecular behaviour as a thin film of CBAMINE was investigated using Langmuir-Blodgett (LB) thin film method. Gas sensing behaviour of LB thin films of CBAMINE was investigated using SPR. This article also reports the results obtained from UV-visible

IOP Conf. Series: Journal of Physics: Conf. Series **1186** (2019) 012011 doi:10.1088/1742-6596/1186/1/012011

spectroscopy and atomic force microscopy (AFM) measurements for the characterization of CMAMINE LB films.

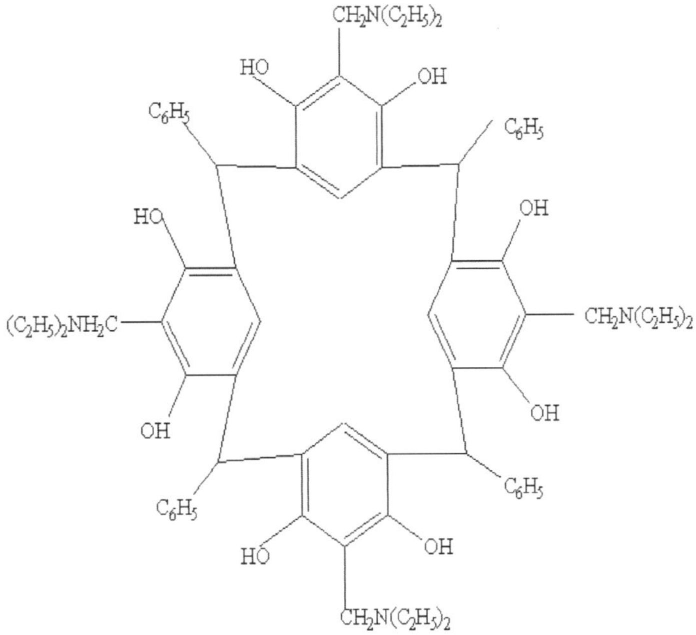

Figure 1. Chemical structure of a CBAMINE molecule.

2. Experimental details

LB films of CBAMINE were deposited by a NIMA 622 alternate LB trough with automated surface balance system. The substrates were ultrasonically cleaned microscopic glass slides for depositing process. The surface pressure is controlled using a Wilhemly plate method [7].

AFM measurement was carried out in air in tapping mode. A standard silicon nitride tip was used to obtain the 2D and 3D images and the spring constant of the cantilever was 40 nm. The AFM image was taken in an area of 4μm x 4μm. A glass slide was ultrasonically cleaned to prepared thin film and then15 layers LB films of CBAMINE were transferred onto this substrate.

Surface plasmon resonance (SPR) measurements were performed by A Kretschmann's type (BIOSUPLAR 6 Model) SPR set-up with a resolution of 0.003^0. Biosuplar-Software was used to control the SPR system settings, measurements and data acquisition as well as data presentation. The SPR data were analysed (via least squares algorithm) to obtain film thickness and refractive index using Winspall software for the theoretical fitting of the Fresnel's equations [8]. The response of CBAMINE thin films on exposures to chloroform, benzene, toluene and ethanol vapours were studied using a special a transparent plastic flow cell made in house. The reflection intensity at a constant angle of incidence was monitored as a function of time when the film remained exposed to different vapours for at least 2 minutes followed by allowing the film to recover in clean air. This process was repeated for four times to observe the reproducibility of the CBAMINE thin film as a sensing material. All organic vapour measurements were made using dry air in the small gas cell, thereby eliminating the effect of water vapour on the response properties of CBAMINE LB films [9].

3. Results and Discussion

In this study, a monolayer containing calixarene molecules has been deposited using LB film deposition technique to form multilayer sequences possessing Y-type LB deposition mode. A NIMA 622 alternate LB trough with automated surface balance was utilised to examine the molecular behaviour of CBAMINE at the air–water interface and fabricate LB film multilayer onto the different substrates. The automatically controllable barriers are employed to change molecular arrangement of the monolayers on the water surface by compressing the CBAMINE monolayer with 0.2 mm s^{-1} of barrier speed. Wilhemly plate method was employed to determine the surface pressure. By compressing the barriers, a plot of surface pressure against surface area was obtained as in figure 2.

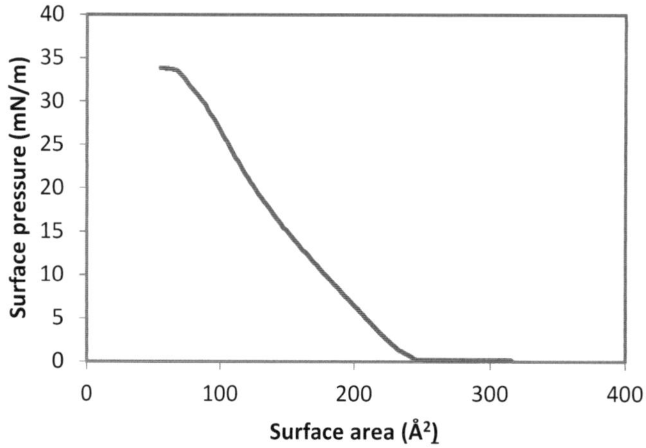

Figure 2. The pressure–area (π–A isotherm) of CBAMINE monolayer on the water surface.

By compressing the barriers, a plot of surface pressure against surface area was obtained.From the pressure–area (π–A) isotherm graph, surface pressure of the solid phase of the floating monolayer on the water surface was obtained as 20 mN/m, which are in good agreement with the reported result [10]; therefore, this surface pressure value was selected for the LB film deposition procedure.

The effectiveness of the monolayer deposition process is described by the transfer ratio (TR). A TR of close to 1.0 implies that the material exhibits good adhesion to the substrate when it is immersed in and then removed from the subphase. However, a low TR value indicates that the material does not adhere to the substrate well, and consequently, the second monolayer deposited on the film might be easily removed, resulting in a low-quality film [11]. The TR value is calculated using the following equation:

$$\tau = A_L/A_s \quad (1)$$

where A_L is the decrease in the area occupied by the molecules at the air/water interface (at solid phase) and A_S is the area of the substrate covered with the monolayer. In this study, for each monolayer, the average reduced area is calculated and the transfer ratio, τ, for a glass substrate is found to be 96 % for LB films of CBAMINE. This value can be used to conclude that steady, reproducible and uniform monolayers of calixarene molecules were deposited from the air–water interface on a glass and quarts substrates.

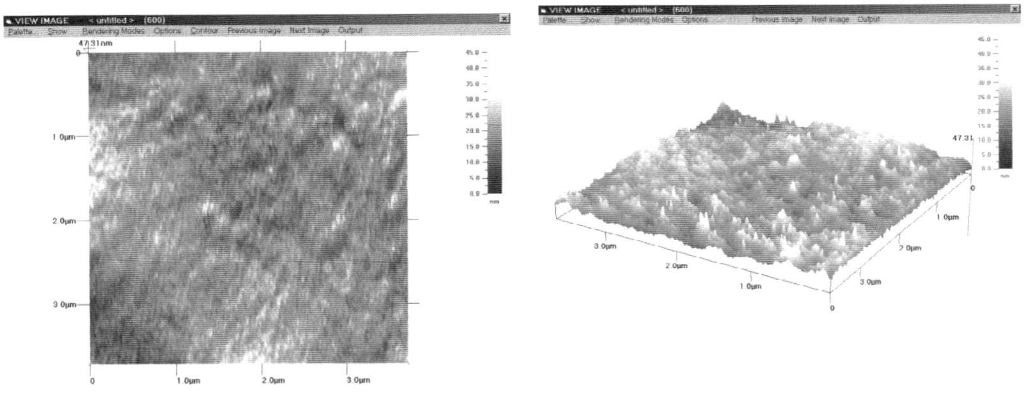

(a) (b)

Figure 3. (a) Two-dimensional and (b) three-dimensional AFM images of CBAMINE LB films.

Morphological examination of 15 thin layers CBAMINE molecules deposited onto an optically flat hydrophilic glass substrates were carried out using AFM tapping mode. Figure 3 demonstrates4 µm x 4 µm AFM image of LB film. Measurements were repeated using different area regions of the sample and finally seen that surface morphologies were very similar to these 2D and 3D AFM pictures. It is seen some big pairs on the surface. These bumps are supposed to be formed by the aggregation of CBAMINE molecules. These hills may be caused by irregularities of glass surface. Although the particles slightly increase the RMS value, results indicate LB films exhibited smooth, compact, uniform and void free morphology with a RMS value 4.6 nm.

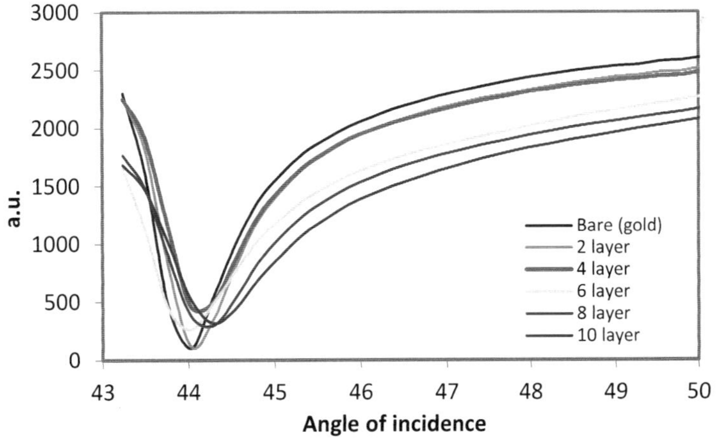

Figure 4. SPR curves of CBAMINE LB films with increase in thickness.

Optical parameters (the refractive index and the thickness) of these LB films were determined using SPR curves with a suitable fitting program (WINSPALL software developed at the Max-Plank-Institute for Polymer Research, Germany). Figure 4 illustrates SPR intensity versus angle curves for the clean gold surface, and for surfaces with deposited layers of LB films. The refractive index values ranging from 1.64 to 1.82 with a approximately thickness value of average 1.14 nm (Table1) were calculated using the fitting program. These results are good agreement with optical index and the thickness of LB films of similar calixarene based derivative [12].

Table 1. The film thickness and refractive index of the deposited LB thin films with respect to number of layers.

Number of layers	Thickness (nm)	Refractive index
2	2.0	1.48
4	4.0	1.64
6	6.4	1.72
8	8.3	1.68
10	11.3	1.90

The adsorption of chloroform, benzene, toluene and ethanol vapours was studied using SPR measurements on gold-coated glass slides with thin films of CBAMINE. The vapour remained periodically turned on and off for about 2 min and this process was repeated four times for same vapour concentration. Typical response on exposure to chloroform, benzene, toluene and ethanol vapour of four different concentrations for 10-layer LB CBAMINE film is presented in figure5. The sensor responses are reversible with recovery being achieved. For all the vapours examined, the thin film response time is fast in seconds, but the thin film response for chloroform is significantly higher when compared to the other three vapours. SPR and different other systems having layers of calixarene derivatives were used to detect chloroform vapour [13-14].

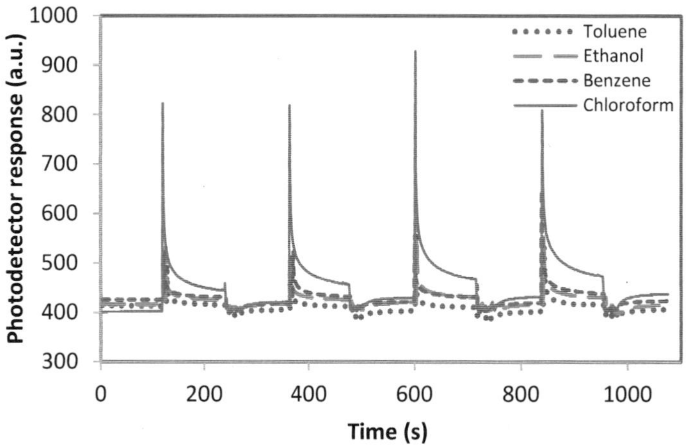

Figure 5. Effect of different VOCs on photodetector response: Toluene, ethanol, benzene, and chloroform.

4. Conclusion

We have studied gas sensing properties of CBAMINE which is calix[4]arene derivative. This molecule have been produced and deposited successfully as LB films on glass and gold-coated glass substrates. The films have been characterized morphologically and optically by means of AFM and SPR respectively. The SPR technique was also used in order to investigate the adsorption properties of CBAMINE LB films. Both steady state and in situ adsorption kinetic characteristics were obtained and LB films of this molecule were determined to be highly sensitive and selective for chloroform vapour.

References

[1] Schwanke A J, Balzer R, Pergher S, 2018 *Handbook of Nanomaterials for Industrial Applications* (Amsterdam: Elsevier Inc) 908-15

[2] Neethirajan S, Jayas D S, Sadistap S, 2009 *Food Bioprocess Technol.* **2** 115–21
[3] Allen M G, 1998 *Meas. Sci. Technol.* **9** 545–62
[4] TsujitaW, YoshinoA, IshidaH, MoriizumiT, 2005 *Sensors and Actuators B* **110** 304–11
[5] Hassan A K, Goy C, Nabok A V, 2008 *Thin Solid Films* **516** 9006–11
[6] Kostyukevych K V, Khristosenko R V, Pavluchenko A S, Vakhula A A, Kazantseva Z I, Koshets I A, Shirshov Y M, 2016 *Sensors and Actuators B* **223** 470–80
[7] Petty M C, 1996 *Langmuir–Blodgett Films* (Chambridge:Chambridge University Press)
[8] Pockrand I, 1978 *Surf. Sci.* **72** 577-88
[9] Erdoğan M, Çapan R, Davis F, 2010 *Sensors and Actuators B* **145** 66–70
[10] Özbek Z, Çapan R, Göktaş H, Şen S, İnce F G, Özel M E, Davis F, 2011 *Sensors and Actuators B* **158** 235–40
[11] Oliveira R F, Barros A, Ferreira M, 2017 *Nanostructured Films: Langmuir–Blodgett (LB) and Layer-by-Layer (LbL) Techniques (Nanostructures)* ed S Holt (Oxford: Elsevier) chapter 4 pp 105–23
[12] Çapan R, Ozbek Z, Goktas H, Sen S, Ince FG, Ozel ME, Stanciu GA, Davis F, 2010 *Sensors and Actuators B* **148** 358–65
[13] Koshets I A, Kazantseva Z I, Shirshov Y M , Cherenok S A, Kalchenko V I, 2005 *Sensors and Actuators B* **106** 177–81
[14] Çapan R, Goktas H, Özbek Z, Sen S, Özel M E, Davis F, 2015 *Applied Surface Science* **350** 129–34

Two interesting scaffolds derived from pyrazolo[3,4-*b*]quinolines: synthesis and applications

K S Danel[1], G Lewińska[2], Z Usatenko[2], T Lemek[1] and AWisła-Świder[1]

[1] Institute of Chemistry, University of Agriculture, 122 Balicka Street, Cracow 31-141, Poland

[2] Institute of Physics, Cracow University of Technology, 1 Podchorążych Street, Cracow 30-084, Poland

rrdanelk@cyf-kr.edu.pl

Abstract. Pyrazolo[3,4-*b*]quinolines are fascinating compounds of great importance. They reveal not only a range of very interesting biological activities, but also have been applied as optical brighteners, in organic light emitting diodes (OLED), in polymer light emitting diodes (PLED), sensors, in nonlinear optics, solar cells and many others. Recently, two interesting scaffolds emerged from the basic ring system, the first group comprises analogues of azafluoranthenes, the second one contains a seven-member ring embedded in the main framework, which can give rise to chiral compounds to be used to the production of diodes emitting circularly polarized light (CPL).

1. Introduction

Pyrazolo[3,4-*b*]quinolines **1** (PQs), formally the *ortho* fusion product between quinoline and pyrazole rings, have been extensively examined compounds since the beginning of the 20th century now. Their story began in 1928 when Tomasik *et al.* synthesized the basic heterocyclic ring system [1], figure 1.

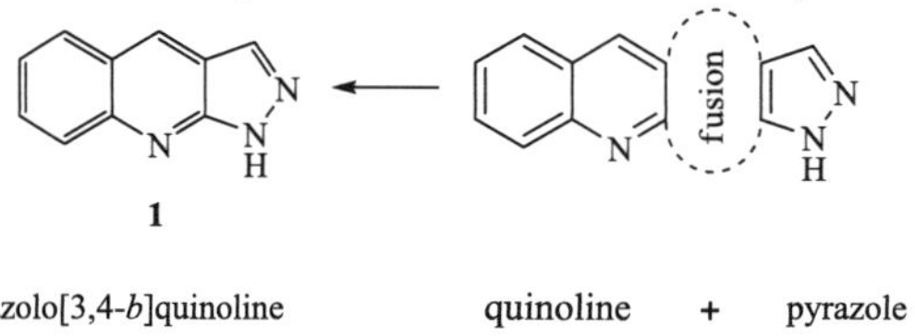

Figure 1. Chemical structures of the basic ring system and its components are presented.

There are a few general methods to decorate **1** with different substituents [2], which determines the physicochemical properties of the compound in case when desired particular application purposes are required. The most straightforward one seems by heating to reflux a mixture of aryl amines, aldehydes and pyrazolones in ethylene glycol, figure 2, or under microwave irradiation, which produces **2** [2].

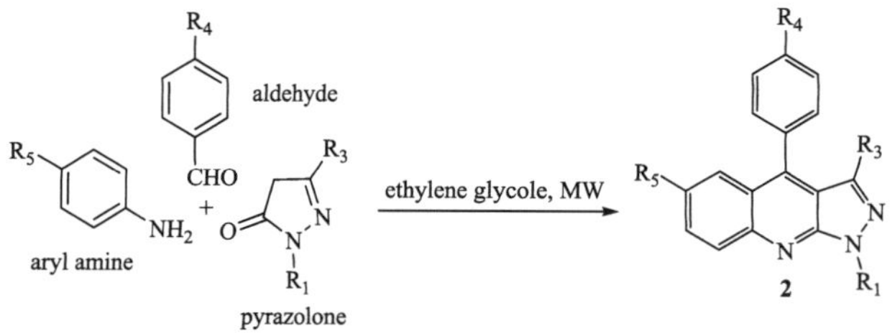

Figure 2. The synthesis of pyrazolo[3,4-*b*]quinolines **2**.

Thus it is a classical example of multicomponent reaction (MCR) in which three or more starting materials react to form a product, where basically all or most of the atoms contribute to the newly formed product [3]. PQs reveal a broad spectrum of pharmacological properties, like antivirals [4], dopaminergic [5], inhibitors of oncogenic Ras [6], interferon inducing agents [7], antimicrobials [8], antimalarials [9] and potent apoptosis inducers [10]. PQs **3** emit mostly in the blue spectral range with high quantum yields close to unity [11]. As OLEDs requires to cover the full spectrum of visible light from blue to red, the batochromic shift can easily be achieved after incorporation within the PQs skeleton some donating moieties, like dialkyl **4** or diaryl amines **5**, figure 3 [12, 13].

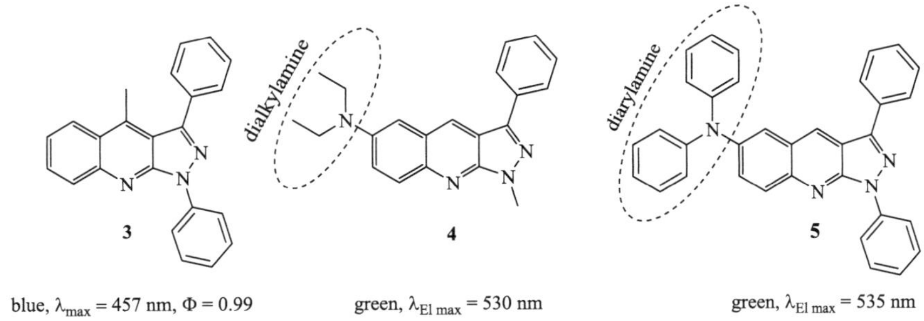

Figure 3. Examples of blue and green emitting PQs.

2. Condensed forms of PQs.

In searching for new derivatives of PQs we paid attention to rubicene derivatives [14, 15]. The title compound is obtained in three consecutive steps, starting from commercial antraquinone **6**, its convertion to diol **7**, which can be reduced to dichloroanthracene **8** and finally, cyclized to afford the target compound **9**, figure 4.

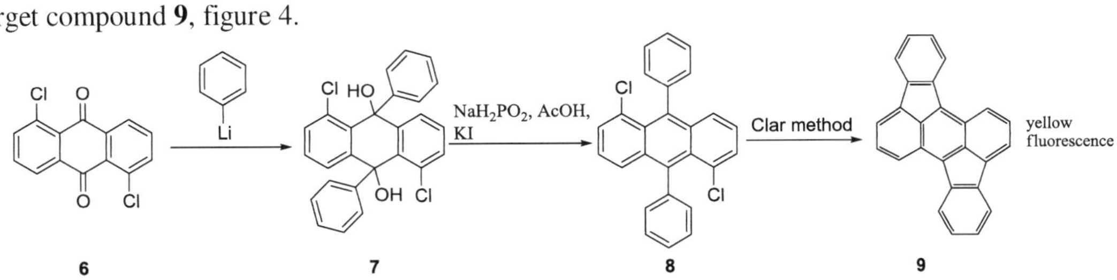

Figure 4. The synthesis of heptacyclic arene, rubicene **6**.

It is worth remembering that the greater the extent of the π electron system, the lower the energy of the low - lying $\pi \rightarrow \pi^*$ transition, and consequently, the longer the wavelength of the corresponding absorption band [16]. When analyzing the basic PQ skeleton (precursor **10**) one can see two probable sites for rings formation, figure 5: connecting phenyl 4 and *peri* position gives indeno-annelated product **11**, whenever phenyls 4 and 3 attacking each other generate a heteroazulene embedded

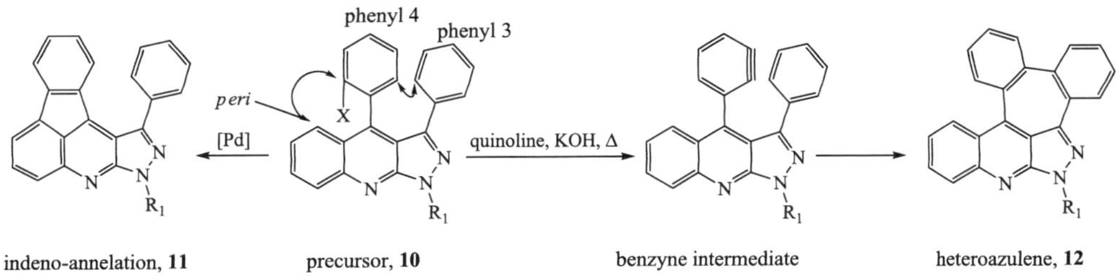

Figure 5. Two possible sites for cyclization produce regioisomers **11** and **12**.

derivative **12** [17]. To bring this task to reality one should use a properly halogenated PQ, the easiest way is when we have the mentioned **10** and X = Cl or Br, then in the presence of boiling quinoline and KOH (so called Clar method) we always get **12**. When X = Br, the indenoannelation is dominating reaction in the presence of palladium catalysts. We believe that vigorous conditions in case of boiling quinoline and KOH lead to a reactive benzyne intermediate, which attacks the neighboring ring 3. The 7-member ring formation by Clar method, originally used for rubicene synthesis [18] is also noteworthy, since up to our knowledge this is the first such case. Similar examples of heptagon formation by means of Scholl reaction, Suzuki coupling and Pd-catalyzed intramolecular C–H arylation can be found in the literature [19–23]. It has been recently stated that in terms of the application of carbon materials to electronic devices and other nano-scale materials, the ability to incorporate seven-membered rings may become a key factor [24]. We also applied this method to produce amines **15** emitting at the longer wavelength [25], figure 6, in two step sequence: (a) firstly, in Hartwig-Buchwald reaction we preferentially reacted the bromine atom with diarylamines, (b) secondly, the chlorine atom was utilized in the ring closure.

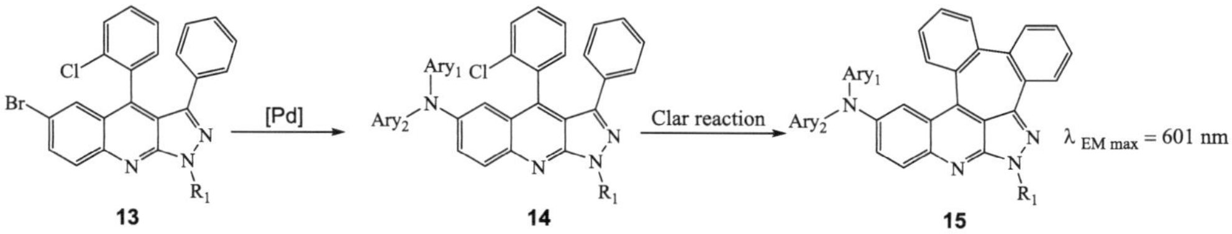

Figure 6. The synthesis of amines **15**.

In 1997, the electroluminescence characteristics of novel materials, among them, 4-phenyl-1,3,5,7-tetramethyl-1,7-dihydrodipyrazolo[3,4-*b*;4',3'-*e*]pyridine **16** were described for the first time [26]. For us it created an opportunity to successfully demonstrate that cyclization of neighboring rings could lead to interesting topologies **18** and **20** [27]. Moreover, the cyclization step could also be accomplished in milder conditions using palladium catalysts. It is worth of mentioning that both classical approaches and palladium catalyzed cyclization reactions delivered analogues of aza *s*-indacenes in short times and high yields, figure 7. One should take a notice of helical arrangements of rings in **18**.

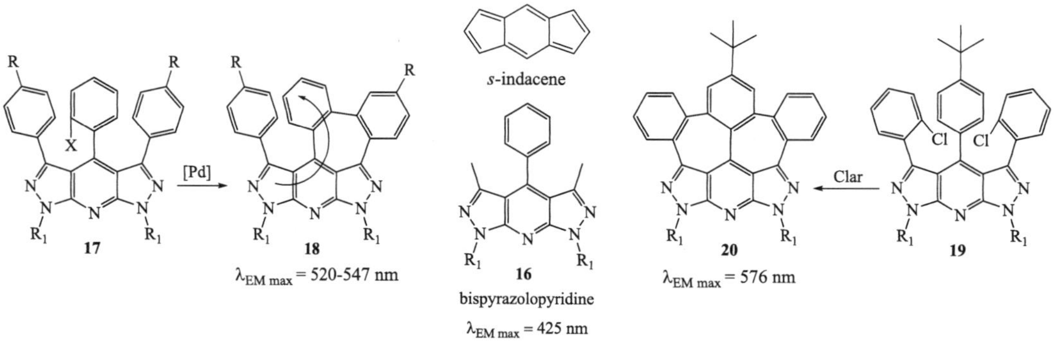

Figure 7. Synthesis of aza *s*-indacenes derivatives **18** and **19**.

3. Applications

In this short review we showed how to exploit effectively halo (Cl and Br) derivatives of PQs. Their use in OLEDs belongs to rarity [28] due to lower quantum yields (heavy atom effect) and performance, except for fluorine atoms, which can enhance device stability [29, 30]. Thus a bromine atom is preferably exchanged for other side radicals (e.g. CN), later converted to oxadiazoles [31], or utilized in C-N coupling to produce emissive hole-transporters (HTL) [13, 25, 32], as well as by a nucleophilic displacement of bromomethyl PQ derivatives to afford useful sensors [33]. It seems that intramolecular annulation reactions performed on PQs widen this interesting library due to their potential applications in medicine, electroluminescent cells [34], non-linear optics [35] and sensing devices [33, 36]. In fact, one can produce two regioisomers **11** and **12** from a common starting material **10**, separately, not as a mixture. Two groups of indenoannulated derivatives (A-D and F-E), figure 8, were tested in diodes [37, 38]. The second group was also tested in bulk-heterojunction type solar cells [39]. The configuration of the emitting device was as the follows: ITO/PEDOT:PSS/PVK:DYE/Ca/Al and that

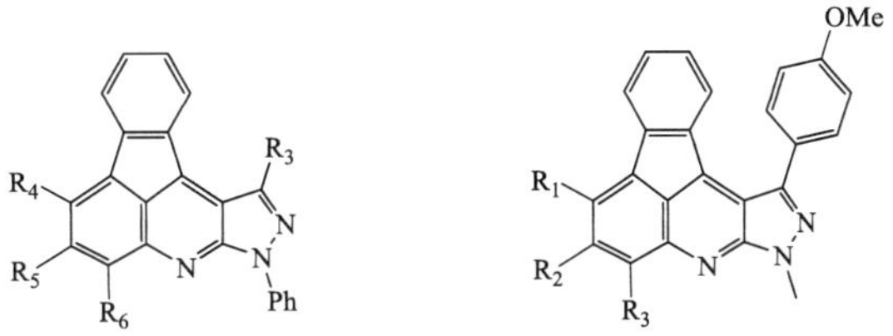

A: R_3 = Ph, R_4 = R_5 = R_6 = H
B: R_3 = Ph, R_4 = R_6 = H, R_5 = Me
C: R_3 = Me, R_4 = H, R_5 = R_6 = Me
D: R_3 = *n*-Pr, R_4 = OMe, R_5 = R_6 = H

E: R_1 = H, R_2 = R_3 = Me
F: R_2 = H, R_1 = R_3 = Me
G: R_3 = H, R_1 = R_2 = Me

Figure 8. Indenopyrazoloquinolines A-D and E-F.

of the light harvesting one: ITO/PEDOT:PSS/organic active layer/Al. Both groups of compounds were emitting green light, table 1. One can clearly see the influence of *p*-methoxyphenyl moiety on the emission wavelength of the dyes (E-G) 527-537 nm in relative to the first group 510-515 nm. In case of solar cells the authors used a ternary mixture as an active layer consisting of P3HT (donor 1), dyes (donor 2) and finally the acceptor molecule, phenyl-C71-butyric acid methyl ester (PCBM). The highest efficiency for that configuration reached 3.71%. The heterocyclic part of indenopyrazoloquinoline derivatives 11 is planar, which have been shown by quantum chemical calculations [40]. This fact, in addition to their luminescent properties, makes them potential candidates as intercalating molecules. Moreover, a series of interesting biological properties concerning azafluoranthene alkaloids have been described [41].

Table 1. Selected data for indenopyrazoloquinoline based OLEDs.

Compound	$\lambda_{EL\ max}$	CIE (1931)	
	(nm)	X	Y
A	515	0.1516	0.6938
B	510	0.1342	0.6843
C	512	0.1312	0.6737
D	511	0.1371	0.6950
E	537	0.3422	0.5929
F	531	0.3201	0.5845
G	527	0.3009	0.5877

Contrary to **11**, their isomers **12**, containing a seven-membered ring are twisted. A steric hindrance exerted by protons at the bay and fjord regions impose helicity on these molecules, figure 9. This distortion is reminiscent to that of encountered in either dibenzo[*a*,*l*]pyrene [42] or benzo[*c*]phenanthrene alias [4]helicene [43]. The helical structure of azulenes can be clearly seen from their crystal structure [44], however they easily interconvert through a planar transition state. The calculated energy barrier for this process is 25 kcal/mol [45]. Nevertheless, twisted skeleton of these

IOP Conf. Series: Journal of Physics: Conf. Series **1186** (2019) 012012 doi:10.1088/1742-6596/1186/1/012012

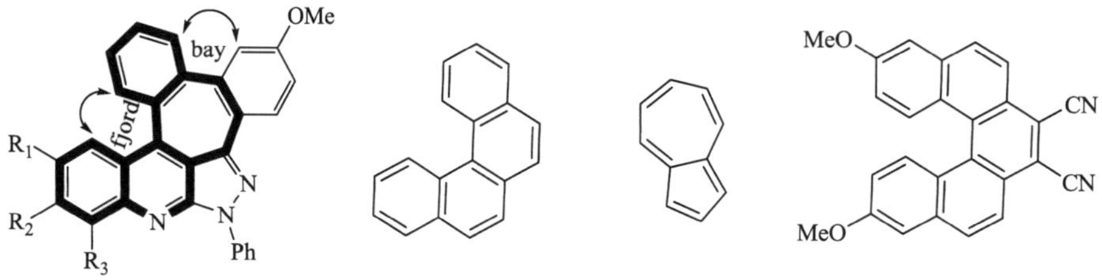

12a: R_1 = H, R_2 = R_3 = Me
12b: R_2 = H, R_1 = R_3 = Me
12c: R_3 = H, R_1 = R_2 = Me

benzo[*c*]phenanthrene or [4]helicene

azulene

[5]helicene

Figure 9. Azulene aza[4]helicenes **12**.

molecules should prevent destructive π...π stacking in the bulk, thus enhancing the lifetime of EL devices, what has been demonstrated by application of 3,12-dimethoxy-7,8-dicyano[5]helicene (DDH) as a novel emissive material [46]. The helical dyes were applied as a direct light emitting layer in OLED cells, table 2. The configuration of the devices: ITO/PEDOT:PSS/azulene dyes/Ca/Al, emitting greenish-yellow light with CIE coordinates (0.44, 0.54) was reported. The helical molecules were also tested in heterojunction solar cells of the following configuration: ITO/PEDOT:PSS/organic active layer/aluminum, in which an active layer was two- or three-component organic compounds mixture [47]. Conversion efficiency of light energy achieved for helicene and P3HT active layer was about 1%, whereas for P3HT:PCBM and helicene about 2%. Moreover, the helical molecules have been used to induce circularly polarized electroluminescence from an achiral light-emitting polymer (LEP) via a chiral small-molecule dopant [48].

Table 2. Selected data for helical molecules in OLED devices.

	$\lambda_{max\ abs}$ (nm)	$\lambda_{max\ em}$ (nm)	Φ_{fl}	HOMO (eV)	LUMO (eV)	T_m (°C)	T_g (°C)
2a	460	555	0.27	5.65	3.25	201	98
2b	465	555	0.38	5.65	3.21	195	100
2c	460	555	0.31	5.66	3.22	223	113

Acknowledgments
The work was supported from the basic funding of the University of Agriculture in Krakow DS-3711/ICh/17

References
[1] Musierowicz A, Niementowski S and Tomasik Z 1928 *Roczn. Chem.* **8** 325
[2] Nizioł J, Danel A, Gondek E, Armatys P, Sanetra J and Boiteux G 2004 *Macromol. Symp.* **212** 473
[3] Ugi I, Domling A and Horl W 1994 *Endeavor* **18** 115
[4] Siminoff P, Bernard A M, Hursky V S and Price K E 1973 *Antimicrob. Agents Chemother.* **3** 742
[5] Sarre S, Vandeneede D, Ebinger G, Michotte Y 1998 *J. Neurochem.* **70** 1730
[6] Wolin R, Wang D, Kelly J, Afonso A, James L, Kirschmeier P and McPhail A T 1996 *Bioorg. Med. Chem. Lett.* **6** 195
[7] Crenshaw R R, Luke G M and Siminoff P 1976 *J. Med. Chem.* **19** 262
[8] El-Sayed O A and Aboul-Enein H Y 2001 *Arch. Pharm. Pharm. Med. Chem.* **334** 117
[9] Stein R G, Biel J H and Singh T 1970 *J. Med.Chem.* **13** 15

[10] Zhang H Z, Claassen G, Crogran-Grundy C, Tseng B, Drewe J and Cai S X 2008 *Bioorg. Med. Chem.* **16** 222
[11] He Z, Milburn G HW, Baldwin K J, Smith D A, Danel A, Tomasik P 2000 *J. Lumin.* **86** 1
[12] Tao Y T, Balasubramaniam E, Danel A, Jarosz B, Tomasik P 2000 *Appl. Phys. Lett.* **77** 1575
[13] Danel A, Chacaga Ł, Uchacz T, Pokladko-Kowar M, Gondek E, Karasiński P and Sahraoui B 2015 *Adv. Device Mater.* **1** 17
[14] Smet M, Shukla R, Fülöp L and Dehaen W 1998 *Eur. J. Org. Chem.* **12** 2769
[15] Huang T H, Lin J T, Tao Y T and Chuen C H 2003 *Chem. Mat.* **15** 4854
[16] Valeur B and Berberan-Santos M N 2012 *Molecular Fluorescence* (Wiley-VCH Verlag GmbH & Co. KGaA) p 38
[17] Danel K S, Wisła A and Uchacz T 2009 *Arkivoc* **x** 71
[18] Clar E and Willicks 1958 *J. Chem. Soc.* **0** 942
[19] Pradhan A, Dechambenoit P, Bock H and Durola F 2013 *J. Org. Chem.***78** 2266
[20] Mughal E U and Kuck D 2012 *Chem. Commun.* **48** 8880
[21] Li X, Han J-W, Zhang Y-X and Wong H N C 2017 *Asian J. Org. Chem.* **6** 1876
[22] Kawai K, Kato K, Peng L, Segawa Y, Scott L T and Itami K 2018 *Org. Lett.* **20** 1932
[23] Kawasumi K, Zhang Q, Segawa Y, Scott L T and Itami K 2013 *Nature Chem.* **5** 739
[24] Sato K 2014 *TCIMAL* 162
[25] Gąsiorski P, Danel K S, Matusiewicz M, Uchacz T and Kityk A V 2010 *J. Lumin.* **130** 2460
[26] He Z, Milburn G H W, Danel A, Puchala A, Tomasik P and Rasala D 1997 *J. Mater. Chem.* **7** 2323
[27] Danel K S, Uchacz T and Karelus M 2011 *Arkivoc* **ix** 272
[28] Tao Y T, Balasubramaniam E, Danel A,Wisla A and Tomasik P 2001 *J. Mater. Chem.* **11** 768
[29] Strukelj M, Papadimitrakopoulos F, Miller T M and Rothberg L J 1995 *Science* **267** 1969
[30] Szlachcic P and Uchacz T 2018 *J. Lumin.* **194** 579
[31] Fuks-Janczarek I, Danel K, Kwiecień B and Kityk I V 2008 *Mat. Chem. Phys.* **111** 69
[32] Danel K, Gondek E, Pokładko M and Kityk I V 2009 *J. Mater. Sci.: Mater. Electron.* **20** 171
[33] Mac M, Uchacz T, Danel A, Danel K, Kolek P and Kulig E 2011 *J. Fluoresc.* **21** 375
[34] Danel A nad Danel K 2010 *Chemik* **64** 511
[35] Makowska-Janusik M, Gondek E, Kityk I V, Wisła A, Sanetra J and Danel A 2004 *Chem. Phys.* **306** 265
[36] Rurack K, Danel A, Rotkiewicz K, Grabka D, Spieles M and Rettig W 2002 *Org. Lett.* **4** 4647
[37] Gąsiorski P, Gondek E, Danel K S, Pokładko-Kowar M, Armatys P and Kityk A V 2013 *Opt. Mater.* **35** 681
[38] Gąsiorski P, Gondek E, Pokładko-Kowar M, Danel K S, Matusiewicz M, Kuźnik W and Kityk A V 2011 *Opt. Mater.* **34** 317
[39] Lewińska G, Danel K S, Łukaszewska I, Lewiński G, Niemiec W and Sanetra J 2018 *J. Mater. Sci.: Mater. Electron.* **29** 17809
[40] Gąsiorski P, Danel K S, Matusiewicz M, Uchacz T, Vlokh R, Kityk A V 2011 *J. Fluoresc.* **21** 443
[41] Ponnala S and Harding W W 2013 *Eur. J. Org. Chem.* **6** 1107
[42] Katz A K, Carrell H L and Glusker J P 1998 *Carcinogenesis* **19** 1641
[43] Brulé C, Laali K K, Okazaki T and Lakshman M K. 2007 *J. Org. Chem.* **72** 3232
[44] Szlachcic P, Danel K S, Gryl M, Stadnicka K, Usatenko Z, Nosidlak N, Lewińska G, Sanetra J and Kuźnik W 2015 *Dyes Pigment* **114** 184
[45] Danel K S and Uchacz T 2001 *Intertech proceedings* 268
[46] Sahasithiwat S, Mophuang T, Menbangpung L, Kamtonwong S and Sooksimuang T 2010 *Synth. Met.* **160** 148
[47] Lewińska G, Danel K S and Sanetra J 2016 *Solar Energy* **135** 848
[48] Yang Y, Correa da Costa R, Smilgies D M, Campbell A J and Fuchter M J 2013 *Adv. Mater.* **25** 2624

IOP Conf. Series: Journal of Physics: Conf. Series **1186** (2019) 012013 doi:10.1088/1742-6596/1186/1/012013

Electret and Dielectric Properties of Lyophilized Polymer Films

Ivan Bodurov, Asya Viraneva and Temenuzhka Yovcheva

Faculty of Physics and Technology, University of Plovdiv "Paisii Hilendarski", 24 Tzar Assen str., 4000 Plovdiv, Bulgaria

bodurov@uni-plovdiv.net

Abstract. Polymer electrets are one of the most important types of electrets, which are widely used both in academic research and industrial applications. The effect of trapped charges on dielectric properties of the polymer electrets is crucial for more intelligent utilization of these materials. The aim of the present paper is to investigate the lyophilization effect on electret and dielectric properties of polymer films. The properties of one synthetic polymer (polystyrene – PS) and two biopolymers (polylactic acid – PDLA and poly-ε-caprolactone – PεC) were investigated and compared. The samples were analyzed by means of dielectric relaxation spectroscopy in the range of 20 Hz – 1 MHz using QuadTech 1910 Inductance Analyzer. The frequency dependences of the impedance magnitude and the phase angle were measured. Regarding the electret properties, the samples were charged in a conventional three-electrode corona system. Positive and negative 5 kV voltage was applied to the corona electrode and 600 V voltage with the same polarity was applied to the grid as to the corona electrode. The results showed that the lyophilization increases time relaxation of the trapped charges. It was established that the samples charged in a positive corona are more stable than those charged in a negative corona.

1. Introduction

The increased industrial interest on the dielectric and electret properties of polymers reflects the increasing use of these materials in electronic devices, optoelectronic switches, printed circuit boards, microwave radar systems, batteries, fuel cells and many others. Polymer electrets are one of the most important types of dielectrics, which are widely used in scientific research and industrial applications. The effect of trapped charges on dielectric properties of the polymer electrets is important for more intelligent utilization of these materials. The electret properties of polymer dielectrics have been investigated in many literary sources [1-4].

In the recent years, the interest in the fundamental and the applied aspects of dielectric spectroscopy of polymer material has increased considerably since dielectric properties of the materials provide information on how the substance interact with electromagnetic energy. Detailed information of the basic aspects of the dielectric behavior of a polymeric materials can be found in many articles, books and reviews [5-9]. The information of these articles, however, refers almost exclusively to systems that do not change over time or are valid under certain conditions. The systematic studies with dialectric spectroscopy of systems with a time evolution of the structure have a more recent origin. The enormous potential of the dielectric spectroscopy of polymer materials is extensively motivated by G. Williams in [10]. The examples include testing systems that are chemically or physical changed as a result of chemical reaction, crystallization, phase separation, time, temperature and others.

The rapid development of modern technologies requires the research of new materials, polymer mixtures (blends) and composites based on polymer materials, design of dielectric materials and its modification at different voltages and temperatures.

The aim of the present paper is to investigate the lyophilization effect on electret and dielectric properties of different polymer films.

2. Experimental

2.1. Materials and samples preparation
Polystyrene (PS), polylactic acid (PDLA) and poly-ε-caprolactone (PεC) purchased from Lactel Absorbable Polymers (USA) were used for the preparation of different films. The two film types (non-lyophilized and lyophilized) were prepared in the following way: 1) non-lyophilized films of PS, PDLA and PεC – by 1 g polymer in 50 ml dioxane. The solution was poured in petri dishes and dried until the solvent evaporation; 2) lyophilized films of PS, PDLA and PεC – by dissolving of 1 g polymer in 50 ml dioxane. The solution was placed in a vacuum dryer at temperature –50 °C and dried until the total solvent evaporation. After that the obtained films were kept in an exicator at room temperatures and 0 % relative humidity (RH). After that two types samples of 30 mm diameter were cut from obtained films.

2.2. Corona charging and surface potential measurement
The charging of the samples in a corona discharge was carried out by means of a conventional corona triode system consisting of a corona electrode (needle), a ground plate electrode and a grid placed between them. The distance between the corona electrode and the grid was 10 mm and the distance between the grid and the grounded plate electrode was 3 mm. The experimental setup is shown on figure 1. The samples were charged under room conditions (T = 21-23 °C and RH = 50-60 %) for 1 minute. Positive or negative 5 kV voltage was applied to the corona electrode. Voltage of 600 V of the same polarity as that of the corona electrode were applied to the grid. Introducing a grid between the corona electrode and sample limits the electret surface potential to that of the grid and produces a more uniform distribution of charge on the electret surface. The electrets surface potential of the charged samples was measured by the vibrating electrode method with compensation and the estimated error was less than 5 %. The normalized surface potentials V/V_0 were calculated, as the value V_0 was the initial surface potential measured about 1 minute just after charging the electrets.

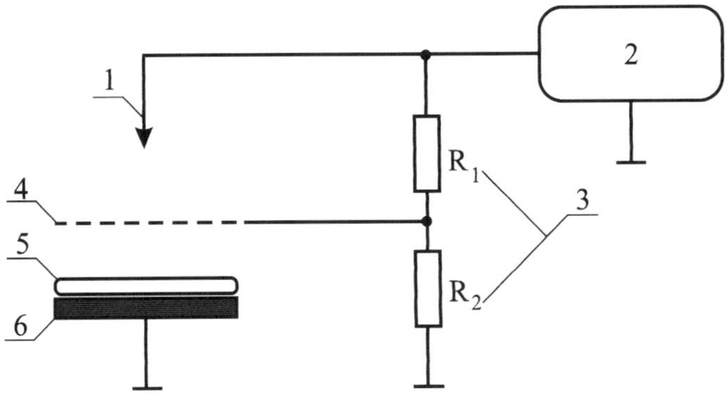

Figure 1. Conventional three electrode system:

1 – corona electrode (needle);
2 – high voltage power supply;
3 – voltage divider;
4 – grid;
5 – sample;
6 – ground plate electrode.

2.3. Dielectric spectroscopy
The measurement of the impedance spectra – impedance magnitude and phase angle in the frequency range from 20 Hz up to 1 MHz was done using QuadTech 1910 Inductance Analyzer. The magnitude of the test signal was 1 V. The tested sample was placed between two flat electrodes fixed to a Teflon holder. A spring was used to ensure good contact between the electrodes and the sample and to

IOP Conf. Series: Journal of Physics: Conf. Series **1186** (2019) 012013 doi:10.1088/1742-6596/1186/1/012013

provide constant tension force. The system of the sample and the electrodes was placed in an insulated chamber.

3. Results and discussion

3.1. Time storage influence on electrets surface potential decay

The dependences of the normalized surface potentials on the storage time for positively and negatively charged PS, PDLA and PεC non-lyophilized and lyophilized films have been studied for 360 minutes. The surface potential was measured once for 2 minutes except for the first 10 minutes because the change was rapidly decaying. After the period of 360 minutes steady state values of the surface potential were established for all investigated samples. Time dependences of the normalized surface potential for positively and negatively charged PS, PDLA and PεC non-lyophilized films are presented in figure 2a and figure 2b respectively. Such dependences for positively and negatively charged PS, PDLA and PεC lyophilized films are presented in figure 3a and figure 3b, respectively. In figure 4 the steady state values (after 360 minutes) of the normalized surface potential for all investigated samples is presented. Each point in the figure is a mean value from 4 samples. The calculated standard deviation was better than 5 % from the mean value with confidence level 0.95.

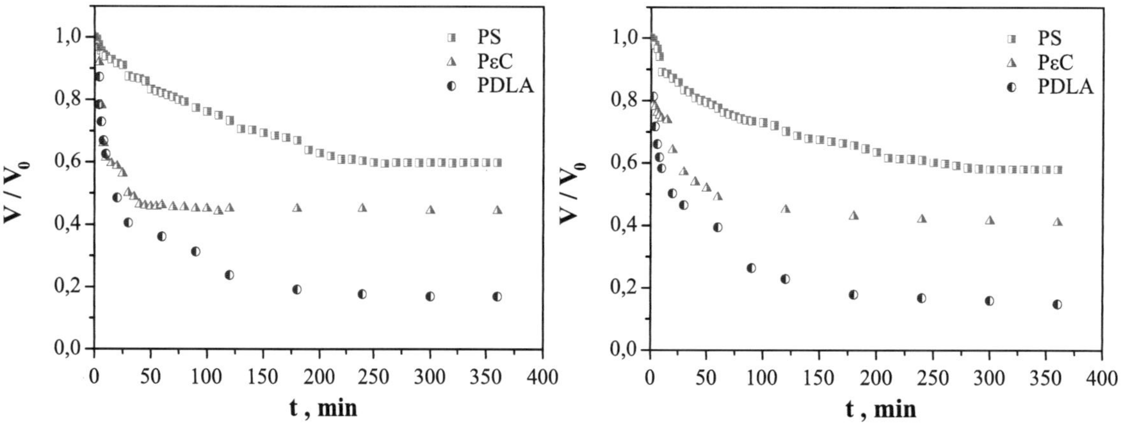

a) positive corona b) negative corona

Figure 2. Time dependences of the normalized surface potential for PS, PDLA and PεC non-lyophilized films.

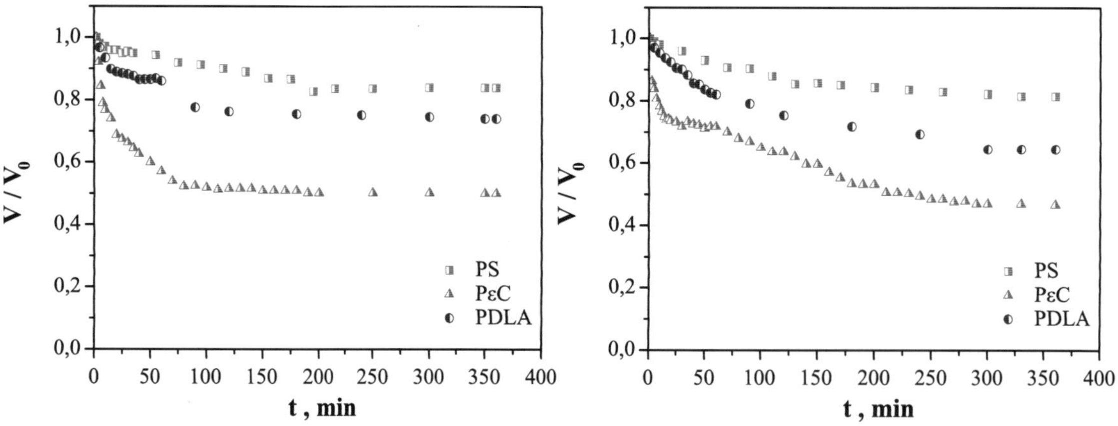

a) positive corona b) negative corona

Figure 3. Time dependences of the normalized surface potential for PS, PDLA and PεC lyophilized films.

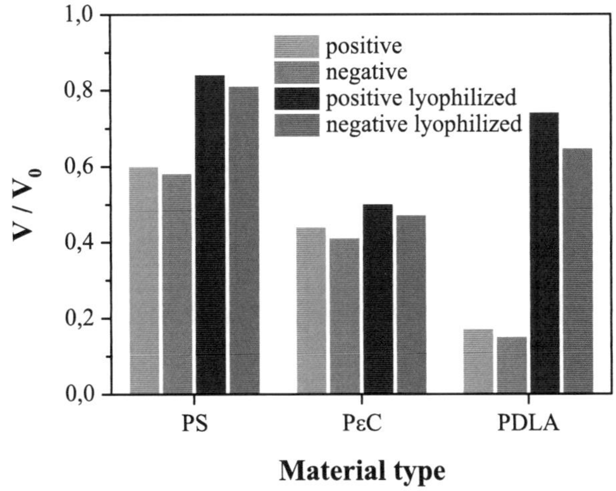

Figure 4. The dependences of the steady state value of the normalized surface potential for all investigated samples.

The results presented in figure 2-4 show that:

- For all investigated samples the values of the normalized surface potential are initially decaying exponentially for the first 50 minutes and then are slowly decreasing and are practically stabilized to the 360th minute.

The values of the surface potential of electrets depends on the amount of trapped charges in the different localized surface states of the samples. In the initial period of time after the corona charging, the surface potential rapidly decreases due to the release of the weakly captured charges from the shallow energy states. Then the surface potential stabilizes to a steady state value caused by the tightly captured charges in the deep energy traps. A similar behavior of exponential decay in electrets is observed in [10].

- The steady state values of the normalized surface potential for the samples charged in a positive corona are higher than those for the samples charged in a negative corona for all investigated samples.

It was established in [11-13] that during the corona discharge in the air, at atmospheric pressure, different types of ions are deposited on a sample, since charging in a corona discharge depending on the corona polarity. In the case of a positive corona the ions are mainly $H^+(H_2O)_n$ and the ones for the negative corona – CO^{3-}. Those ions are bound in traps of various depths and are released from them depending on the surrounding conditions.

- The values of the normalized surface potential of non-lyophilized films decay faster than those of the lyophilized independently of the corona polarity and the material type.
- The highest steady state values of the normalized surface potential and the best stability were observed for all investigated lyophilized films.

We assume that the lyophilization leads to the formation of porous structure and this significantly improves the ability to store charge and to a better stability with time. This was explained by investigating the morphology with scanning electron microscopy (SEM) in [14].

3.2. Dielectric spectroscopy

In figure 5, the frequency dependences of impedance magnitude Z for one biopolymer PεC and one synthetic polymer PS samples are presented. In figure 5, the frequency dependences of phase angle P for the same two PεC and PS samples are presented. Since the uncertainty did not exceed 4 %, the confidence intervals are not presented in the figures.

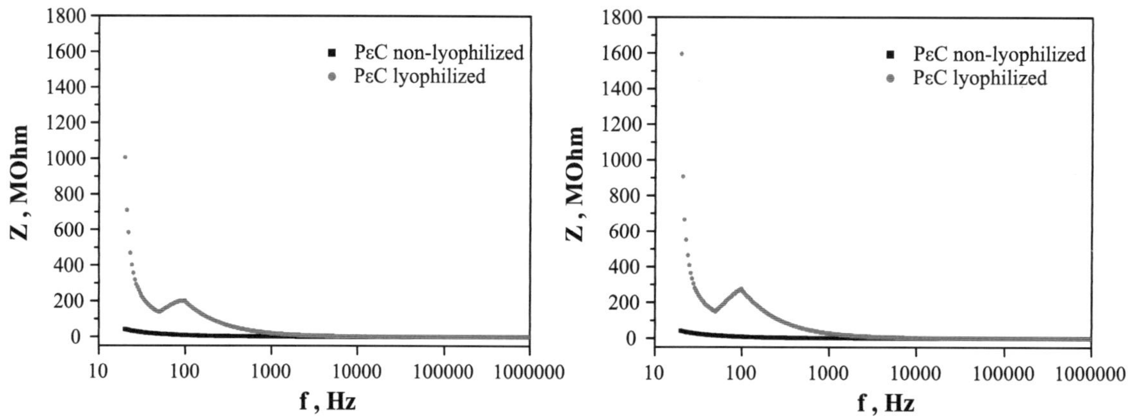

Figure 5. Impedance magnitude dependences on frequencies of investigated polymers.

As it can be seen in figure 5 and figure 6, the impedance magnitude values Z and phase angle P for lyophilized samples show considerable differences in the low frequency interval and the curves merge in the high frequency interval. This is supposed to be related to the porous structure of lyophilized samples. According of these results, frequency dependences of magnitude values Z and phase angle P are the dielectric spectroscopy parameters, sensitive to the uniformity of the film structure. Since the main differences were observed in the low frequency interval (up to 400 Hz) one may choose to monitor the behavior of Z and P at this low frequency range.

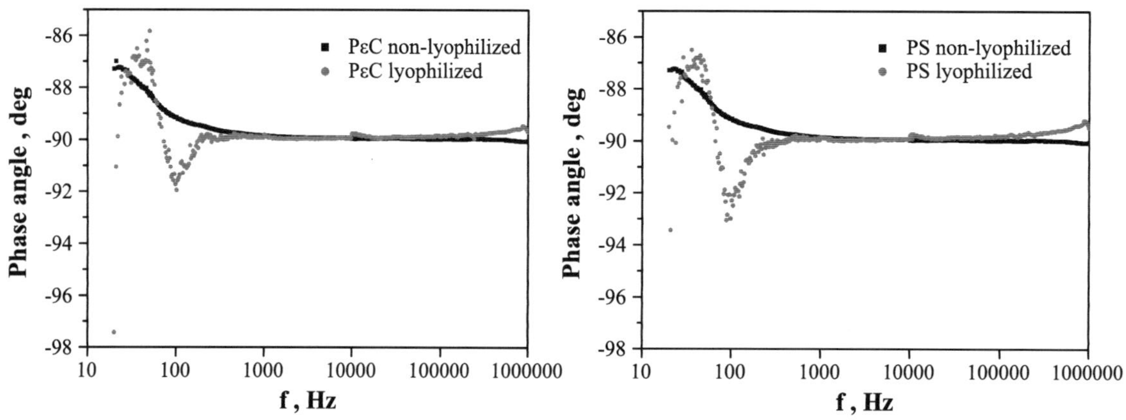

Figure 6. Phase angle dependences on frequencies of investigated polymers.

4. Conclusion

In this paper the lyophilization effect on electret and dielectric properties of different polymer films was investigated. The experimental results showed that the lyophilization influences on electret and dielectric properties of all investigated polymer films (biopolymer and synthetic polymer). The highest steady state values of the normalized surface potential and the best stability were observed for all investigated lyophilized polymer films. Therefore, the lyophilization effect lead to significantly improvement of the ability to store electrical charges and obtaining a stable electrets, which could be used in various applications. It was found that the frequency dependences of the dielectric spectroscopy parameters (magnitude values and phase angle) for lyophilized samples show considerable difference in the low frequency interval (up to 400 Hz). In this interval they were more sensitive to the uniformity of the film structure.

Acknowledgments
The research was supported by University of Plovdiv "Paisii Hilendarski", Department of Scientific Research, under the project FP17-FF-010.

References

[1] Sessler G M 1980 *Electrets* (Berlin-Heidelberg-New York:Springer-Verlag)
[2] Yovcheva T A 2010 *Corona charging of synthetic polymer films* (New York: Nova Science Publishers Inc)
[3] Cheng S-W, Chien Yu-H C, Huang T-Y, Liu Ch-L and Liou G-S 2018 *Polymer* **148** 382
[4] Dang Zh-M 2018 Dielectric Polymer Materials for High Density Energy Storage (Oxford: Elsevier)
[5] McCrum N G, Read B and Williams G 1967 *Anelastic and Dielectric Effects in Polymeric Solids* (New York: John Wiley & Sons Ltd)
[6] Karasz F E 1972 *Dielectric Properties of Polymers* (New York: Plenum Press)
[7] Jonscher A K 1983 *Dielectric Relaxation in Solids* (London: Chelsea Dielectric Press)
[8] Ku C C and Liepins R 1987 *Electrical Properties of Polymers* (Munich: Hanser)
[9] Riende E and Saiz E 1992 *Dipole Moments and Birefrengence of Polymers* (New York: Prentice Hall)
[10] Williams G 1997 *Keynote Lectures in Selected Topics of Polymer Science* (Madrid: CSIC)
[11] Sessler G M 2001 *J Electrostat* **51-52** 137
[12] Giacometti J A, Fedosov S and Costa M 1999 *Brazilian J Phys* **29** 269
[13] Giacometti J A and Olivera O N 1992 *IEEE Trans Electr Insul* **27** 924
[14] Yovcheva T A, Marudova M G, Viraneva A P, Sotirov S I, Rusev S H, Bodurov I P, Pilicheva B A, Uzunova Y I, Exner G K, Grancharova Ts Ts and Vlaeva I Y 2016 *AIP Conf Proc* **1722** 220026

IOP Conf. Series: Journal of Physics: Conf. Series **1186** (2019) 012014 doi:10.1088/1742-6596/1186/1/012014

Phase diagrams of Heisenberg chains with different cell spins: role of the tree-site exchange interactions

Nedko Ivanov[1,2] and Jürgen Schnack[1]

[1]Fakultät für Physik, Universität Bielefeld, Postfach 100131, D-33501 Bielefeld, Germany
[2]ISSP, Bulgarian Academy of Sciences, Tzarigradsko chaussee 72, 1784 Sofia, Bulgaria

E-mail: `ivanov@physik.uni-bielefeld.de`

Abstract. Additional three-site exchange interactions, naturally arising, e.g., in the strong-coupling expansion of the two-band Hubbard model and in various effective spin Hamiltonians, may stabilize various exotic spin phases which are not typical for the pure Heisenberg models on frustrated lattices and/or with extra biquadratic exchange terms. The reason is hidden in some specific features of the isotropic three-site exchange interactions, such as the promotion of collinear spin configurations as well as the reinforcing tendency towards clustering of the quantum spins on the shortest exchange bonds. As a result, systems with integer and half-integer composite spins in the unit cell might be expected to support different quantum phase diagrams. In this work, we study this phenomenon by comparing the quantum phase diagrams of two Heisenberg chains with extra three-site exchange terms and two different alternating pairs of site spins S and σ [$(S,\sigma) = (1,\frac{1}{2})$ and $(\frac{3}{2},\frac{1}{2})$], resulting in half-integer and integer cell spins $S+\sigma$, respectively.

1. Introduction

Quantum spin systems with competing exchange interactions present an excellent background for studying novel quantum states of matter such as the quantum spin ice, various spin nematic states, as well as different types of quantum spin liquids. Most of these systems are currently considered as real candidates for future applications in quantum computing and information technologies. The Heisenberg spin systems with three-site exchange couplings discussed below, which constitute a special class of magnetically frustrated systems, have been remaining scarcely explored [1, 2, 3, 4, 5, 6, 7], as opposed to those with extra biquadratic terms [8]. The large number of newly synthesized Heisenberg spin chains with alternating S and σ site spins $S > \sigma$ in the unit cell (for a review, see, e.g., Ref. [9]) offer very suitable platform where the effects of the frustrating three-site exchange interaction may be explored even in the extreme quantum case with site spins $\sigma = \frac{1}{2}$. Clearly, this is not the case with the standard uniform-spin models, since the square of the spin-$\frac{1}{2}$ matrix is a constant. Based on extended numerical density-matrix renormalization group (DMRG) and exact diagonalization (ED) calculations [5, 7], below we compare the quantum phase diagrams of this model for half-integer and integer cell spins. The Heisenberg chain with extra three-site exchange terms (see figure 1) is defined by the following Hamiltonian

$$\mathcal{H} = \sum_{n=1}^{L} J_1 \boldsymbol{S}_{2n}\cdot(\boldsymbol{\sigma}_{2n-1}+\boldsymbol{\sigma}_{2n+1}) + J_3\left[(\boldsymbol{S}_{2n}\cdot\boldsymbol{\sigma}_{2n-1})\,(\boldsymbol{S}_{2n}\cdot\boldsymbol{\sigma}_{2n+1})+h.c.\right], \tag{1}$$

IOP Conf. Series: Journal of Physics: Conf. Series **1186** (2019) 012014 doi:10.1088/1742-6596/1186/1/012014

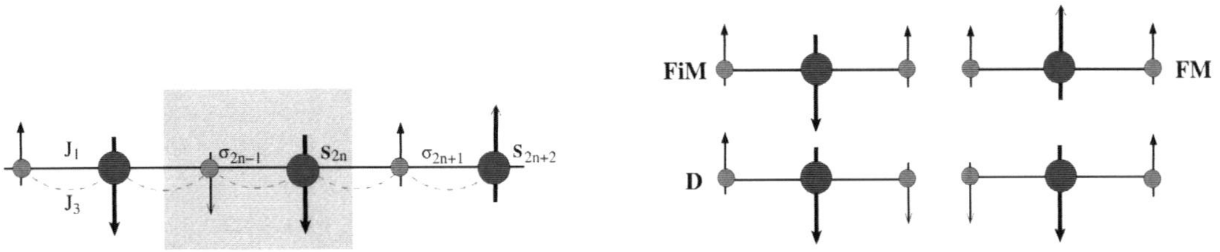

Figure 1. (left) The considered Heisenberg chains and (right) the optimal cluster states used as building blocks of the classical spin phases presented in figure 2.

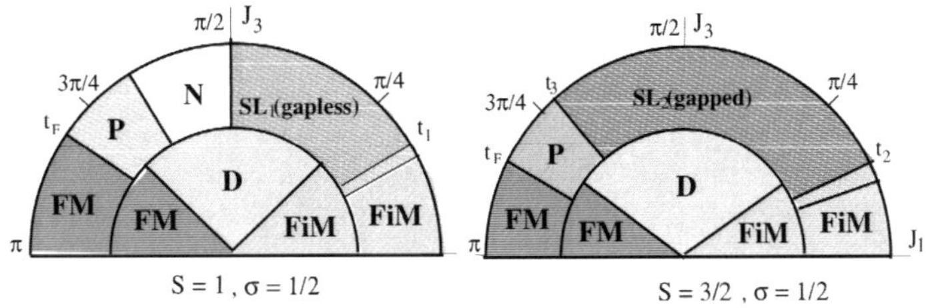

Figure 2. Classical (inner circles) and quantum (outer circles) phase diagrams of the models with half-integer (left) and integer (right) cell spins $S+\sigma$. FM and FiM stand for, respectively, the ferromagnetic and ferrimagnetic phases. $\boldsymbol{P}$ is a partially polarized magnetic state with a spatially modulated magnetization. $\boldsymbol{D}$ denotes a 2^L-fold degenerate classical phase composed of two types of collinear cluster spin configurations shown in figure 1. This phase contains both non-magnetic and magnetic ground-state configurations. $\boldsymbol{SL_1}$ and $\boldsymbol{SL_2}$ denote, respectively, gapless and gapped spin-liquid (SL) states characteristic, respectively, for the models with half-integer and integer cell spins. $\boldsymbol{N}$ stands for a quasi-nematic state with short-range nematic order and spin-2 lowest-lying excitations. Here, $t_1 \approx 30^\circ$, $t_2 = 25.5^\circ$, and $t_3 = 132^\circ$. $t_F = \pi - \arctan\frac{1}{\sigma(2S+1)}$ is the exact FM phase boundary for arbitrary site spins S and σ.

where L stands for the number of elementary cells, each containing two different site spins ($S > \sigma$). In what follows we use the standard parameterization of the coupling constants $J_1 = J\cos t$ and $J_3 = J\sin t$ ($0 \leq t < 2\pi$), where $J = 1$ is used as an energy unit. Note that for extreme quantum systems with $\sigma = \frac{1}{2}$ and arbitrary S, the well-known biquadratic spin interactions $(\boldsymbol{S}_i \cdot \boldsymbol{\sigma}_j)^2$ here reduce to bilinear Heisenberg terms, so that the extra three-site term in equation (1) actually is a general form for the isotropic nearest-neighbor exchange interactions which include up to tree lattice sites.

2. Classical phase diagram

First, let us briefly comment on the classical phase diagram of equation (1), which can be constructed by using the ground-state (GS) cluster configurations shown in figure 1. By fitting the directions of the sharing σ spins of the FM and FiM three-spin clusters, we can obtain the macroscopic non-degenerate (up to a global rotation of spins) FM and FiM configurations corresponding to local minima of the classical energy (FM and FiM phases in figure 2). Using the two cluster configurations D, the same procedure now leads to a 2^L-fold degenerate classical phase, denoted by D in figure 2. At a classical level, the phase diagrams of both models are qualitatively indistinguishable. The established classical diagram has been independently confirmed by classical Monte Carlo simulations.

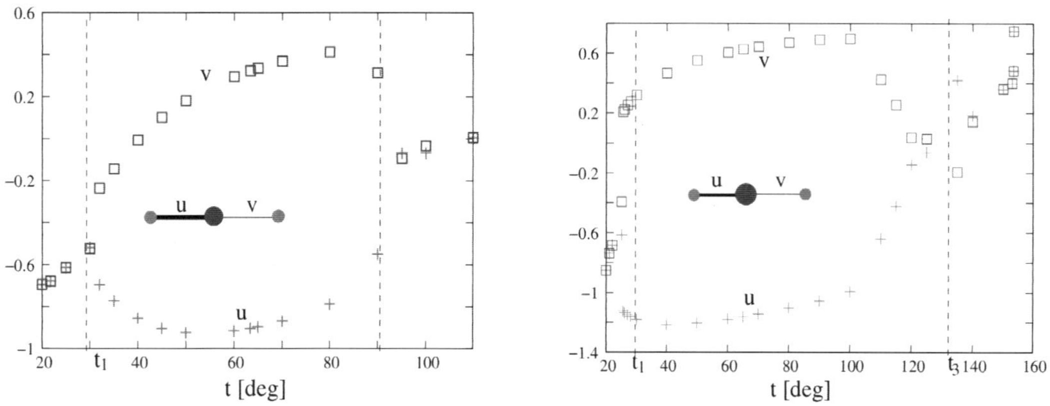

Figure 3. DMRG results for the nearest-neighbor spin-spin correlations $u = \langle \boldsymbol{S}_{2n} \cdot \boldsymbol{\sigma}_{2n-1} \rangle$ and $v = \langle \boldsymbol{S}_{2n} \cdot \boldsymbol{\sigma}_{2n+1} \rangle$ ($u \neq v$) in the $(S, \sigma) = (1, \frac{1}{2})$ (left) and $(\frac{3}{2}, \frac{1}{2})$ (right) chains *vs* t in the regions occupied by the non-magnetic doubly degenerate SL states.

3. Comparison of the quantum phase diagrams

3.1. Modulated non-Lieb-Mattis type magnetic states

The established partially-polarized magnetic states (see, e.g., the P sectors of the diagrams in figure 2) in both models do not exist in the classical phase diagram, so that their appearance is a pure quantum effect. Due to the frustration effect of the tree-site interaction, the Lieb-Mattis theorem [10] is not applicable, so that the so-called quantization of the unit-cell magnetic moment is destroyed, i.e., the magnetic moment per cell can take arbitrary values. This means, in particular, that in such systems the magnetic moment may change continuously to zero near the transition from magnetic to non-magnetic states (t_1 and t_2 points in figure 2), or to be spatially modulated, like in the P sectors of both models, which are close to the ferromagnetic phase boundary t_F. In the latter case, the local magnetization clearly shows an incommensurate modulation with long-distance periodicity depending on the distance from the boundary. As a matter of fact, these modulated states follow the rule $q(S + \sigma - m_0) =$ integer, where q is the period of the modulated structure, and m_0 is the magnetic moment per unit cell [5, 7]. For $t \lesssim t_F$, the modulated structure is characterized by $(q, m_0) = (8, \frac{9}{8})$ and $(3, \frac{5}{3})$ for the $(1, \frac{1}{2})$ and $(\frac{3}{2}, \frac{1}{2})$ models, respectively. Skipping further discussions on this interesting modulated ferrimagnetic states, we only mention that similar partially-polarized non-Lieb-Mattis-type phases have been earlier identified [11, 12] and studied [13] in other magnetically frustrated one-dimensional systems.

3.2. Doubly degenerated non-magnetic states

The DMRG analysis of the short-range correlations in open chains reveals regions of the parameter space in both models where in the GS a regular alternating-bond structure characterized by different values of the correlators $\langle \boldsymbol{\sigma}_{2n-1} \cdot \boldsymbol{S}_{2n} \rangle = u$ and $\langle \boldsymbol{S}_{2n} \cdot \boldsymbol{\sigma}_{2n+1} \rangle = v$ ($u < v$) appears. The uv (vu) "dimerized" GS $|\Psi_L\rangle$ ($|\Psi_R\rangle$) is stabilized in open chains with a σ spin on the left (right) end of the chain and corresponds to a uv (vu) dimerization of the local Hamiltonians h_n (see figure 3). The established uv structure of the GS is strongly pronounced in the middle of the SL regions in both models, see figure 2, where the values of u and v indicate the formation of nearly pure spin-$\frac{1}{2}$ and spin-2 (or spin-3) states of the nearest-neighbor spins in the systems with half-integer and integer cell spins $S + \sigma$, respectively. Very recently, it was also

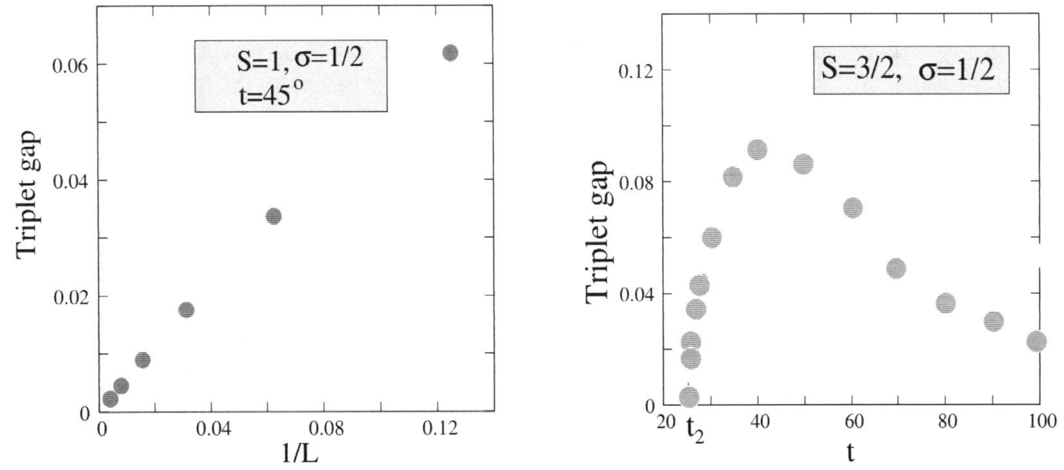

Figure 4. (left) Finite-size scaling of the lowest triplet excitation in the gapless SL_1 phase of the $(S,\sigma)=(1,\frac{1}{2})$ chain (DMRG results for $t=45°$). (right) Extrapolated triplet gap *vs* t in the gapped SL_2 phase of the $(S,\sigma)=(\frac{3}{2},\frac{1}{2})$ chain (DMRG results). Irrespective of the gap structure of the low-energy spectrum, the non-magnetic states in both models are doubly degenerate as a result of the dimerization.

demonstrated that similar clustering effects, but in uniform-spin Heisenberg chains with extra isotropic three-site interactions, stabilize exact Majumdar-Ghosh type states at some special points of the parameter space [1, 4]. Since a given $\boldsymbol{S}_{2n}$ spin can form such dimer states in two ways (with the left or right nearest-neighbor σ spin), all states in the SL regions are doubly degenerate, irrespective of the low-energy structure of the spectrum (gapped or gapless) of the non-magnetic phase. As an example, in figure 4 we show DMRG results for the gaps of the SL phases of both models with $(S,\sigma)=(1,\frac{1}{2})$ and $(\frac{3}{2},\frac{1}{2})$. The finite-size scaling of the lowest triplet excitation in the first model suggests a gapless non-magnetic state, as may be expected from the dimerized structure of the GS, which is characterized by local spin-$\frac{1}{2}$ dimers. On the other hand, the dimerized SL state of the $(\frac{3}{2},\frac{1}{2})$ chain suggests the formation of either spin-1 (smaller t), or spin-2 (larger t) local dimers, which leads to the formation of a gapped SL state in the whole non-magnetic region.

3.3. Nematic-like phase

The three-site spin interaction term in equation (1) for every unit cell can be represented in the following symmetric form

$$V_n^{(3)}=\frac{1}{2}\left(S_{2n}^{\alpha}S_{2n}^{\beta}+S_{2n}^{\beta}S_{2n}^{\alpha}\right)\left(\sigma_{2n-1}^{\alpha}\sigma_{2n+1}^{\beta}+\sigma_{2n-1}^{\beta}\sigma_{2n+1}^{\alpha}\right), \tag{2}$$

where $\alpha,\beta=x,y,$ and z. The two symmetric terms in the parentheses may be considered as tensor order parameters (up to some normalization factors) for on-site and bond nematic phases constructed from S and σ types spins, respectively. This representation of $V_n^{(3)}$ means that in systems with dominant isotropic three-site spin interactions nematic orders of different types could be stabilized. As a matter of fact, the numerical ED calculations concerning the low-lying excitations in periodic $L=8$ chains, described by the Hamiltonian (1), point towards the formation of such a short-ranged nematic order in the subsystem of S spins [6], the N sector in figure 2. Information about the non-magnetic N state in the $(1,\frac{1}{2})$ model can be extracted

IOP Conf. Series: Journal of Physics: Conf. Series **1186** (2019) 012014 doi:10.1088/1742-6596/1186/1/012014

form the lowest-laying states in different total-spin (S_t) sectors. The established tower of well-separated lowest multiplets containing only even S_t sectors is a special feature of the spectrum of this quasi-quadrupolar spin state. This spectrum structure is known as a fingerprint of the spin quadrupolar (i.e., nematic) order, unlike the Anderson tower of states [14] containing all S_t sectors. The latter is a characteristic of the Néel order. In fact, Anderson towers of states have been observed even in some finite isotropic spin-S chains and magnetic molecules [15, 16, 17]. Interestingly, such regions with enhanced quadrupolar fluctuations do not appear on the phase diagram of the $(\frac{3}{2}, \frac{1}{2})$ model.

4. Summary

In the present work, based on extensive DMRG and ED numerical calculations, we compared the quantum phase diagrams of two Heisenberg chains with extra three-site spin exchange interactions and different (half-integer and integer) cell spins $S + \sigma$. We demonstrated that in the regions with dominating three-site spin interactions the models with half-integer and integer cell spins exhibit different non-magnetic phases. Nevertheless, the stabilization of these phases is closely related to some peculiarities of the three-site spin interaction, such as the support of collinear spin states, the pronounced tendency towards dimerizations of the nearest-neighbor exchange bonds, and the enhancement of quadrupolar spin fluctuations. We believe that the presented results will encourage future studies on real systems with dominating isotropic three-site exchange interactions.

Acknowledgments

We are grateful to Jörg Ummethum for adapting his DMRG code systems with three-site spin interactions. Support by the Bulgarian Science Foundation, Grant DN0818/14.12.2016, is gratefully acknowledged.

References

[1] Michaud F, Vernay F , Manmana S R and Mila F 2012 *Phys. Rev. Lett.* **108** 127202
[2] Michaud F, Manmana S R and Mila F 2013 *Phys. Rev.* B **87** 140404(R)
[3] Michaud F and Mila F 2013 *Phys. Rev.* B **88** 094435
[4] Wang Z -Y, Furuya S C, Nakamura M and Komakura R 2013 *Phys. Rev.* B **88** 224419
[5] Ivanov N B, Ummethum J. and Schnack J 2014 *Eur. Phys. J.* B **87** 226
[6] Ivanov N B and Schnack J 2014 *J. Phys.: Conf. Series* **558** 012015
[7] Ivanov N B, Petrova S I and Schnack J 2016 *Eur. Phys. J.* B **89** 121
[8] Lhuillier C and Misguich G, *Introduction to frustrated magnetism Materials, Experiments, Theory* 2011 (Springer Series in Solid State Sciences vol 164) ed C Lacroix, P Mendels and F Mila (Berlin Heidelberg: Springer Verlag) chapter 2 pp 23-41
[9] Ivanov N B and Schollwöck U, *Quantum Heisenberg ferrimagnetic chains Some Contemporary Problems of Condensed Matter Physics* 2001 ed Vlaev S J and Gaggero-Sager L M (Nova Science Publishers Inc., Hungtington, New York) chapter 5 pp 145-176
[10] Lieb E H and Mattis D C 1962 *J. Math. Phys.* **3** 749
[11] Ivanov N B, Richter J 2004 *Phys. Rev.* B **69** 214420
[12] Shimokawa T, Nakano H, 2013 *J. Kor. Phys. Soc. (SI)* **63** 591
[13] Furuya Sh S, Giamarchi Th 2014 *Phys. Rev.* B **89** 205131
[14] Anderson P W 1952 *Phys. Rev.* B **86** 694
[15] Schnack J and Luban M 2000 *Phys. Rev.* B **63** 014418
[16] Waldmann O 2001 *Phys. Rev.* B **65** 024424
[17] Machens A, Konstantinidis N P, Waldmann O, Schneider I and Eggert S 2013 *Phys. Rev.* B **87** 144409

Entropy and specific heat of a critical quantum system with long-range interaction

Ekaterina S Pisanova

Faculty of Physics and Technology, University of Plovdiv "Paisii Hilendarski", 24 Tzar Assen Str., 4000 Plovdiv, Bulgaria

E-mail: katia_pisanova@abv.bg

Abstract. The entropy and the specific heat of a quantum spherical model with a long-range interaction (decreasing at large distances r as $r^{-d-\sigma}$ where d is the space dimensionality and $0 < \sigma \leq 2$) are studied in the quantum critical region at the upper quantum critical dimension $d = 3\sigma/2$. The problem of obtaining the temperature-dependent corrections to the ground state free energy involves the solution of a transcendental equation, the exact solution of which is expressed in terms of the Lambert W-function. The free energy, the entropy and the specific heat in the quantum critical region are derived in terms of the Lambert W-function. For systems in real physical dimensions (chains, thin layers, i.e. films and three-dimensional systems) the exact results for the entropy and the specific heat obtained in terms of the Lambert W-function and the leading asymptotic ones are compared on the basis of the calculated relative errors. It can be concluded that for a class of quantum models at the upper critical dimension the Lambert W-function is a very effective tool for an exact computation of low-temperature critical properties, especially in the finite-temperature quantum critical region.

1. Introduction

The aim of this study is to obtain the entropy and the specific heat of a critical quantum system in the finite-temperature quantum critical region at the upper quantum critical dimension.

The model considered here is a quantum version of the spherical model or more precisely the "spherical quantum rotors" model (SQRM) [1]. Its equivalence with the quantum nonlinear $O(n)$ sigma model (QNLσM) [2–5], widely used in the exploration of the critical behavior of quantum magnets, in the limit $n \to \infty$ is shown in [1]. The mathematical analogy between the SQRM and a modification of the φ^4-lattice model used extensively in studying the structural phase transitions (see e.g. [6] and references therein) is discussed in [7]. In [1] the critical exponents of the SQRM for both the zero-temperature quantum critical point and the finite-temperature classical one as functions of the dimensionality have been obtained. Due to its exact solvability for each dimensionality the SQRM is very useful for analytical research of the scaling properties of quantum critical phenomena and of finite-size effects [7–9].

In the framework of the model of structural phase transitions, mentioned above, it has been shown that the Lambert W-function [10] is a very effective tool for an exact computation of non-universal critical properties in the neighborhoods of both the classical and the quantum critical points at the corresponding upper critical dimensions [11, 12]. In [13] the specific heat of the classical mean-spherical model has been exactly calculated in terms of the Lambert W-function.

In this paper, using the properties of the Lambert W-function we study the entropy and the specific heat of the SQRM in the quantum critical region at the upper quantum critical dimension.

2. The model

The Hamiltonian of the model is [1]

$$H = \frac{1}{2} g \sum_l P_l^2 - \frac{1}{2} \sum_{l,l'} J_{l,l'} S_l S_{l'} + \frac{1}{2} \mu \sum_l S_l^2 - h \sum_l S_l, \tag{1}$$

where S_l are spin operators at site l. The operators P_l play the role of "conjugated" momenta (i.e. $[S_l, S_{l'}] = 0$, $[P_l, P_{l'}] = 0$, and $[P_l, S_{l'}] = i\delta_{ll'}$, with $\hbar = 1$). The coupling constant g measures the strength of the quantum fluctuations (below, it will be called quantum parameter), $J_{l,l'} = J(r)$ is the interaction matrix between spins at sites l and l', h is an ordering magnetic field, and the spherical field μ is introduced so as to ensure the constraint

$$\sum_l \langle S_l^2 \rangle = N. \tag{2}$$

Here N is the total number of quantum spins located at sites "l" of a d-dimensional hypercubical lattice and $\langle \ldots \rangle$ denotes the standard thermodynamic average taken with the Hamiltonian (1). The commutation relations for the operators S_l and P_l together with the kinetic term in the Hamiltonian (1) do not describe quantum Heisenberg-Dirac spins but quantum rotors as is pointed out in [1].

The considered here long-range interaction potential $J(r) \sim r^{-d-\sigma}$, where $\sigma > 0$ and $r \to \infty$, enters the exact expression for the free energy of the model only through its Fourier transform [14]. The long-wavelength (small $k \equiv |\mathbf{k}|$) leading asymptotic of its Fourier transform is $U(k) \cong J\,|\mathbf{k}|^{\sigma}$, $0 < \sigma \leq 2$, where the energy scale has been fixed so that $U(0) = 0$. The case $\sigma \geq 2$ corresponds to short-range interaction, i.e. the universality class then does not depend on σ. Values satisfying $0 < \sigma < 2$ correspond to long-range interactions and the critical behavior depends on σ.

The normalized free energy density $\widetilde{f} = f/J$ of the model (1) in the thermodynamic limit $N \to \infty$ at $h = 0$ has the form [9]

$$\widetilde{f}(t, \lambda | d, \sigma) = \sup_{\phi} \left\{ t k_d \int_0^{x_D} x^{d-1} \ln \left[2 \sinh \left(\frac{\lambda}{2t} \sqrt{\phi + x^{\sigma}} \right) \right] dx - \frac{\phi}{2} \right\} - d, \tag{3}$$

where the supremum in the equation (3) is attained at the solutions of the mean-spherical constraint, equation (2), that is obtained by differentiating the above expression with respect to ϕ

$$1 = \frac{\lambda}{2} k_d \int_0^{x_D} \frac{x^{d-1}}{\sqrt{\phi + x^{\sigma}}} \coth \left(\frac{\lambda}{2t} \sqrt{\phi + x^{\sigma}} \right) dx. \tag{4}$$

In equations (3) and (4) we have introduced the notations: $\lambda = \sqrt{g/J}$ is the normalized quantum parameter, $t = T/J$ - the normalized temperature, $\phi = \mu/J$ - the scaled spherical field, $k_d^{-1} = \frac{1}{2}(4\pi)^{\frac{d}{2}}\,\Gamma(d/2)$ (Γ is the Euler gamma function) and $x_D = 2\pi(d/S_d)^{1/d}$ is the radius of the sphericalized Brillouin zone, $S_d = 2\pi^{d/2}/\Gamma(d/2)$. From the condition that in the quantum critical point ($\lambda = \lambda_c, t = 0$) the solution of (4) is $\phi = 0$, one obtains the critical value λ_c, $\lambda_c = (2 - \sigma/d) x_D^{\sigma/2}$.

Equations (3) and (4) provide the basis for studying the bulk critical behavior of the model.

In this paper we shall consider the entropy and the specific heat capacity in the quantum critical region ($\lambda = \lambda_c$) at the upper quantum critical dimension $d = 3\sigma/2$. In the neighborhood of the quantum critical point, by using the exact solution of the spherical field equation (4) in terms of the Lambert W-function, we establish exact analytical expressions for the free energy density, the entropy and the specific heat capacity. When the upper critical dimension coincides with real physical dimensions (chains, thin layers, i.e. films and three-dimensional systems), the obtained exact results for the specific heat are compared with the leading asymptotic ones on the basis of the calculated relative errors.

3. Temperature-dependent corrections to the free energy density at the quantum critical point

In the low-temperature limit ($t << 1$) replacing $\frac{\lambda}{t}x_D^{\sigma/2}$ by infinity at $d = 3\sigma/2$ the above expression for the free energy density (3) takes the form

$$\widetilde{f}(t,\lambda|\sigma) = \sup_{\phi}\left\{\frac{1}{x_D^{\sigma}}\frac{\lambda}{\lambda_c}\left[\frac{2}{3}x_D^{3\sigma/2}\left(\phi + x_D^{\sigma}\right)^{1/2}F\left(-\frac{1}{2},1;\frac{5}{2};\frac{x_D^{\sigma}}{x_D^{\sigma}+\phi}\right) + 4\left(\frac{t}{\lambda}\right)^4\Phi(\phi)\right] - \frac{\phi}{2}\right\} - \frac{3}{2}\sigma, \tag{5}$$

where $F(a,b;c;z)$ is the hypergeometric function and

$$\Phi(\phi) = \int_0^{\infty} x^2 \ln\left[1 - \exp\left(-\sqrt{\frac{\lambda^2}{t^2}\phi + x^2}\right)\right]dx. \tag{6}$$

Sufficiently close to the quantum critical point ($\phi << 1$) the integral (6) can be evaluated as $\Phi(\phi) = \Phi(0) + \Phi'(0)\phi$ and retaining the leading terms, including $\phi^2\ln\phi$-term, after some algebra, we can write (5) in the form

$$\widetilde{f}(t,\lambda|\sigma) = \widetilde{f}_c + \frac{1}{2}\left(\frac{\lambda}{\lambda_c} - 1\right)\phi + A\phi^2\ln\phi + Bt^4 + Ct^2\phi, \tag{7}$$

where $\widetilde{f}_c = x_D^{\sigma}/2 - 3\sigma/2$ is the free energy density at the quantum critical point,

$$A = \frac{1}{\left[6(4\pi)^{\frac{3}{4}\sigma}\sigma\Gamma\left(\frac{3}{4}\sigma\right)\right]^{2/3}}, \tag{8}$$

$$B = -\frac{18\zeta(4)}{\left[2(4\pi)^{\frac{3}{4}\sigma}\sigma\Gamma\left(\frac{3}{4}\sigma\right)\right]^{2}} \tag{9}$$

and

$$C = \frac{3\zeta(2)}{\left[\sqrt[4]{6}(4\pi)^{\frac{3}{4}\sigma}\sigma\Gamma\left(\frac{3}{4}\sigma\right)\right]^{4/3}}, \tag{10}$$

$\zeta(x)$ is the Riemann zeta function. In (7) ϕ is the solution of the corresponding spherical field equation, that reads

$$\frac{1}{2}\left(\frac{\lambda}{\lambda_c} - 1\right) + 2A\phi\ln\phi + A\phi + Ct^2 = 0. \tag{11}$$

At $\lambda = \lambda_c$, i.e. in the quantum critical region, (11) has the form

$$\phi\ln\phi + \frac{1}{2}\phi = -\frac{C}{2A}t^2. \tag{12}$$

By the substitution $\ln\phi + 1/2 = W(x)$, the equation (12) takes the form of the defining equation for the Lambert W-function [10],

$$W(x)\exp(W(x)) = x. \tag{13}$$

When x is a real variable, then for $-1/e \leq x < 0$ there are two branches with real values of $W(x)$: the branch $W_0(x)$ satisfying $-1 \leq W_0(x)$ and the branch $W_{-1}(x) \leq -1$, as $\lim_{x\to 0^-} W_{-1}(x) = -\infty$. Thus, in terms of the Lambert W-function, the exact solution of (12) is

$$\phi = \exp\left[W_{-1}\left(-\epsilon_t\right) - 1/2\right], \tag{14}$$

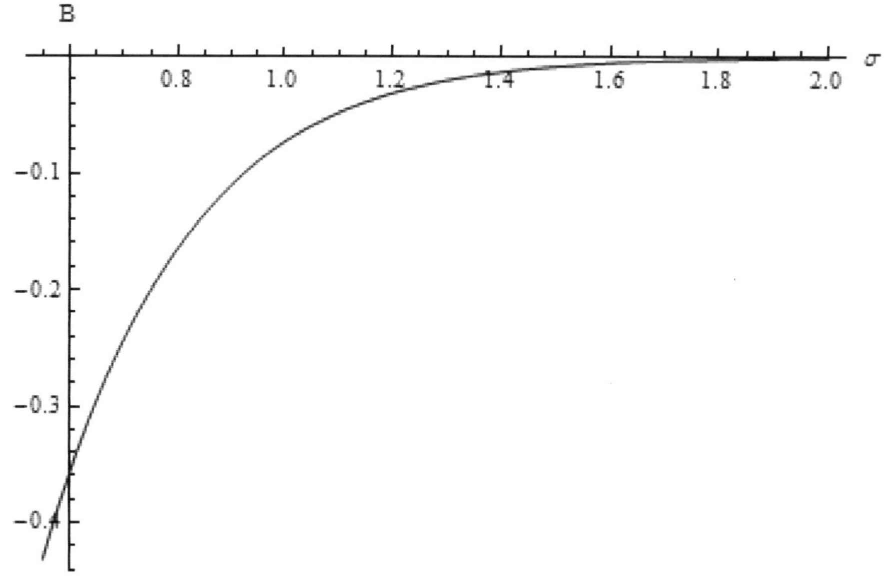

Figure 1. The coefficient B as a function of σ, i.e. of the dimensionality of the system.

where $\epsilon_t = (C/2A)\sqrt{e}t^2$. The choice of the solution $W_{-1}(x)$ corresponds to the fact that at the quantum critical point ($\lambda = \lambda_c, t = 0$) the scaled spherical field ϕ vanishes.

By using the convergent series for the branch $W_{-1}(-x)$ at small $x > 0$ [10] and retaining the leading two terms, we get the asymptotic expression for ϕ,

$$\phi \approx -\frac{C}{4A}\frac{t^2}{\ln t}. \tag{15}$$

Let us note that for a wide class of models the logarithmic behavior of the susceptibility $\chi = \phi^{-1}$ is well known in the theory of phase transitions (see, e.g. [6]).

At $\lambda = \lambda_c$ near to the quantum critical point, by substituting (14) into (7), we get the following exact expression for the temperature-dependent corrections to the free energy density at the quantum critical point

$$\tilde{f}(t, \lambda_c|\sigma) - \tilde{f}_c = Bt^4\left[1 + \frac{5}{4}\frac{1}{W_{-1}(-\epsilon_t)} + \frac{5}{8}\frac{1}{W_{-1}^2(-\epsilon_t)}\right]. \tag{16}$$

The asymptotic expression for the temperature-dependent corrections to the free energy density at the quantum critical point can be obtained by substituting (15) into (7). It is easy to see that the coefficient B, given by (9), characterizes the leading t-corrections to the ground state free energy of the system at its quantum critical point. The coefficient B as a function of σ is graphically presented on figure 1.

4. Derivatives of the free energy with respect to the temperature - entropy and specific heat

By using (7) and taking into account the spherical field equation (12), for the entropy, S, and the specific heat, c, per spin in the quantum critical region ($\lambda = \lambda_c$) near to the quantum critical point we have

$$S = -\frac{\partial \tilde{f}}{\partial t} = -4Bt^3 - 2Ct\phi \tag{17}$$

and

$$c = t\frac{\partial S}{\partial t} = -12Bt^3 - 2Ct\phi - 2Ct^2\frac{d\phi}{dt}. \tag{18}$$

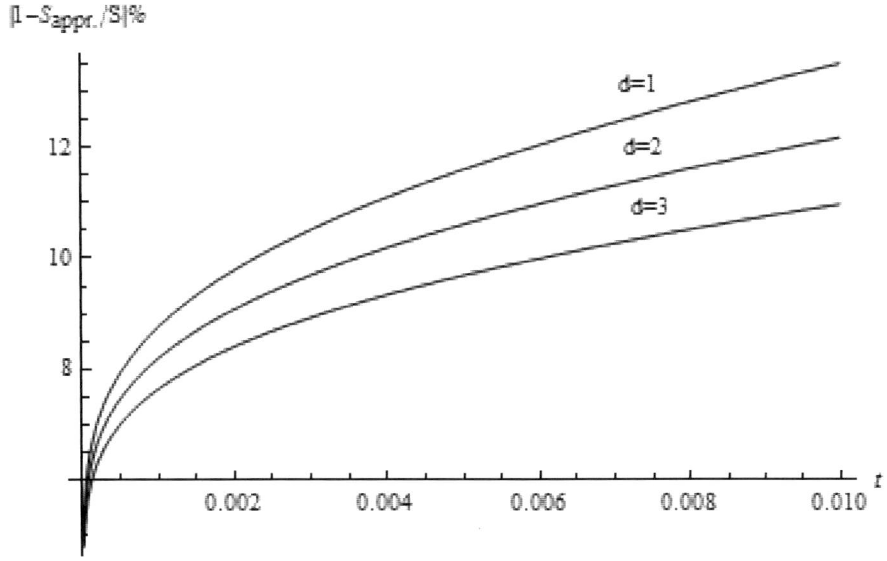

Figure 2. The percent relative error $\left|1 - S_{appr.}/S\right|$.

Substituting in (17) and (18) the exact solution for ϕ, given by (14), we obtain the following exact expressions

$$S = -4Bt^3\left[1 + \frac{5}{4}\frac{1}{W_{-1}\left(-\epsilon_t\right)}\right] \tag{19}$$

and

$$c = -12Bt^3\left[1 + \frac{5}{12}\frac{1}{W_{-1}\left(-\epsilon_t\right)} + \frac{5}{6}\frac{1}{\left(1 + W_{-1}\left(-\epsilon_t\right)\right)}\right]. \tag{20}$$

From (19) and (20), by using the series for the Lambert W-function [10] we get the leading asymptotic behavior of the entropy and the specific heat as follows

$$S_{appr.} = -4Bt^3\left[1 + O\left(\frac{5}{8\ln t}\right)\right] \tag{21}$$

and

$$c_{appr.} = -12Bt^3\left[1 + O\left(\frac{5}{8\ln t}\right)\right]. \tag{22}$$

For systems in real physical dimensions (chains ($\sigma = 2/3$), films ($\sigma = 4/3$) and three-dimensional systems ($\sigma = 2$)) the percent relative errors $\left|1 - S_{appr.}/S\right|$ and $\left|1 - c_{appr.}/c\right|$ as functions of the normalized temperature t and the dimensionality of the system d are graphically presented on figure 2 and figure 3, respectively.

5. Conclusions

In the quantum critical region at the upper quantum critical dimension $d = 3\sigma/2$ exact expressions for the free energy density (16), the entropy (19) and the specific heat (20) of the model (1) are obtained in the Lambert W-function terms. The coefficient B, characterizing the leading temperature-dependent corrections to the free energy density at the quantum critical point, is an increasing function of the long-range exponent σ, i.e. of the dimensionality d of the system (see figure 1).

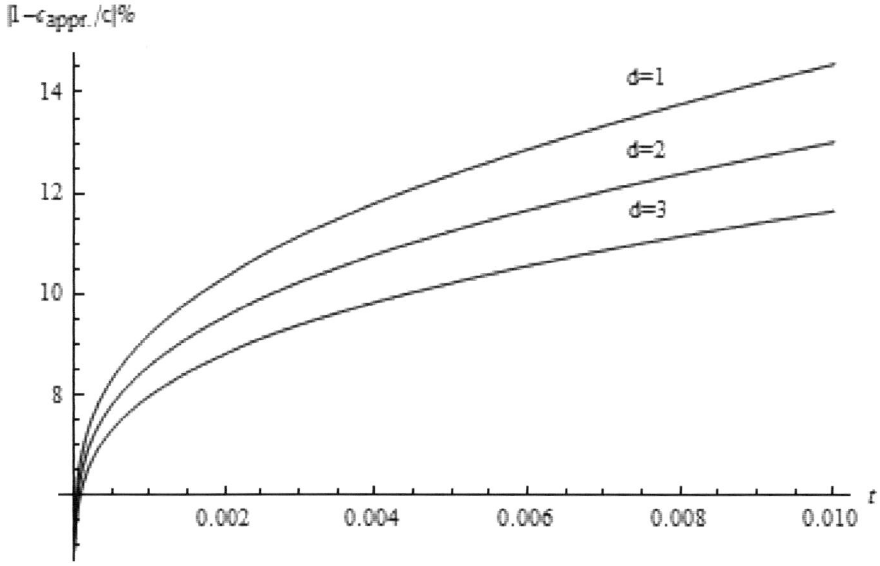

Figure 3. The percent relative error $\left|1 - c_{appr.}/c\right|$.

Sufficiently close to the quantum critical point the asymptotic results for the entropy, (21), and the specific heat capacity, (22), are applicable. As one can see the critical amplitudes of the entropy and the specific heat capacity are decreasing functions of σ, i.e. of the dimensionality d of the system. It is not surprising that the critical exponent α, characterizing the leading critical behavior of the specific heat, at the upper critical dimension is not depends on σ and for the specific heat the Debye law holds, $c \sim t^3$.

From figure 2 and figure 3 it is easy to see that the obtained exact results for the entropy and the specific heat are especially applicable to low-dimensional systems in the finite-temperature quantum critical region.

Acknowledgments
This study was supported by Project FP17-FF-010 at the University of Plovdiv (Bulgaria).

References

[1] Vojta T 1996 *Phys. Rev.* B **53** 710
[2] Rosenstein B, Warr B J and Park S H 1990 *Nucl. Phys.* B **336** 435
[3] Castro Neto A H and Fradkin E 1993 *Nucl. Phys.* B **400** 525
[4] Chubukov A V, Sachdev S and Ye J 1994 *Phys. Rev.* B **49** 11919
[5] Sachdev S 2011 *Quantum Phase Transitions* (Cambridge: Cambridge University Press)
[6] Brankov J G, Danchev D M and Tonchev N S 2000 Theory of Critical Phenomena in Finite-Size Systems: Scaling and Quantum Effects *Series on Modern Condensed Matter Physics* vol 9 (Singapore: World Scientific) 105
[7] Chamati H, Danchev D M, Pisanova E S and Tonchev N 1997 *Preprint* IC/97/82 (*Preprint* cond-mat/9707280)
[8] Chamati H, Pisanova E S and Tonchev N 1998 *Phys. Rev.* B **57** 5798
[9] Chamati H, Danchev D and Tonchev N 2000 *Eur. Phys. J.* B **14** 307 (*Preprint* cond-mat/9809315v3)
[10] Corles R M, Gonnet G H, Hare D E G, Jeffrey D J and Knuth D E 1996 *Adv. Comput. Math.* **5** 329
[11] Pisanova E S, Tonchev N S and Kisov Hr T 2009 *AIP Conf. Proc.* **1203** 255
[12] Pisanova E S and Ivanov S 2013 *Bulg. J. Phys.* **40** 159
[13] Pisanova E S and Ivanov S 2015 *Bulg. Chem. Commun., Spec. Issue* B **47** 269
[14] Joyce G S 1972 *Phase Transitions and Critical Phenomena* Vol. 2, ed C. Domb and M. S. Green (London: Academic Press) 375

IOP Conf. Series: Journal of Physics: Conf. Series **1186** (2019) 012016 doi:10.1088/1742-6596/1186/1/012016

Interaction of solitons with a qubit in an anisotropic Heisenberg spin chain

S K Varbev, R S Kamburova and M T Primatarowa

Georgi Nadjakov Institute of Solid State Physics, Bulgarian Academy of Sciences, 1784 Sofia, Bulgaria

E-mail: `stanislavvarbev@issp.bas.bg`

Abstract. The interaction between solitons, propagating in a magnetic chain, with a qubit is studied. The spin chain is described by the Heisenberg model in a nearest-neighbour approximation with anisotropy. The quantum bit (qubit) is modeled by a two level quantum system. The equation for the evolution of the qubit is obtained. The role of the parameters of the soliton for the quantum bit dynamics is investigated. The results are analyzed using a geometrical representation for the qubit as a Bloch sphere.

1. Introduction

Quantum bit in a quantum computer can be realised in many ways. In general, we could think of it as a two-level quantum system. A possible realization for a qubit is its implementation as an electron spin in Ref. [1]. Another proposed realization for a quantum bit is as a composed particle or a molecular cluster with magnetic moment and spin in Ref. [2].

The propagation and interaction of solitons in one-dimensional magnetic systems were studied in Refs. [3,4]. Propagation of soliton-like excitations along spin chains has been proposed as a possible way for transmitting both classical and quantum information between two distant parts of the system with negligible dispersion and dissipation in Refs. [5-7]. This gives the possibility to control the qubit levels, varying the parameters of the soliton.

In the present paper we study the interaction between a soliton, propagating in a magnetic chain, and a single-spin $\frac{\bar{h}}{2}$ qubit. The soliton is a solution of semiclassical equations of motion for the spin chain described by the Heisenberg model. The evolution of the qubit is described with the use of density operator formalism. The density operator for a qubit can be written in terms of the Bloch vector. The evolutionary equations for the components of the Bloch vector are determined. The system of equations is solved numerically for different values of the parameters. The dependence of the quantum bit dynamics on the parameters of the soliton is investigated. The results are analysed geometrically onto the Bloch sphere.

2. The model of the system

We consider the interaction of a ferromagnetic Heisenberg chain of N spins with magnitude S described in the nearest-neighbor approximation and a quantum bit by the following Hamiltonian

$$\hat{H} = -J\sum_{n=1}^{N} \hat{\vec{S}}_n \cdot \hat{\vec{S}}_{n+1} - A\sum_{n=1}^{N} (\hat{S}_n^z)^2 - \mu H_0 \sum_{n=1}^{N} \hat{S}_n^z - \nu H_0 \hat{\sigma}_j^z + d_{xy}(\hat{S}_j^x \hat{\sigma}_j^x + \hat{S}_j^y \hat{\sigma}_j^y) + d_z \hat{S}_j^z \hat{\sigma}_j^z. \quad (1)$$

IOP Conf. Series: Journal of Physics: Conf. Series **1186** (2019) 012016 doi:10.1088/1742-6596/1186/1/012016

The identification of the quantities is as follows. $\hat{\vec{S}}_n$ are the spin operators for the spin chain, $\hat{\vec{\sigma}}_j$ is the spin operator for the qubit placed at an arbitrary site j. $J > 0$ is the exchange integral, and A is the on-site anisotropy constant for the chain which can be positive (easy axis) or negative (easy plane). H_0 is the external magnetic field applied along the z-axis, so that in the ground state of the system all spins are aligned in the z-direction, μ is the magnetic moment per spin in the chain, ν is the magnetic moment per spin for the qubit and d_{xy} and d_z characterize the coupling interaction between the qubit and the spin from the chain labeled by the index j.

We use for the scalar products of any spin vector operators in (1) the rule ($\hbar = 1$)

$$\hat{\vec{S}}_n \cdot \hat{\sigma}_j = \frac{\hat{S}_n{}^{+} \sigma_j^{-} + \hat{S}_n{}^{-} \hat{\sigma}_j{}^{+}}{2} + \hat{S}_n{}^{z} \hat{\sigma}_j{}^{z}. \tag{2}$$

The relevant commutators are

$$[\hat{\sigma}_n^{\pm}, \hat{\sigma}_j^{z}] = \mp\hat{\sigma}_n^{\pm}\delta_{nj}, \qquad [\hat{\sigma}_n^{+}, \hat{\sigma}_j^{-}] = 2\hat{\sigma}_n^{z}\delta_{nj} \tag{3}$$

and similarly for $\hat{\vec{S}}_n$. All the other commutators give zero.

The Heisenberg equations of motion for $\hat{S}_n^{\pm}$ are

$$i\dot{\hat{S}}_n = [\hat{S}_n, \hat{H}], \tag{4}$$

which in this case yield

$$\begin{aligned} \pm \ & i\dot{\hat{S}}_n^{\pm} = -J[\hat{S}_n^{z}(\hat{S}_{n-1}^{\pm} + \hat{S}_{n+1}^{\pm}) - \hat{S}_n^{\pm}(\hat{S}_{n-1}^{z} + \hat{S}_{n+1}^{z})] \\ + \ & A(\hat{S}_n^{z}\hat{S}_n^{\pm} + \hat{S}_n^{\pm}\hat{S}_n^{z}) + d_{xy}\delta_{jn}\hat{S}_j^{z}\hat{\sigma}_j^{\pm} - d_z\delta_{jn}\hat{S}_j^{\pm}\hat{\sigma}_j^{z} + \mu H_0\hat{S}_n^{\pm}. \end{aligned} \tag{5}$$

For now on, we assume that S is large enough for the semiclassical approximation to be valid, and hence, using $\alpha_n = \hat{S}_n^{+}/S$, $\alpha_n^* = \hat{S}_n^{-}/S$, $\hat{S}_n^{z}/S = \sqrt{1 - |\alpha_n|^2}$, we obtain

$$\begin{aligned} i\frac{\partial \alpha_n}{\partial t} = & - JS\left[(\alpha_{n+1} + \alpha_{n-1})\sqrt{1 - |\alpha_n|^2} - \alpha_n\left(\sqrt{1 - |\alpha_{n+1}|^2} + \sqrt{1 - |\alpha_{n-1}|^2}\right)\right] \\ & + 2AS\alpha_n\sqrt{1 - |\alpha_n|^2} + \mu H_0\alpha_n. \end{aligned} \tag{6}$$

3. Soliton solutions

We shall look for solutions in the form of amplitude-modulated waves

$$\alpha_n(t) = \varphi_n(t)e^{i(kn - \omega t)}, \tag{7}$$

where k and ω are the wave number and the frequency of the carrier waves (the lattice constant equals unity) and the envelopes $\varphi_n(t)$ are slowly varying functions of the position and time. In the continuum limit equations (6) transform into the following modified Nonlinear Schrödinger equation for the envelopes in Ref. [8]

$$i\left(\frac{1}{S}\frac{\partial \varphi}{\partial t} + 2J\sin k\frac{\partial \varphi}{\partial x}\right) = (\varepsilon - \omega S^{-1})\varphi - J\cos k\frac{\partial^2 \varphi}{\partial x^2} + g|\varphi|^2\varphi, \tag{8}$$

where

$$\varepsilon = \mu H_0 S^{-1} - 2g\,, \quad g = J(\cos k - 1 - A/J)\,. \tag{9}$$

For a bright soliton $g\cos k < 0$, equation (8) has the following solution

$$\varphi(x,t) = \varphi_0 \mathrm{sech}\frac{x-vt}{L},$$

$$\varphi_0^2 = -\frac{2J\cos k}{gL^2}, \quad \omega = \varepsilon S - \frac{JS\cos k}{L^2}, \quad v = 2JS\sin k\,, \tag{10}$$

where the parameters L and v are the width and the velocity of the soliton. The region for the k-values where this soliton solution exists in the case of easy axis anisotropy ($A > 0$) is $0 \le k < \pi/2$.

4. Qubit dynamics

The equation of the evolution for the qubit is

$$i\frac{\partial\hat{\rho}}{\partial t} = [\hat{H}, \hat{\rho}], \tag{11}$$

where the reduced density operator for the qubit is given via

$$\hat{\rho} = \frac{1}{2}(\hat{1} + \vec{a}\cdot\hat{\vec{\sigma}}_j), \tag{12}$$

$\hat{1}$ is the identity operator, and

$$\vec{a} = \mathrm{Tr}(\rho\hat{\vec{\sigma}}_j) \tag{13}$$

is the expectation value of the spin operator for the qubit (the Bloch vector). Evidently, $\hat{H}$ is the Hamiltonian of the system, after averaging over states of the spin chain.

From the Liouville equation (11) we find the following system of ordinary differential equations for the components of the Bloch vector

$$\begin{aligned} i\partial_t(a^-) &= (d_z S\sqrt{1-\varphi^2} - \nu H_0)a^- - d_{xy}S\varphi^* e^{-i(kj-\omega t)}a^z, \\ i\partial_t(a^+) &= -(d_z S\sqrt{1-\varphi^2} - \nu H_0)a^+ + d_{xy}S\varphi e^{i(kj-\omega t)}a^z, \\ i\partial_t(a^z) &= \frac{d_{xy}}{2}(S\varphi^* e^{-i(kj-\omega t)}a^+ - S\varphi e^{i(kj-\omega t)}a^-). \end{aligned} \tag{14}$$

Using $a^+ = (a^-)^*$ and $\vec{a}^2 = 1$ we obtain for a^+ and a^- the following solutions

$$a^+ = R(t)e^{ig(t)}, \qquad a^- = R(t)e^{-ig(t)}, \tag{15}$$

where $R(t)$ and $g(t)$ are real functions. Finally, for $a^z(t)$ we find

$$a^z(t) = \pm\sqrt{1-R^2(t)}. \tag{16}$$

The dynamics of the qubit is governed by two real functions. One of these functions $R(t)$ is explicitly related to $a^z(t)$ component of the Bloch vector, which is the expectation value for the spin z component. The other function $g(t)$ represents the azimuth angle of the Bloch sphere.

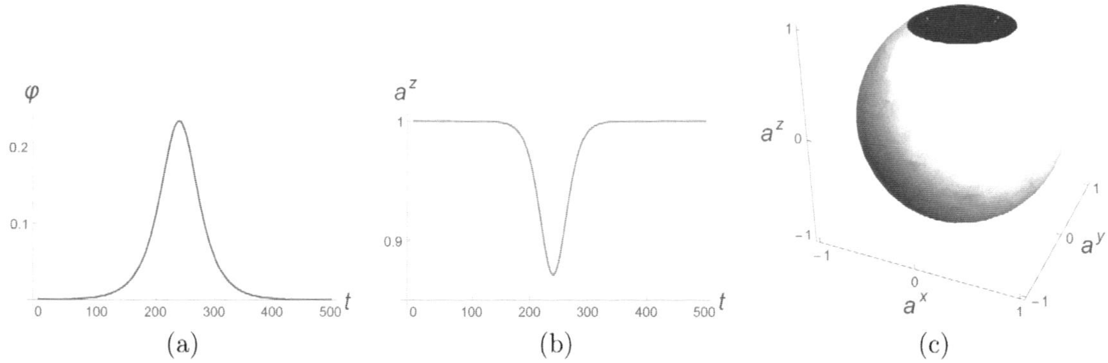

Figure 1. Soliton-qubit interaction of passing near the qubit soliton with parameters $A = 1$, $d_{xy} = 20$, $d_z = 5$, $x = 50$, $k = \frac{\pi}{30}$. (a) soliton solution $\varphi(t)$ for $x = 50$, (b) z-component of the Bloch vector and (c) Bloch sphere.

5. Numerical results

In this section we employ numerical study of (14). We choose the initial condition for the Bloch vector to be $\vec{a} = (0, 0, 1)$ - spin up. We define the energy scale with $J = 1$, the spin of the chain in terms of $\hbar$ is $S = 1$, and the width of soliton ($L = 6$) is taken to be larger than the lattice constant, which is unity. The lattice constant defines the scale of the distance. The terms with the external field are taken to be of the order of the exchange interaction in the spin chain $\mu H_0 = 0.5$, $\nu H_0 = 1.5$, where we supposed that the magnetic moment per spin for the qubit is larger than the magnetic moment per spin in the chain. The site j where the qubit is situated and is coupled with the spin S_j from the chain is arbitrary. x stands for the distance between the peak of the soliton and the site j at the initial moment. The result does not depend from j up to translation, which is equivalent to renumeration of the sites in the chain. For this reason we can take $j = 10$.

First, let us assume that the soliton is far away from the qubit in the initial moment, after approaching it the interaction takes place, and finally, the soliton is going away from the qubit. In this case $x = 50$ means that in the initial moment the soliton is far away from the qubit. The soliton as a function of time is shown in [figure 1(a)]. When the soliton is in the vicinity of the qubit, the interaction takes place. This interaction does not change the initial state much and one cannot switch the qubit spin from up to down $\vec{a} = (0, 0, -1)$. This can be seen in the time evolution of the z-component of the Bloch vector which is shown in [figure 1(b)]. We see that when the soliton retrieves, the qubit restores its initial state. The path of the Bloch vector onto the Bloch sphere is shown [figure 1(c)].

Second, we consider the case where the soliton resides onto the j-th spin at the initial moment [figure 2(a)], i.e. the soliton peaks onto the j-th site. It is seen that the interaction changes the z-component of the Bloch vector, which is a measure how close we are to spin up or spin down state. This is shown for different values of the final time [figures 2(b) - 2(d)]. We also plot the respective Bloch vector path onto the Bloch sphere [figures 2(e) - 2(g)]. For long enough time $a^z(t)$ tends to asymptotic value between 0.8 and 1, which implies non-negligible probability to find the qubit in spin down state.

With an increase of the interaction between the soliton and the qubit in the $x - y$ plane, the expectation value $a^z(t)$ oscillates reaching smaller values, see [figures 3(a) - 3(c)]. For long enough time $a^z(t)$ tends to an asymptotic value with higher probability to find the qubit in spin down state.

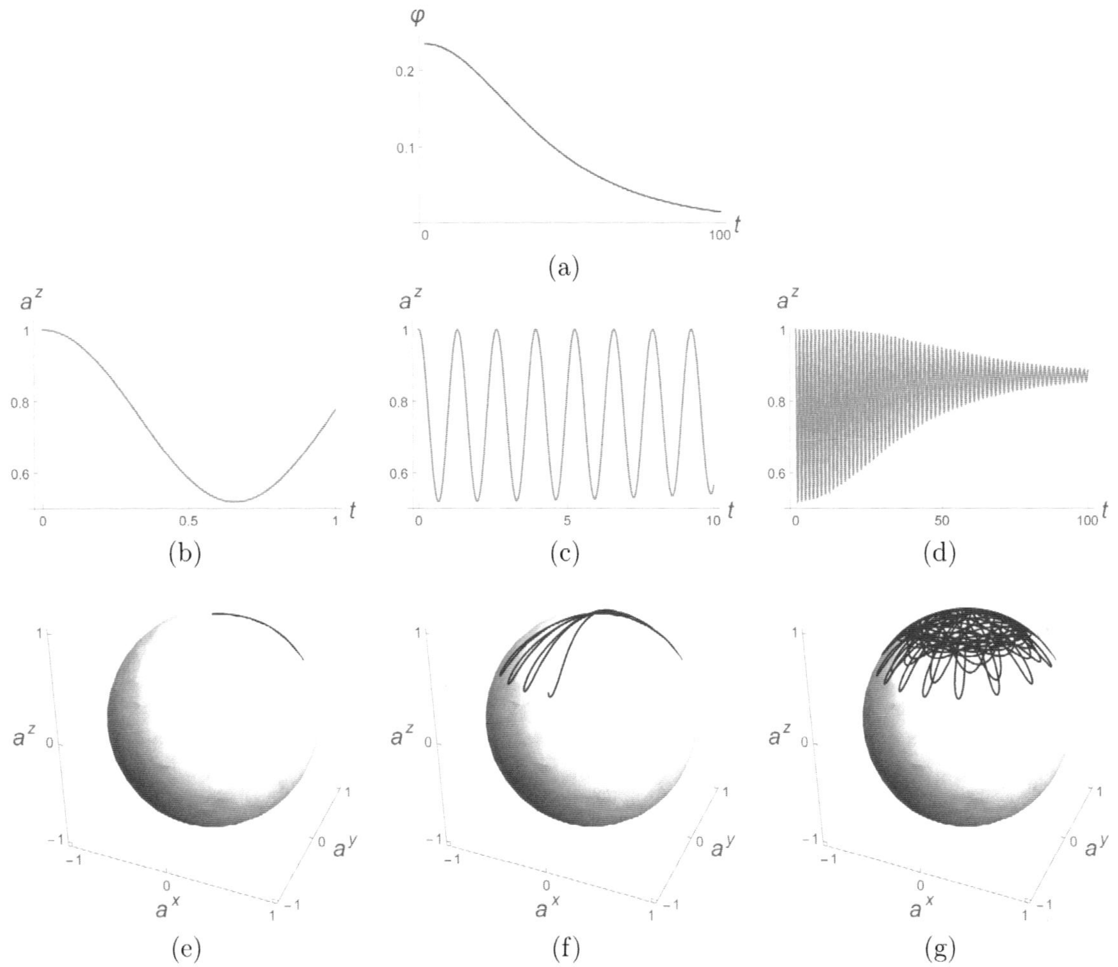

Figure 2. Soliton-qubit evolution for parameters $A = 1$, $d_{xy} = 20$, $d_z = 5$, $x = 0$, $k = \frac{\pi}{30}$. (a) soliton solution $\varphi(t)$ for fixed $x = 0$. Pairs of frames representing the z-component of the Bloch vector and the Bloch sphere are plotted for different final time t: (b) and (e) $t = 1$, (c) and (f) $t = 10$, (d) and (g) $t = 100$.

In the case of small interaction [figures 3(d) and 3(e)], the Bloch vector covers only a small part of the sphere around the "north pole". For larger interaction it covers almost all the states onto the sphere [figure 3(f)], which means that almost arbitrary superpositions are possible.

6. Conclusion

We have studied the interaction between a quantum bit and a soliton in a magnetic chain. The magnetic chain was described by a Heisenberg model in a nearest-neighbour approximation with anisotropy. The evolutionary differential equation for the density operator for the qubit is used to describe its dynamics. It is shown that for large enough values of the interaction constant between the soliton and the qubit in the $x - y$ plane d_{xy} and long enough time the Bloch vector covers almost all states onto the sphere. We showed that the spin switching and creation of superpositions are in principle possible.

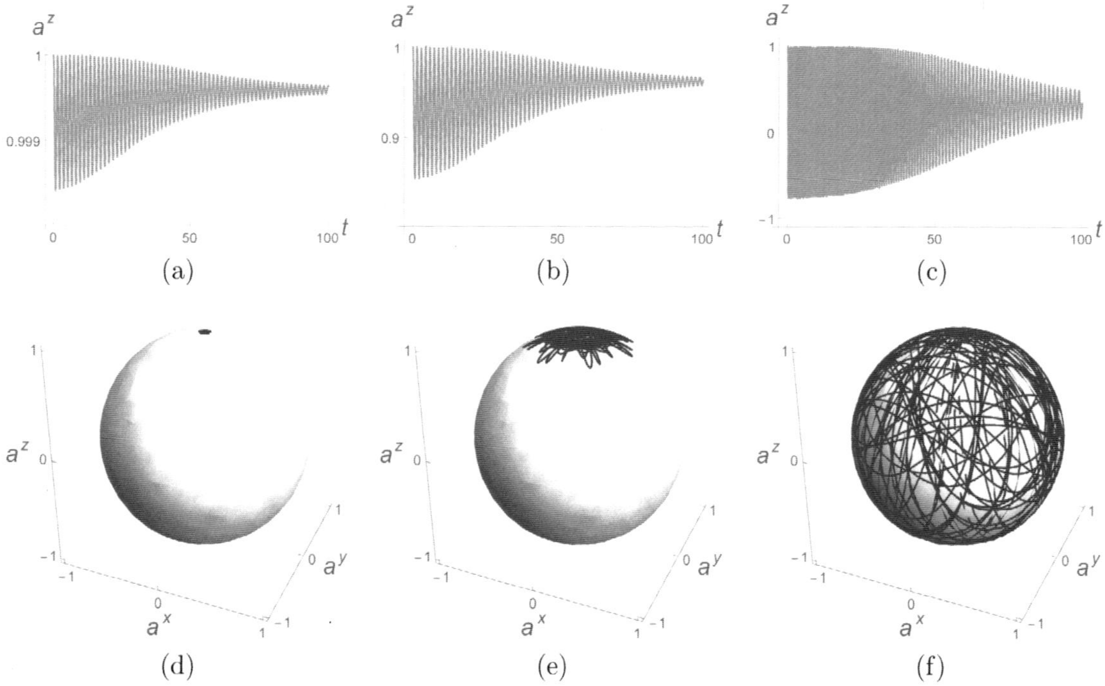

Figure 3. Soliton-qubit system for parameters $A = 1$, $d_z = 5$, $x = 0$, $k = \frac{\pi}{30}$. Pairs of frames representing the z-component of the Bloch vector and the Bloch sphere are plotted for different values of the interaction d_{xy}: (a) and (d) 1, (b) and (e) 10, (c) and (f) 100.

Acknowledgments

This work is supported by the Program for career development of young scientists and PhD students, BAS (Grant agreement no: DFNP-17-11/24.07.2017). The authors would like to express their great appreciation to I. Boradjiev for the many helpful discussions.

References

[1] Engel H-A, Kouwenhoven L P, Loss D and Marcus C M 2004 *Quantum Inf. Process.* **3** 115
[2] Tejada J, Chudnovsky E M, del Barco E, Hernandez J M and Spiller T P 2001 *Nanotechnology* **12** 181
[3] Kosevich A M, Ivanov B A and Kovalev A S 1990 *Phys. Reports* **194** 117
[4] Lai R, Sievers A J 1999 *Physics Reports* **314** 147
[5] Cuccoli A, Nuzzi D, Vaia R, Verrucchi P 2014 *Journal of Applied Physics* **115** 17B302
[6] Cuccoli A, Nuzzi D, Vaia R, Verrucchi P 2014 *Int. J. Quantum Inf.* **12** 2 1461013
[7] Cuccoli A, Nuzzi D, Vaia R, Verrucchi P 2015 *New J. Phys.* **17** 083053
[8] Varbev S K, Primatarowa M T, Kamburova R S 2017 *Journal of Physics: Conference Series* **794** 1 012027

Soliton-impurity interaction in two coupled ferromagnetic chains

R S Kamburova, M T Primatarowa and S K Varbev

Georgi Nadjakov Institute of Solid State Physics, Bulgarian Academy of Sciences, 1784 Sofia, Bulgaria

E-mail: krad@issp.bas.bg

Abstract. The propagation of soliton excitations in a system of two magnetic chains with impurities is studied. The inhomogeneous chains coupled through ferromagnetic interaction between the opposite spins are described by a Heisenberg model in the nearest-neighbor approximation. The presence of the impurities leads to linear perturbing terms in the evolutionary equations. We investigated numerically the influence of the soliton parameters, the interchain coupling and the defect strength on the soliton dynamics. The conditions of perfect soliton switching are obtained.

1. Introduction

The existence of solitons and their properties in magnetic systems have been subject of investigation for many years [1-7]. The most considered models are based on the one-dimensional spin Heisenberg chains. In the last years growing interest is devoted to the study of coupled spin chains (spin ladders) with different exchange interactions between them [8-11] as model to explain the magnetic and electronic structure of different experimentally discovered magnetic materials [12,13]. In a previous paper we have studied the dynamics of bright solitons in two coupled ferromagnetic chains [14]. We have obtained that the coupling between the two anisotropic ferromagnetic chains leads to linear and nonlinear terms in the dynamics equations. The condition for a perfect soliton switching in such a system was derived. We have observed that in contrast to the linear interchain coupling for the nonlinear coupling not only its value but also its sign plays a crucial role for the soliton dynamics.

Another interesting topic of research with practical importance is the dynamics of solitons in coupled parallel inhomogeneous chains [15-20]. In the present paper we study the interaction of bright solitons with impurities in two coupled anisotropic ferromagnetic chains.

2. Hamiltonian of the system

We consider two ferromagnetic Heisenberg chains of N spins with magnitude S described in the nearest-neighbor approximation by the following Hamiltonian:

$$\begin{aligned} \hat{H} = & - J\sum_{n=1}^{N}(\hat{\vec{S}}_n \cdot \hat{\vec{S}}_{n+1} + \hat{\vec{\sigma}}_n \cdot \hat{\vec{\sigma}}_{n+1}) - A\sum_{n=1}^{N}[(\hat{S}_n^z)^2 + (\hat{\sigma}_n^z)^2] \\ & - \sum_{n=1}^{N}(\mu H_0 + \epsilon\delta_{n,n_0})(\hat{S}_n^z + \hat{\sigma}_n^z) - d\sum_{n=1}^{N}\hat{\vec{S}}_n \cdot \hat{\vec{\sigma}}_n, \end{aligned} \quad (1)$$

where $J > 0$ is the exchange integral and A is the on-site anisotropy constant which can be positive (easy axis) or negative (easy plane). H_0 is the external magnetic field applied along the z-axis, so that in the ground state of the system all spins $\hat{\vec{S}}_n$ and $\hat{\vec{\sigma}}_n$ corresponding respectively to the first and the second chains are aligned in the z-direction, μ is the magnetic moment per spin. d characterizes the coupling interaction between the two chains which can be ferromagnetic ($d > 0$) or antiferromagnetic ($d < 0$). The constant ϵ determines the strength of the impurity on site n_0, which can be attractive or repulsive. It corresponds to a change in the external magnetic field or in the magnetic moment.

We use for the scalar products in (1) the rule

$$\hat{\vec{a}} \cdot \hat{\vec{b}} = \frac{\hat{a}^+\hat{b}^- + \hat{a}^-\hat{b}^+}{2} + \hat{a}^z\hat{b}^z, \tag{2}$$

$$[\hat{S}_i^{\pm}, \hat{S}_j^z] = \mp\hat{S}_i^{\pm}\delta_{ij}, \quad [\hat{S}_i^+, \hat{S}_j^-] = 2\hat{S}_i^z\delta_{ij}, \quad [\hat{\sigma}_i^{\pm}, \hat{\sigma}_j^z] = \mp\hat{\sigma}_i^{\pm}\delta_{ij}, \quad [\hat{\sigma}_i^+, \hat{\sigma}_j^-] = 2\hat{\sigma}_i^z\delta_{ij}. \tag{3}$$

All the other commutators are zero.

The equations of motion $i\dot{\hat{a}} = [\hat{a}, \hat{H}]$ for $\hat{S}_i^{\pm}$ and $\hat{\sigma}_i^{\pm}$ ($\hbar = 1$) yield

$$\begin{aligned} \pm i\dot{\hat{S}}_i^{\pm} &= - J[\hat{S}_i^z(\hat{S}_{i-1}^{\pm} + \hat{S}_{i+1}^{\pm}) - \hat{S}_i^{\pm}(\hat{S}_{i-1}^z + \hat{S}_{i+1}^z)] \\ &+ A(\hat{S}_i^z\hat{S}_i^{\pm} + \hat{S}_i^{\pm}\hat{S}_i^z) - d(\hat{S}_i^z\hat{\sigma}_i^{\pm} - \hat{S}_i^{\pm}\hat{\sigma}_i^z) + (\mu H_0 + \epsilon\delta_{in_0})\hat{S}_i^{\pm}, \\ \pm i\dot{\hat{\sigma}}_i^{\pm} &= - J[\hat{\sigma}_i^z(\hat{\sigma}_{i-1}^{\pm} + \hat{\sigma}_{i+1}^{\pm}) - \hat{\sigma}_i^{\pm}(\hat{\sigma}_{i-1}^z + \hat{\sigma}_{i+1}^z)] \\ &+ A(\hat{\sigma}_i^z\hat{\sigma}_i^{\pm} + \hat{\sigma}_i^{\pm}\hat{\sigma}_i^z) + d(\hat{S}_i^z\hat{\sigma}_i^{\pm} - \hat{S}_i^{\pm}\hat{\sigma}_i^z) + (\mu H_0 + \epsilon\delta_{in_0})\hat{\sigma}_i^{\pm}. \end{aligned} \tag{4}$$

In the semiclassical approximation valid for large values of S the components of the spin operators are complex amplitudes, $\alpha_n = \hat{S}_n^+/S$, $\alpha_n^* = \hat{S}_n^-/S$, $\hat{S}_n^z/S = \sqrt{1-|\alpha_n|^2}$, $\beta_n = \hat{\sigma}_n^+/S$, $\beta_n^* = \hat{\sigma}_n^-/S$ and $\hat{\sigma}_n^z/S = \sqrt{1-|\beta_n|^2}$. Then we have

$$\begin{aligned} i\frac{\partial\alpha_n}{\partial t} &= -JS\left[(\alpha_{n+1} + \alpha_{n-1})\sqrt{1-|\alpha_n|^2} - \alpha_n\left(\sqrt{1-|\alpha_{n+1}|^2} + \sqrt{1-|\alpha_{n-1}|^2}\right)\right] \\ &+ 2AS\alpha_n\sqrt{1-|\alpha_n|^2} - dS\left(\beta_n\sqrt{1-|\alpha_n|^2} - \alpha_n\sqrt{1-|\beta_n|^2}\right) + (\mu H_0 + \epsilon\delta_{nn_0})\alpha_n, \\ i\frac{\partial\beta_n}{\partial t} &= -JS\left[(\beta_{n+1} + \beta_{n-1})\sqrt{1-|\beta_n|^2} - \beta_n\left(\sqrt{1-|\beta_{n+1}|^2} + \sqrt{1-|\beta_{n-1}|^2}\right)\right] \\ &+ 2AS\beta_n\sqrt{1-|\beta_n|^2} - dS\left(\alpha_n\sqrt{1-|\beta_n|^2} - \beta_n\sqrt{1-|\alpha_n|^2}\right) + (\mu H_0 + \epsilon\delta_{nn_0})\beta_n. \end{aligned} \tag{5}$$

The set of differential equations (5) describes our system.

3. Soliton solutions

We shall look for solutions in the form of amplitude-modulated waves

$$\alpha_n(t) = \varphi_n(t)e^{i(kn - \omega t)}, \quad \beta_n(t) = \psi_n(t)e^{i(kn - \omega t)}, \tag{6}$$

where k and ω are the wave number and the frequency of the carrier waves (the lattice constant equals unity) and the envelopes $\varphi_n(t)$, $\psi_n(t)$ are slowly varying functions of the position and time. In the continuum limit and for $\varphi^2 \ll 1$, $\psi^2 \ll 1$ equations (5) transform into the following coupled modified nonlinear Schrödinger equations for the envelopes:

$$\begin{aligned} i\left(\frac{1}{S}\frac{\partial\varphi}{\partial t} + 2J\sin k\frac{\partial\varphi}{\partial x}\right) &= (\omega_0 + \epsilon\delta(x-x_0)S^{-1} - \omega S^{-1})\varphi \\ &- J\cos k\frac{\partial^2\varphi}{\partial x^2} + g|\varphi|^2\varphi - \frac{d}{2}(2\psi - |\varphi|^2\psi + |\psi|^2\varphi), \\ i\left(\frac{1}{S}\frac{\partial\psi}{\partial t} + 2J\sin k\frac{\partial\psi}{\partial x}\right) &= (\omega_0 + \epsilon\delta(x-x_0)S^{-1} - \omega S^{-1})\psi \\ &- J\cos k\frac{\partial^2\psi}{\partial x^2} + g|\psi|^2\psi - \frac{d}{2}(2\varphi - |\psi|^2\varphi + |\varphi|^2\psi), \end{aligned} \tag{7}$$

where

$$\omega_0 = \mu H_0 S^{-1} - 2g + d, \quad g = J(\cos k - 1 - A/J). \tag{8}$$

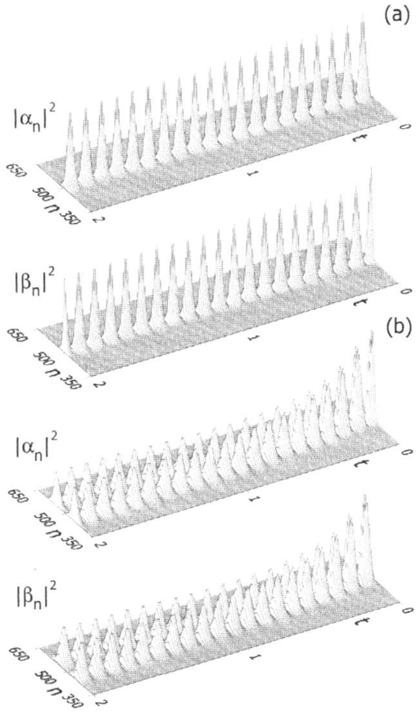

Figure 1. Switching of the bound soliton-impurity solution (11) with $J = S = 1$, $A = 1$, $L = 10$ and $|d| = 0.0314$ for $\epsilon = -0.08$ (a) and $\epsilon = 0.08$ (b). The time is in units of $10^3/JS$.

Further we shall consider the case $d = 0$. For $\epsilon = 0$ the uncoupled homogeneous nonlinear equations (7) possess bright-soliton solutions of the form

$$\varphi(x,t) = \varphi_0 \mathrm{sech}\frac{x - vt}{L}, \quad \psi(x,t) = \psi_0 \mathrm{sech}\frac{x - vt}{L} \tag{9}$$

with

$$\varphi_0^2 = \psi_0^2 = -\frac{2J\cos k}{gL^2}, \quad \omega = \omega_0 S - \frac{JS\cos k}{L^2}, \quad v = 2JS\sin k, \tag{10}$$

which appear for $gJ\cos k < 0$. L and v are soliton's width and velocity, respectively.

For the static inhomogeneous case ($k = v = 0$ and $\epsilon \neq 0$) the uncoupled nonlinear equations (7) possess bound soliton-impurity solutions of the form

$$\varphi(x) = \psi(x) = \frac{2J}{AL^2}\mathrm{sech}\left(\frac{|x - x_0|}{L} + \Delta\right), \quad \Delta = \mathrm{Arth}\left(\frac{-\epsilon L}{2SJ}\right), \tag{11}$$

which exists only for $A > 0$.

For $d \neq 0$ we shall investigate the evolution of a soliton which at the initial time $t = 0$ is launched in one of the chains at the position $n_0 = 500$. Figure 1 shows the evolution of a bound soliton-impurity state (11). We have obtained that the single peak solution ($\Delta > 0$, $\epsilon < 0$, attractive impurity) is stable and is transferred from one chain to the other and back [figure 1(a)]. The process is close to a perfect switch of the soliton in the linear case with a period $t_0 = \pi/|d|$. The double peak solution which corresponds to a repulsive impurity ($\Delta < 0$, $\epsilon > 0$) is not stable. It is split in two parts which can oscillate around the impurity or propagate with opposite velocities during the transfer process [figure 1(b)].

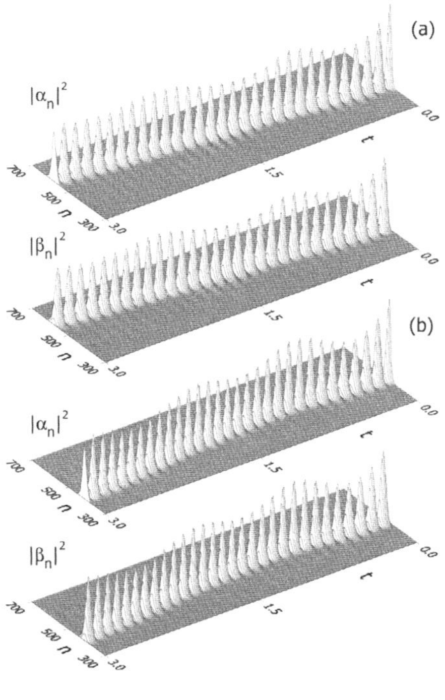

Figure 2. Scattering of a soliton α_n launched at $n_s = 450$ for $k = 0.025$ from a repulsive defect with $\epsilon = 0.01$ (a) and $\epsilon = 0.03$ (b). All other parameters are the same as in figure 1.

4. Soliton scattering on the linear impurities

We shall investigate the evolution of a soliton with the form (9) which at the initial time $t = 0$ is launched in one of the chains at the position n_s far enough from the impurity position n_0

$$\alpha_n(0) = \frac{1}{L}\sqrt{\frac{2\cos k}{A/J + 1 - \cos k}}\,\mathrm{sech}\frac{n - n_s}{L}e^{ikn}, \quad \beta_n(0) = 0 \tag{12}$$

solving numerically the system (5). The simulations are carried out for large enough chains compared to the soliton width L and periodic boundary conditions. We consider anisotropy values for which $\varphi_0^2 \ll 1$. Then, the linear coupling between the chains is dominant in the system (7) and perfect soliton switching with the period t_0 can occur when the condition

$$\left|\frac{J\cos k}{2L^2 d}\right| \ll 1 \qquad (13)$$

is fulfilled [14].

Now, when the soliton is perfectly transferred from one chain to the other, its scattering on the defects is like the soliton dynamics in one inhomogeneous chain. The evolution depends strongly on the initial soliton amplitude (φ_0), velocity ($v \sim k$) and the strength of the defect ϵ. First we consider slow solitons. In the case of repulsion $\epsilon > 0$ the soliton is either transmitted [figure 2(a)] or reflected [figure 2(b)].

The evolution is more complex in the case of attractive defects ($\epsilon < 0$, figure 3). For a fixed small initial velocities we observe between the regions of transmission and reflection regions where the soliton is trapped [figure 3(a)] or split in a trapped and a reflected part [figure 3(b)] depending on the defect strength.

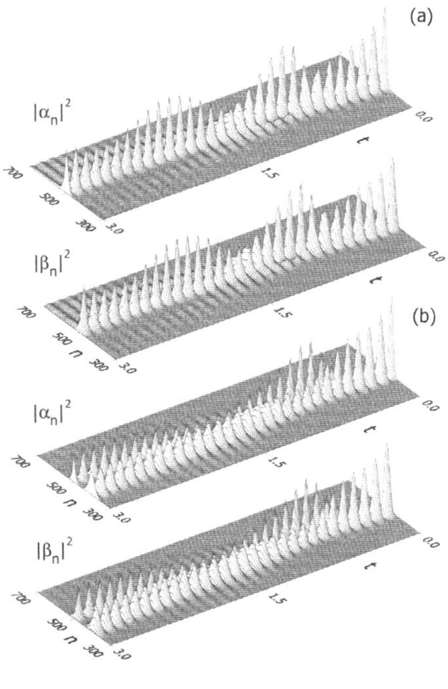

Figure 3. Scattering of a soliton α_n launched at $n_s = 450$ for $k = 0.025$ from an attractive defect with $\epsilon = -0.04$ (a) and $\epsilon = -0.09$ (b). All other parameters are the same as in figure 1.

Further we consider fast solitons. For large values of the velocity the soliton behaves like a wave and it is split in a transmitted and a reflected part. Their size depends on the defect strength and not on its sign (figure 4). We observe transmission [figure 4(a)], splitting [figure 4(b)] or reflection [figure 4(c)] of the initial soliton which in the same time is switched from one chain to the other. In all these cases the soliton-impurity interaction is not influenced by the coupling constant between the two chains.

5. Conclusion

Soliton dynamics in two inhomogeneous Heisenberg spin chains with ferromagnetic interaction between them is studied. We obtained that when the condition for perfect soliton switching is

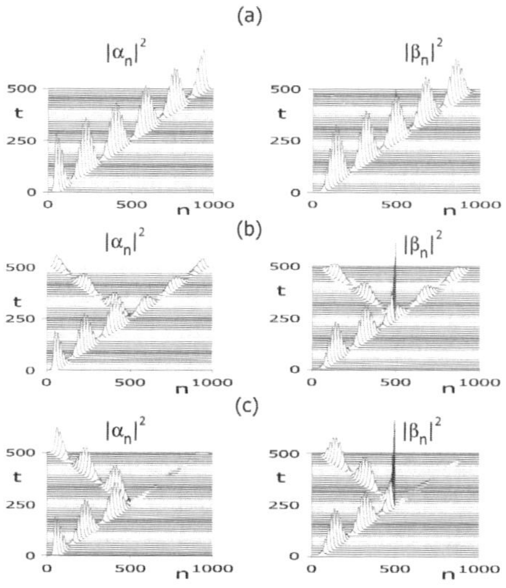

Figure 4. Scattering of a soliton α_n launched at $n_s = 50$ for $k = 1.1$ from a defect with $|\epsilon| = 0.1$ (a), $|\epsilon| = 2$ (b) and $|\epsilon| = 10$ (c). The time is in units of $1/JS$. All other parameters are the same as in figure 1.

fulfilled the soliton-impurity interaction is not influenced by the coupling constant between the chains. The numerical investigation of the stability of bound soliton-defect solutions showed that the single-peak solution is stable, while the double-peak solution is unstable and easily destroyed.

Acknowledgments

This work is supported by the Bulgarian National Science Fund under Grant No. DN 08/18 from 14.12.2016.

References

[1] Mikeska H-J and Steiner M 1991 *Adv. Phys.* **40** 191
[2] Wallis R F, Mills D L and Boardman A D 1995 *Phys. Rev.* B **52** R3828
[3] Rakhmanova S and Mills D L 1998 *Phys. Rev.* B **58** 11458
[4] Slavin A N, Kivshar Yu S, Ostrovskaya E A and Benner H 1999 *Phys. Rev. Lett.* **82** 2583
[5] Zaspel C E, Mantha J H, Rapoport Yu G and Grimalsky V V 2001 *Phys. Rev.* B **64** 064416
[6] Bischof B, Slavin A N, Benner H and Kivshar Yu S 2005 *Phys. Rev.* B **71** 104424
[7] Ivanov B A, Galkin A Yu, Khymyn R S and Merkulov A Yu 2008 *Phys. Rev.* B **77** 064402
[8] Ivanov N B and Richter J 2004 *Phys. Rev.* B **69** 214420
[9] Znidaric M 2013 *Phys. Rev.* B **88** 205135
[10] Kavitha L et al 2015 *Appl. Math. Comp.* **251** 643
[11] Tandon K, Lal S, Pati SK et al 1999 *Phys. Rev. B.* **59** 396
[12] Eccleston RS, Barnes T, Brody J et al 1994 *Phys. Rev. Lett.* **73** 2626
[13] Azuma M, Hiroi Z, Takano M et al 1994 *Phys. Rev. Lett.* **73** 3463
[14] Varbev S K, Primatarowa M T and Kamburova R S 2017 *J. Phys.: Conf. Series* **794** 012027
[15] Królikowski W and Kivshar Yu S 1996 *J. Opt. Soc. Am.* B **13** 876
[16] Longhi S 2006 *Phys. Rev.* E **74** 026602(9)
[17] Morales-Molina L and Vicencio R A 2006 *Optics Lett.* **31** 966
[18] Goodman R H and Weinstein M I 2008 *Physica* D **237** 2731
[19] Belmonte-Beitia J, Pérez-García V M and Torres P J 2009 *Nonlin. Science* **19** 437
[20] Kamburova R S and Primatarowa M T 2010 *AIP Conf. Proc.* **1203** 261

A different optical composition for a broadband linear polarization rotator

Hristina Hristova[1], Stefano Ognyanski[1], Andon Rangelov[2] and Emiliya Dimova[1]

[1] Institute of Solid State Physics, Bulgarian Academy of Sciences, 72 Tsarigradsko Chaussée Blvd., 1784 Sofia, Bulgaria

[2] Department of Physics, Sofia University, James Bourchier 5 blvd., 1164 Sofia, Bulgaria

E-mail: hhristova@issp.bas.bg, e.dimova@issp.bas.bg

Abstract. We suggest and experimentally demonstrate a broadband polarization rotator. Our device consist of a Fresnel's rhomb retarder and a composite broadband half-wave plate with relative rotation α between their fast polarization axis. Such a optical device rotate the polarization plane of a linearly-polarized light at angle 2α over a wide range of wavelengths.

1. Introduction

Being fundamental property of light [1–5], the polarization became an important tool for many optical measurement techniques used in stress measurements, ellipsometry, physics, chemistry, biology, astronomy and others [6–8]. The ability to manipulate light polarization is highly desirable for practical applications. For example, display and telecommunications technologies rely on the ability for controlled rotation of the light polarization [9]. For the purpose optical polarization rotators, which rotate the light polarization plane at a desired angle [1–5], ordinary are used.

Two types of polarization rotators are commercially available. First of them exploits the effect of circular birefringence. The main advantages of this type are independence of the angle of light polarization rotation via rotation of the rotator around its own optical axis, and low price. On the other hand, they can be used only for a limited range of wavelengths. The second type of rotators are those for which polarization plane rotation is achieved through consecutive reflections, for example, Fresnel rhombs [1–5]. They can operate over a wide range of wavelengths but are quite expensive. Due to their features both of the types are used for measurements [10].

There are also schemes to enlarge the spectral width of a light polarization plane rotation - for several fixed wavelengths simultaneously using up to four half-wave plates in a series, suggested by Koester [11], and using two wave plates of the same material, shown by Kim and Chang [12].

The approach of Kim and Chang [12] can be extend in combination with the Koester idea [11] allowing to design an arbitrary broadband polarization rotator [13, 14]. The presented scheme consists of two crossed half-wave plates, where the angle between their fast axes is half the angle of polarization rotation. Due to the the analogy with the composite phase gates [15], the usage of composite half-wave plates ensures broadband and stability with respect to wavelength variations.

In this paper, we propose an alternative design for a broadband polarization rotator and demonstrate the device working. The main difference from [13,14] consists in using of a Fresnel's rhomb instead of one of the both composite half-wave plates.

2. Theory

Using the Jones calculus, any reversible polarization transformation can be considered as a composition of a retarder $\mathfrak{J}$ and a rotator $\mathfrak{R}$ Jones matrices [16], represented in the horizontal-vertical basis as,

$$\mathfrak{R}(\theta) = \begin{bmatrix} \cos\theta & \sin\theta \\ -\sin\theta & \cos\theta \end{bmatrix}, \mathfrak{J}(\varphi) = \begin{bmatrix} e^{i\varphi/2} & 0 \\ 0 & e^{-i\varphi/2} \end{bmatrix}, \tag{1}$$

where θ is the angle of rotation, the phase shift is $\varphi = 2\pi L(n_f - n_s)/\lambda$, with λ being the vacuum wavelength, n_f and n_s the refractive indices along the fast and slow axes, correspondingly, and L the thickness of the retarder. The Fresnel's rhombs and the wave plates are used as retarders. The performance of the wave plates is usually limited to a narrow range of wavelengths around λ. A broadband rotator can be constructed as a sequence of two broadband half-wave plates with an additional rotation between them. In references [13, 14, 17] the main steps in derivation of the Jones matrices for the composite half-wave plate(CHWP) are shown and broadband rotator composed of two CHWPs. But there is another possibility for creation of a broadband rotator, namely, to replace one of the CHWPs in the sequence with a half-wave Fresnel's rhomb (HWFR).

For a single half-wave plate rotated at angle θ, the Jones matrix $\mathfrak{J}$ has the form

$$\mathfrak{J}_\theta(\pi) = \mathfrak{R}(-\theta)\mathfrak{J}(\pi)\mathfrak{R}(\theta) = \begin{bmatrix} i\cos(2\theta) & -i\sin(2\theta) \\ -i\sin(2\theta) & -i\cos(2\theta) \end{bmatrix}. \tag{2}$$

Considering a simple sequence of two ordinary half-wave plates rotated at angles θ_1 and θ_2 respectively, multiplication of the Jones matrices gives the total propagator

$$\mathfrak{J}_{\theta_1}(\pi)\mathfrak{J}_{\theta_2}(\pi) = -\begin{bmatrix} \cos(2\alpha) & \sin(2\alpha) \\ -\sin(2\alpha) & \cos(2\alpha) \end{bmatrix}, \tag{3}$$

where $\alpha = \theta_1 - \theta_2$ denoted the relative rotation between the fast axes of the two waveplates. The last matrix represents a Jones matrix for a rotator up to unimportant π phase (minus sing), which is not broadband. For our purpose to extend the range of operation of the rotator, we replace the two ordinary half-wave plates with a HWFR, described again by equation 2, and a CHWP, respectively. With a relative rotation of $\alpha/2$ between them, the final propagator is described again by equation3.

3. Experimental setup

Using commercially available optical elements, a broadband linear polarization rotator is developed.

The constructed rotator is composed by two types optical components. First is a half-wave Fresnel's rhomb retarder (FR600HM, ThorLabs) and the second one is an ultra-broadband CHWP. The last one is developed by using the quantum-optical analogies. The composite half-wave plate is built as a stack of ordinary achromatic half-wave plates (WRM053-mica, 700-1100nm, aperture 20mm, MellesGriot) which respective optical axes are adjusted according to the procedure described in [17]. In such a way it plays as half-wave plate for ultra-broadband spectral range, covering whole visual and near IR spectral region.

The experimental setup, shown on the figure 1, is similar to those in [14,17]. For visualization of the investigated effect we have used white light from a lamp TUNGSRAM powered by a 6V d.c. power supply. The light was collimated in a beam (2 mm diameter for a distance of up to

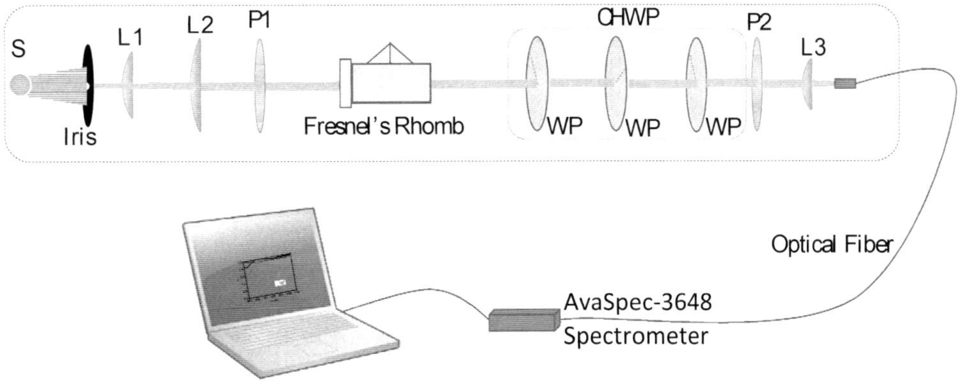

Figure 1. Experimental setup. A Lamp *S*, an iris *I*, lenses *L1* and *L2*, and a polariser *P1* produce a collimated beam of white polarized light. After passing through the investigated composite rotator, the light beam is focused by an analyser *P2* and a lens *L3* onto the entrance *F* of an optical fibre connected to a spectrometer.

2 m) by a set of an iris and two lenses, L1 (f=35 mm) and L2 (f=150 mm). Good collimation, as well as beam horizontality, are critical to HWFR's work. The polarisers, P1 and P2, were borrowed from a Lambda-950 spectrometer, Perkin Elmer and were of GlanTayler type with a wide spectral range of 210-1100 nm. The light beam was polarized in the horizontal plane by the polariser P1. P2 was used to analyse the polarization of the output light. The objects under investigation were placed between them. The light after P2 was sent by set of a lens and a fibre to a spectrometer AvaSpec 3648 Fiber Optic Spectrometer, controlled by the proprietary software AvaSoft version 7.5.

The investigating polarization rotator is a composition of a HWFR and a half-wave plate. We investigated three cases. In the first one we used combination between HWFR and a achromatic half-wave plate (WRM053-mica, 700-1100nm, aperture 20mm, MellesGriot), and the other case two half-wave plates, the same type, have been used. The taken transmittance spectra from this two compositions we used for comparison with the one of the developed new construction. The rotator of interest is constructed by HWFR and a ultra - broadband CHWP made by tree achromatic half-wave plates. Each of the optical elements is mounted on a separate RSP1 rotation mount (Thorlabs Inc.), which can be used for 360^0 rotation. The scale marked at 2^0 increments allowed for precise, repeatable positioning and fine angular adjustment. The mounted half-wave plates were slightly tilted around their vertical axes in order to avoid the back reflection of the light.

4. Measurement procedure and results

The experimental properties of the composite rotator have been studied by analyzing the polarization of the transmitted light through the investigated object. With the upper explained setup we can achieve a reliable data in the range 500-1000 nm, because of the properties of the light source.

The measurement procedure was similar to one described in the articles [14, 17]. All experiments started with measurement of the dark and reference spectra. The dark spectrum, taken with the light path blocked, is further used to automatically correct for hardware offsets. The reference spectrum is usually taken with the light source on and using a blank sample rather than the sample under test. In our case, however, we measured the transmission spectrum of the already assembled composite polarization rotator, with the axes of the polarizer P1 and the analyzer P2 and the fast axis of the single wave plates and HWFR all set to be parallel. The measured light spectrum was used as a reference for the subsequent measurements.

First step of our measurement was to built the CHWP. For this experiment we stopped on the CHWP assembled by three ordinary half-wave plates, [17], and by turning the fast axes of the half-wave plates respectively on angles: first half-wave plate: 60^0; second half-wave plate: 120^0 and third half-wave plate: 60^0. As a next step, we set the composite rotator in two parts similar to the one reported in the [14]. Instead of a half-wave plate, we used HWFR here and the second component is CHWP, which we have already built. The composed linear polarization rotator was tested at randomly selected angles. For more clear representation of the results, we choose angles 40^0, 60^0, 90^0 and 120^0.

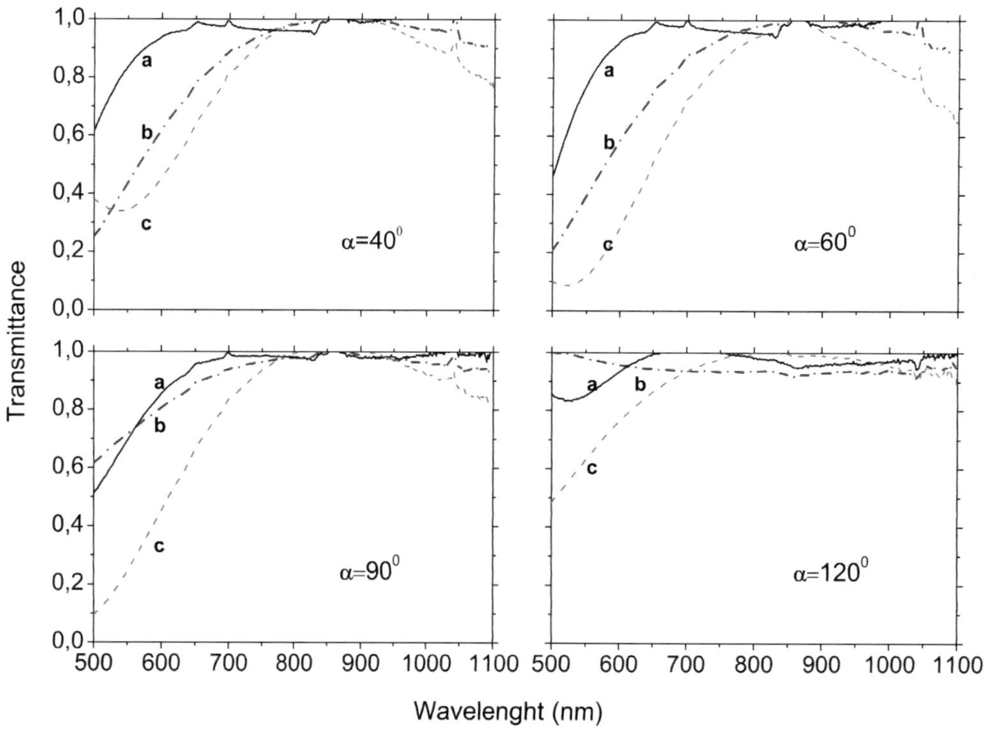

Figure 2. Experimental results for ultra-broadband spectral bandwidth linear polarization rotator built by half-wave Fresnel's rhomb and composite achromatic half-wave plates, black solid line. Three single half-wave plates were used to assemble the composite half-wave plate and their respective optical axes were to the respective rotation angles in accordance with angles in table 1 in the article [17]. As reference data the spectra of two types of rotators are also shown. First on is built by HWFR and one half-wave plate (b), blue dash dot line, while the other one is constructed by two half-wave plates (c), red dashed line.

On the figure 2 the spectra of the composed broad-bandwidth rotator at the chosen angles are presented. Each curve is taken with integration time of 10 ms and average of 1000. The comparison with the reference spectra of rotator, composed by a HWFR and a half-wave plate, as well as rotator, composed of two half-wave plates, shows the broadening effect of proposed by us optical construction. We expect that in case of the composite half-wave plates stacked by more number of elements (5, 7, 9), the linear polarization rotator will have effect even at broader spectrum. As the principle of the construction of this type of linear polarisation rotator is universal on the type of the half-wave plates, we can expect a broader application of it.

5. Conclusion

We have presented an approach to construct broadband composite polarization rotator which operates over a wide range of wavelengths. Such a rotator is comprised of a Fresnel's rhomb and a composite broadband half-wave plates with a relative rotation of $\alpha/2$ between them. Furthermore we prove experimentally that such a polarization rotator is indeed broadband and operate over a wide range of wavelengths, using a double Fresnel's rhomb and a broadband half-wave plate.

Acknowledgments

H. Hristova acknowledges for the financial support by Bulgarian Academy of Science Programme for young scientists and PhD students 2017, DFNP 17-121. A. Rangelov, E. Dimova and S. Ognyanski acknowledge for the financial support by the Bulgarian Science Fund Grant DN18/14.

References

[1] Hecht E 2002 *Optics* (Addison Wesley, San Francisco)
[2] Born M and Wolf E 2005 *Principles of Optics* (Cambridge University Press, Cambridge)
[3] Azzam M A and Bashara N M 1977 *Ellipsometry and Polarized Light* (North Holland, Amsterdam)
[4] Goldstein D 2003 *Polarized Light* (Marcel Dekker, New York)
[5] Duarte F J 2003 *Tunable Laser Optics* (Elsevier Academic, New York)
[6] Pye J D 2001 *Polarised Light in Science and Nature* (Institute of Physics, Bristol)
[7] Damask J N 2005 *Polarization Optics in Telecommunications* (Springer, New York)
[8] Degl'innocenti E L and Landolfi M 2004 *Polarization in Spectral Lines* (Kluwer Academic Publishers, Dordrecht)
[9] Matioli E, Brinkley S, Kelchner K M, Hu Y L, Nakamura S, DenBaars S P, Speck J S and Weisbuch C 2012 *Light: Sci. Appl.* **1** e22
[10] Kulish M R and Virko S V 2013 *Semicond. Physics, Quantum Electron. Optoelectron.* **16** 64
[11] Koester C J 1959 *J. Opt. Soc. Am.* **49** 405
[12] Kim J S and Chang J K 2006 *J. Korean Phys. Soc.* **48** 51
[13] Rangelov A and Kyoseva E 2014 *Opt. Communications* **338** 574
[14] Dimova E, Rangelov A and Kyoseva E 2015 *Photon. Res.* **3** 177
[15] Torosov B T and Vitanov N V 2014 *Phys.Rev.* A **90** 012341
[16] Hurvitz H and Jones R C 1941 *J. Opt. Soc. Am.* **31** 493
[17] Dimova E, Huang W, Popkirov G, Rangelov A and Kyoseva E 2016 *Opt. Communications* **366** 382

IOP Conf. Series: Journal of Physics: Conf. Series **1186** (2019) 012019 doi:10.1088/1742-6596/1186/1/012019

Comparative analysis of red wines on the base of optical and chemical characteristics by statistical methods

St Minkova[1], M Milev[2], Kr Nikolova[1*], Ir Ivanova[3], N Hristova-Avakumuva[4], V Hadjimitova[4], M Kakalova[1], V Vladev[2]

[1]Medical University, 55 Marin Drinov Blvd., 9000 Varna, Bulgaria
[2] Univerity of Food Technologies, 26 Maritsa Blvd., 4000 Plovdiv, Bulgaria
[3] Agricultural University, 12 Mendeleev Blvd., 4000 Plovdiv, Bulgaria
[4] Medical University, Str. Zdrave 5., Sofia, Bulgaria

*E-mail: kr.nikolova@abv.bg

Abstract. Thirty – five samples of red wines from different geographic regions have been characterised according to 14 basic optical and chemical characteristics - colour parameters, content of anthocyanins and antioxidant activity. The goals of this study is to compare the tested wines using different statistical methods. Conducting hierarchical cluster analysis the samples have been divided into 4 clusters according to their difference and similarity. The result is presented graphically by a dendrogram. It could be concluded that the optical parameters and related chemical indicators are more decisive for the division of wines based on the region of production than on the basis of the predominant grape variety. The total number of studied parameters was reduced to 4 factors by factor analysis using Principle Component Analysis (PCA). These new factors explained 86.75% of the entire variation. The first factor contained chemical parameters and color coefficients in SIELab colorimetric system. The second one was connected with the quality characteristics of wine, which were important for consumers such as brilliance of red, color intensity. The third and the fourth factors were related with the Hues angles, color parameter *b* and saturation for the red wines.

1. Introduction

The red wine is very rich in polyphenols (catechin and epicatechin), flavanols (quercetin, rutin, myristin), phenolic acids (gallium, caffeic and p-cumaric acid). Phenolic compounds play a significant role in the sensory properties of wines such as flavor, color and taste and have a positive effect on health [1]. These components are important because of their proven bactericidal effect [2]. The content of polyphenols in grapes is different for various types of grapes in different years. The quality of grape is different from year to year, because the rainfall, sunshine-exposure and average temperature of each harvest year are not the same. These factors affect the polyphenol content and color parameters of wine samples as reported in the literature [3-7].

The purpose of the present study is to group the examined red wines examined into clusters on the basis of the degree of similarity between them, to assess the influence of factors such as grape variety, geographic region or year of production on the modelling of the samples, as well as by the method of principle component analysis (PCA) to reduce the huge number of examined variables to several new artificial factors having a determining significance for the respective groups.

2. Materials and methods

2.1. Samples

The database includes 35 types of red wines from 4 countries - Bulgaria, Italy, Chile and Sought Africa. The samples are purchased from the Bulgarian and foreign supermarkets or obtained directly from winery. Details for the geographical region and grape variety for the investigated samples are presented in table 1.

Table 1. The wine collection.

Country	Code	Region	Variety grapes	Year	Country	Code	Region	Variety grapes	Year
Bulgaria	1	Harmanly	Merlot	2012	Bulgaria	19	Montana	Cabernet cauvignon	2013
Bulgaria	2	Harmanly	Merlot	2013	Bulgaria	20	Tracian Plain	Cabernet cauvignon	2013
Bulgaria	3	Harmanly	Merlot	2014	Bulgaria	21	Tracian Plain	Cabernet cauvignon	2013
Bulgaria	4	Harmanly	Merlot	2015	Bulgaria	22	Harmanly	Cabernet cauvignon	2016
Bulgaria	5	Harmanly	Merlot	2016	Bulgaria	23	Stara Zagpra	Cabernet cauvignon	2013
Serbia	6	Pleven	Merlot	2013	Bulgaria	24	Montana	Cabernet cauvignon + Merlot	2013
Bulgaria	7	Tracian Plain	Merlot	2013	Bulgaria	25	Vidin	Cabernet cauvignon +Syrah	2013
Bulgaria	8	Yambol	Merlot	2013	Bulgaria	26	Sakar Mountain	Cabernet cauvignon +Syrah	2013
Holland	9	Asenovgrad	Merlot	2013	Bulgaria	27	Sakar Mountain	Cabernet cauvignon +Syrah+ Merlot	2013
Bulgaria	10	Stara Zagora	Merlot	2013	Bulgaria	28	Stara Zagora	Cabernet franc	2013
Bulgaria	11	Veliki Preslav	Merlot	2013	Bulgaria	29	Harmanly	Cabernet cauvignon+Syrah+Pamid+ Merlo	2016
Bulgaria	12	Danube Plain	Merlot	2013	Bulgaria	30	Svilengrad	Merlot +Malbec	2014
South Africa	13		Merlot	2013	Bulgaria	31	Tracian Plain	Merlo+Maller	2014
Chile	14		Merlot	2014	Bulgaria	32	Harmanly	Syrah	2016
Bulgaria	15	Harmanly	Cabernet cauvignon	2016	Bulgaria	33	Harmanly	Malbec	2016
Bulgaria	16	Suhindol	Cabernet cauvignon	2014	Bulgaria	34	Trakian Plain	Beaujolais	2015
Bulgaria	17	Stara Zagora	Cabernet cauvignon	2013	Italy	35		Barbera dAsti	2013
Bulgaria	18	Tracian Plain	Cabernet cauvignon	2013					

2.2. Used methods

Color measuring: CIELab coordinates [8] have been measured by using spectrophotometer (Lovibond Trintometer PFX 195, UK). The content of carotene is determined by special software. The samples were used without any dilution, poured into a 10 mm cuvette.

Absorption spectra: The absorption spectra for investigated samples are obtained by spectrometer Brolight, using optical scheme Czerny-Turner. The parameters such as color intensity, tint, saturation, logarithmic color density, angle of Hue are used for estimation of quality and ages of wines. The absorbance values (D) at 420 nm, 520 nm and 620 nm [9] are used to obtain numerical values of color intensity etc.

Antioxidant activity**:** Determination of the antioxidant capacity was performed by two complementary methods, evaluating different aspects of the reduction activity of the samples:

ABTS method - The method is carried out according to Re *et al.* [10] DPPH method - The method is performed according to Goupy *et al.* [11].

2.3. Statistical analysis

To compare and group the types of wines on the base of all investigated physico-chemical parameters a hierarchical cluster analysis with a measure of similarity Euclidean distance [12, 13] has been used, applying the method of average linkage between groups. The data have been preliminary standardized with z-standardization in order to avoid the influence of different dimensions. The results have been represented graphically by a dendrogram.

A factor analysis was conducted. Thus the input set of correlating data was transformed into a new set with a smaller number of uncorrelated artificial variables, so-called „factors" or „principal components", which explain the greatest part of the data variation. The reduction in the number of original variables was achieved by grouping the correlated variables in a common factor and the separation of uncorrelated ones in different factors [14-15]. A stronger distinction of the variables to one or another factor is performed by further factor rotation using Varimax method. All data analyses have been performed using the SPSS software program [16].

3. Results and discussion

The results from cluster analysis are presented on the figure 1. The samples from the different from geographical regions which have been investigated are grouped in fourth clusters. The distance between samples is from 1.030 to 100.943. From the dendrogram it can be seen that fourth clusters.

Table 2 shows both - the percentage distribution of the samples into the 4 clusters and the samples participation in each of the 4 clusters. Samples 1, 19, 22, 25, 27, 29, 30, 32, and 35 do not belong to these clusters, because they have some different values for investigated parameters.

Most of the samples in the first cluster belong to the Eastern Thracian Lowland and only 3 of them in the northern Black Sea region. A sample from South Africa, it is the closest to red wine from the Yambol region. The second cluster is joined samples of Western Thracian Lowland and a wine from northwestern Bulgaria with predominant sort Cabernet Sauvignon. Heterogeneity is observed from varieties, regions and years in the third cluster. The fourth includes samples from the Harmanly region 2015 and 2016 year. This can be explained by the high antioxidant activity in that region over the period considered. There are data that SV_{50}=1μl in the same region [17]. The last fact is associated with very high antioxidant activity. Although at a much distance to them, samples 22 and 32 are joined in the same cluster. They are from the same region and year of production but from another grape variety. Samples 1 and 25 are at the longest distance. This is understandable since they are from two climatically opposite regions – the Vidin and the Harmanly. They, as well as samples 30 and 35, could not belong to any of the four clusters. This is natural because sample 35 is Italian wine and the weather conditions there are not similar to ours weather conditions. Sample 30 is from the region of Svilengrad and according to distances from the dendrogram is closest to the case 1, which belongs to Harmanly region.

Therefore, not grape varieties, but the geographical area and the year of production are decisive factors for the classification of red wines according to the physico-chemical characteristics.

The investigated characteristics (total number 14) have been reduced by grouping in different factors. These, which have correlations one with other, have been combined in one factor. The others - uncorrelated have been divided in different factors.

Table 2. Distribution of types wines in clusters.

Variants in clusters	Samples, number	Distribution of the samples of wines,%
6,7, 8, 11,12, 13, 17, 20, 23, 26, 27, 28, 31, 34	14	40
9, 18, 21, 24	4	11.5
2, 3, 10, 14, 16, 29	6	17
4, 5, 15, 33	4	11.5

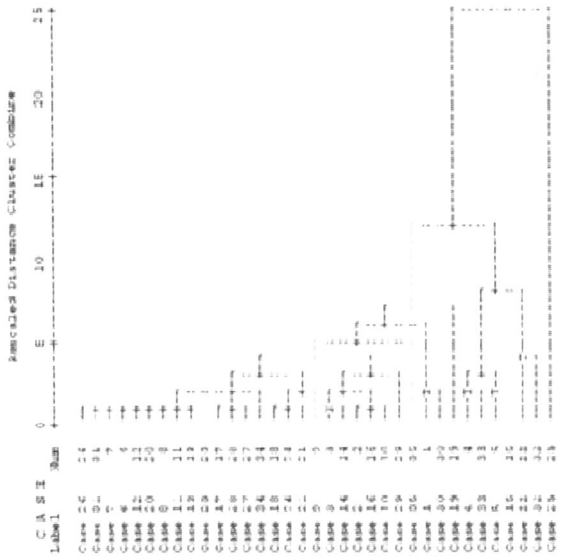

Figure 1. Dendrogram for investigated wines

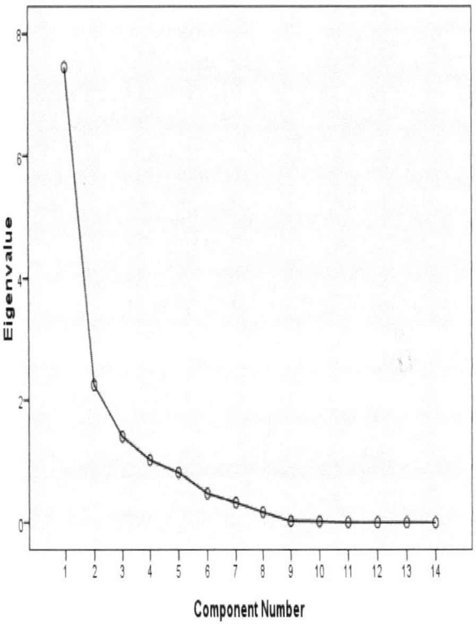

Figure 2. Results from Factor analysis

The determinant of the correlation matrix is different from zero, which allows conducting the factor analysis. The data have to posses normal distribution for obtaining the adequate model. It can be determined with model of Kaizer-Maier- Olkin (KMO test), which value has to be greater than 0.5, in our case this is fulfilled. The Bartlett's test of sphericity is meaningful ($p<0.001$). This gives us reason to perform a factor analysis and that the model will be adequate.

It is determined from the PCA that four factors have values of an own vectors greater than one, which determined the choice of 4 main components, which explained 86.75% of total variation. It is presented on the figure 3. The component plot in a rotated space is shown on the figure 2

The first factor (principle component one - PC 1) explains 42.47% of total variation and strongly correlated with color parameters "a" in color system CIELab, lightness L, Chroma C, anthocyanins, antioxidant activity, determined by ABTS and DPPH methods and nuance. We could summarize that this factor is generalizing, as these parameters have the greatest weight in separating wines into clusters according to consumer sensory assessment. It combines parameters that have correlations between them with a level of significance of 0.1%. The second component summarizes indicators tied with an evaluation of red wine by quality and age–color intensity, brilliance and hue - T), which explains 20.92% of total variation. The third and fourth components are linked to one indicator and explain 13.92% and 9.44% of total variation respectively. So the greatest weight has the parameters that come in the first and second factors.

The variation of the factor weight and distribution between four main components are presented in table 3.

Table 3. Factor matrix, obtained by method PCA

№	Investigated Parameters	Main components (factors) 1	2	3	4
X_1	AOA by ABTS methods	-0.582	-0.210	0.025	-0.246
X_2	AOA by DPPH methods	-0.582	-0.332	-0.022	-0.031
X_3	Anthocyanins 1	0.858	0.277	-0.067	0.208
X_4	Anthocyanins 2	0.859	0.276	-0.069	0.208
X_5	Lightness L	-0.917	-0.170	-0.065	0.290
X_6	"a' color parameter	0.959	0.145	-0.167	-0.032
X_7	"b" color parameter	-0.266	0.536	0.699	0.334
X_8	"C" chroma	0.899	0.329	0.046	0.108
X_9	H angle	-0.768	0.252	0.515	0.238
X_{10}	Saturation	0.242	0.163	0.557	-0.744
X_{11}	Nuance	0.856	0.181	0.166	-0.106
X_{12}	Hue	-0.404	0.781	-0.436	-0.172
X_{13}	Brilliance	-0.733	0.650	-0.178	-0.064
X_{14}	Color Intensity	0.761	-0.587	0.231	0.139
	Total Variation, %				
	Cumulative present from total variation, %				

Table 4. Extraction method: Principal Component analysis

Component	Initial Elgenvalues		Rotation sum of squared loadings		
	% of Variance	Cumula-tive %	total	% of Variance	Cumulative %
X_1	53.226	53.226	5.946	42.468	42.468
X_2	16.095	69.322	2.929	20.923	63.392
X_3	10.075	79.397	1.948	13.917	77.309
X_4	7.355	86.751	1.322	9.443	86.751
X_5	5.883	92.634			
X_6	3.407	96.042			
X_7	2.343	98.384			
X_8	1.223	99.607			
X_9	0.187	99.794			
X_{10}	0.151	99.945			
X_{11}	0.035	99.980			
X_{12}	0.018	99.998			
X_{13}	0.002	100.00			
X_{14}	0,000	100.00			

4. Conclusion

The strongest influence indicates the anthocyanin content, the antioxidant activity and the luminosity that groups the wines into separate clusters. The fourth cluster brings together samples of red wines from southeastern Bulgaria (Harmanli and Svilengrad) with the highest antioxidant activity, anthocyanin content and color indicators. The first one combines the most samples of the most widely consumed red wines from the southern region of Bulgaria with relatively good antioxidant properties.

References

[1] Woraratphoka K, Intarapichet K. 2007 *Food Chem*istry **104** 1485

[2] Singleton V, Esau P 1969 *Phenolic Substance in Grapes and Wine and Their Significance* (Academic Press, New York) p. 125

[3] Lee J E, Hwang G S, Van Den Berg, Lee C H & Hong Y S 2009 *Anal Chim Acta* **648** 71

[4] Lorrain B, Chira K, Teissedre P L 2011 *Food Chemistry* **126** 1991-1999

[5] Pereira G E, Gaudillere J P, Leeuwen C V, Hilbert G, Maucourt M, Deborde C, Rolin D 2006. *Anal Chim Acta* **563** 346

[6] Ferrandino A, Lovisolo C 2014 *Environmental and Experimental Botany* **103** 138

[7] Siegl K, Eder R., Rak G, Abranko L, Koellensperger G, Hann S 2010 *Food Chemistry* **122** 366372

[8] 2004 *C. I. E. Colorimetry* (edited by Central Bureau of the Commision Internationale de LEclairage (Publication C. I. E., Vienna) p.35

[9] Yıldırım H, Elmacı Y, Ova G, Altuğ T and Yücel U 2007 *International Journal of Food Properties* **93** 102

[10] Re R, Pellegrini N, Proteggente A, Pannala A, Yang M and Rice-Evans C 1999 *Radical Biology and Medicine* **26** 1231

[11] Goupy P, Dufour C, Loonis M and Dangles O 2003 *Journal of Agricultural & Food Chemistry* **51** 615

[12] Duran B and Odelle P 1977 *Cluster analysis* (Moscow, Russia: Statistics) p 115

[13] Ward JH 1963 *Journal of the American Statistical Association* **58** 236

[14] Jolliffe I 2002 *Principal Component Analysis* (USA, New York, Springer-Verlag New York) p 76

[15] Vundev D 2003 *Applied Statistic* (Sofia, Bulgaria: SU Kliment Ohridski) p 76

[16] Statistics Toolbox™ User's Guide, R2014a, Copyright 1993–2014 by The MathWorks, Inc

[17] Hristova –Avakumova N G, Atanasova L A, Atanassova S L , Vangelova T V, Hadjimitova V A 2018 *Bulgarian Chemical Communications* **50** 321

IOP Conf. Series: Journal of Physics: Conf. Series **1186** (2019) 012020 doi:10.1088/1742-6596/1186/1/012020

Polymer (PEO)-liquid crystal (LC E8) composites: The effect from the LC inclusion

G B Hadjichristov [1,*], Y G Marinov [1], A G Petrov [1], H K Koduru [2,3] and N Scaramuzza [2,4]

[1] Georgi Nadjakov Institute of Solid State Physics, Bulgarian Academy of Sciences, 72 Tzarigradsko Chaussee Blvd., BG-1784 Sofia, Bulgaria
[2] Dipartimento di Fisica, Università degli Studi della Calabria, Via P. Bucci, Cubo 33B, Rende (CS), IT-87036, Italy
[3] Department of Physics, Madanapalle Institute of Technology & Science, Madanapalle, Andhra Pradesh - 517325, India
[4] CNISM – Unità di Ricerca di Cosenza c/o Dipartimento di Fisica, Università degli Studi della Calabria, Via P. Bucci, Cubo 31C, Rende (CS), IT-87036, Italy

[*]Corresponding author's e-mail address: georgibh@issp.bas.bg

Abstract. The effect from the inclusion of nematic liquid crystal (LC) E8 into the polymer poly(ethylene oxide) (PEO) was inspected by studying of structural and electrical properties of PEO-E8 composites at weight percentage of E8 ranging from 10 % to 50 %. Flexible PEO-E8 films with a thickness of 0.1 mm were structurally characterized by X-ray diffraction and X-ray photo-electron spectroscopy. The results obtained by our analyses indicate that a polymer-LC intermolecular complex is formed by inclusion of E8 LC molecules in the PEO host at a certain their concentration level. The structural properties of PEO-E8 composites were correlated with their electro-conducting properties as depending on the E8 LC concentration. By the amount of the LC fraction one can achieve a controlled modification of the structural and electro-conducting properties of the PEO-E8 material. As compared to PEO, the inclusion of E8 LC in the PEO polymer matrix can lead to a considerably enhanced electrical conductivity of PEO-E8 composites.

1. Introduction

The doping of organic compounds into matrices of organic polymers is a strategic route to improve the performance of material characteristics like structural, optical, non-linear optical, electro-optical, electrical, dielectric, thermodynamical and mechanical properties. This is well exploited in the very quickly developed field of solid polymer electrolytes [1-5]. During the last decade, there are continuous investigations on multifunctional composite materials produced from polymers and liquid crystals (LCs). Such all-organic composites are a class of high-quality advanced materials with a wide spectrum of applications, e.g., in organic electronics, electro-optics and optoelectronics, in mechatronic systems and medicine. In particular, they are of practical interest as electrolytes in rechargeable mini-batteries, organic electronics, photovoltaics and mechatronics [6-9]. Due to the improved electrical properties, various electrolytic polymer-LCs systems based on the polymer poly(ethylene oxide) (PEO) [10-12] have been synthesized and investigated [13-15]. Very recently we have studied PEO-LC composite electrolytes produced of PEO with addition of molecules of the room-temperature nematic cyanophenyl LC with the commercial name E8 [16]. Thin (~ 0.1 mm) films of these flexible blends exhibited enhanced ionic conductivity and dielectric properties. In order to elucidate the effect from the inclusion of E8 LC in PEO, in the present study we have characterized the physico-chemicical structure of PEO-E8 composites (with common methods like X-ray diffraction (XRD) and X-ray photo-electron spectroscopy (XPS)). As depending on the concentration of the LC E8, the structural and ion-conducting properties of PEO-E8 composites were interrelated.

2. Experimental

Samples of our poly(ethylene oxide) (PEO)-based polymer-LC blend electrolyte (flexible free-standing films with a thickness of 0.1 mm) were produced by conventional solvent-cast method (in our case, in methanol at room temperature, as described in [17-19]). By this technique, the structure of the composite film can be controlled by solvent evaporation rate. The structure of PEO is commonly expressed as $[-H_2C-CH_2O-]_n$ [10] (figure 1a). In the present study, high-molecular-weight PEO (M.W. = 8×10^6 g/mol) was used (purchased from Sigma Aldrich). In order to include LC molecules into the PEO polymer network, the powder of PEO monomers were mechanically mixed with the commercially available multicomponent room-temperature nematic LC with the abbreviature E8 (or BL002), supplied from Merck. This LC is an eutectic blend of five cyanoparaphenylene LCs. It contains 45 wt.% of 4-cyano-4'-n-pentyl-biphenyl (5CB), 16 wt.% of 4-n-propoxy-4′-cyanobiphenyl (3OCB), 12 wt.% of 4-n-pentoxy-4'-cyanobiphenyl (5OCB), 16 wt.% of 4-cyano-4'-n-oxyoctylbiphenyl (8OCB) and 11 wt.% of 4-cyano-4"-n-pentyl-p-biphenyl (5CT) (figure 1 b). All these nematic-type LCs have elongated rod-like molecules. The mixture E8 exhibits a single nematic to isotropic transition temperature at ca. 72°C and a glass transition temperature. PEO-E8 blends with 10, 20, 30, 40 and 50 wt.% of E8 were produced. PEO-E8 samples (thin films) were obtained by drying the PEO-E8 blends, and then evaporating the solvent under vacuum. Besides of PEO-E8 composite films, also films of net PEO with the same thickness were prepared by solvent casting.

The X-ray diffractograms of the samples were recorded using a D8 ADVANCE ECO (Bruker-AXS, Bruker Co.) X-ray diffractometer system with Ni-filtered Cu K_α radiation of wavelength λ = 0.154 nm. Angular scans of diffraction in reflection mode were taken in the 2θ range of 10° – 80° with a constant step of 0.1° and at counting time of 35 s. The XPS studies were performed by apparatus described in [20]. XPS spectra were obtained at 45° take-off angle and were referenced to C 1s peak (binding energy (BE) of 285.0 eV) resulting from small traces of adventitious carbon. The acquisition time was 0.6 s, the accuracy of the measured BE was 0.1 eV. The recorded C1s photoelectron lines were corrected by subtracting a Shirley-type background. Peak identification was performed by reference to a database of XPS. The electrical conductivity of the PEO-E8 thin films were measured by means of complex electrical impedance spectroscopy performed by a impedentiometer (SP–200, Biologic) in the frequency range from 0.01 Hz to 2 MHz of the applied electric field. Te PEO-E8 samples were sandwiched between two circular-shaped copper electrodes with diameters of 1 cm. The complex impedance spectra were recorded at a probe voltage of 0.5 V_{RMS} in the sinusoidal waveform. All measurements in this study were carried out at ambient temperature (in our case, 27 – 28 °C).

3. Results and discussion

Optical microscopy studies on produced PEO-E8 films displayed the successful inclusion of nematic LC E8 in the bulk of the films [16]. The structural changes resulting from the dispersed E8 LC molecules into the PEO polymer network were evidenced by vibrational spectroscopy (micro-Raman and FT-IR absorption). The results (will be reported separately) suggest the increase of relative content of the amorphous phase and the flexibility in the examined polymer-LC composite. In addition, XPS studies can provide information about the surface chemistries of PEO-E8 composites (at the top surface and in the very near-surface region of the samples, in our case, within the depth of ~ 5 – 10 nm). In particular, figure 1(c) presents the changes in the core level C1s spectra of PEO polymer that reflect its local structural modification due to the addition of E8 LC.

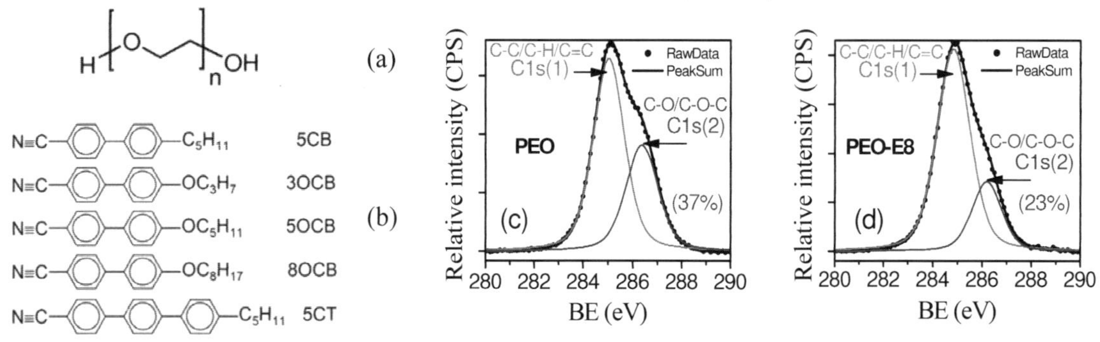

Figure 1. Chemical structures of: (a) PEO monomer; (b) constituents of the LC E8. Deconvoluted XPS spectra (after subtraction of a linear baseline) in the range of the characteristic C1s peak, recorded for pure PEO (c) and for PEO-E8 composite at the 30 wt.% E8 concentration (d) under identical conditions. The resolved components and their envelope (the peak sum) are given with lines, the raw data – with circles. The calculated area percentage of C1s(2) peak is shown in parentheses. The fit-factors (coefficients of determination) are: R^2 = 0.99962028 (c); R^2 = 0.99963553 (d).

The shape of the C1s photoelectron line of PEO (figure 1c) can be deconvoluted with use of two components, labeled as C1s(1) and C1s(2), at BE in vicinity of 285 eV and 286.3 – 286.5 eV, respectively, as referenced to a database of XPS for PEO, e.g. [21,22]. The first one is assigned to the C–C/C–H bonding in the [–CH_2–CH_2–O–] repeat unit of PEO polymer (polymer backbone), and the second one is typically attributed to C–O bonds of carbon atoms in C–O–C moieties of PEO (the alcohol/ether C–OH/C–O–C type carbon component). Figure 1(c) presents the deconvolution of the C1s photoelectron line of PEO as fitted with both components. The curve fitting was performed as described in [20] – the XPS C1s data from the experiment were deconvoluted by least-square fitting; each component was taken as a product of Gaussian and Lorentzian functions (80% : 20%). Because very small, the contributions related to other types of C to O bondings (usually at BE > 288 eV) were ignored in the curve fitting.

On the other hand, the molecular structures of the constituents of E8 LC (figure 1b) suggest that the corresponding C1s XPS spectra should be deconvoluted by using at least 4 peaks. From them, three are related to aromatic carbon, aliphatic carbons CH_2–CH_3, and C–N species (e.g., for 5CB molecule their assignments are at BE of 284.8 eV, 285 eV and 286.7 eV, respectively [23]). A significant additional peak should be the one corresponding to C–O bonding of the cyanoalkoxybiphenyl (nOCB) constituents of E8 LC. Thus, regarding the C1s line in the XPS spectrum of PEO-E8 composite (figure 1d), besides the C–C/C–H bonding there are strong contributions from C=C double bonds (in the phenyl rings of the E8 LC molecules) and C≡N triple bonds, related to the functional groups of the compounds belonging to the mixture E8 (figure 1b). However, the XPS signal attributed to C=C bonds could be hardly distinguished from that associated with hydrocarbon bonds (C–C and C–H), because they are in the same BE region (moreover, for C–C and C=C bond it is difficult to specify the BE because we would need an independent charging reference). The same problem applies to the BE of the carbon atoms bonded to the N. The contribution of C≡N bond in cyanogroups to the XPS C1s core-level spectra strongly overlaps with that of the C–O bond, and partially with C–C/C–H/C=C bonds, hence they could not be readily quantified separately and unambiguously analyzed without introducing of a lot of assumptions. Due to the merging of the peaks corresponding to both PEO and E8 LC, the interpretation of the XPS spectral changes in our blend system using these peaks might not be reasonable. Still, the clear diminishing of the shoulder at BE = 286.3 eV (figure 1d) and corresponding decrease of the percent area of C1s(2) peak with respect to the total C1s peak, is indicative of the reducing concentration of C–O bondings in the PEO-E8 composite, even though in the considered case these bondings are associated also with the functional groups of

IOP Conf. Series: Journal of Physics: Conf. Series **1186** (2019) 012020 doi:10.1088/1742-6596/1186/1/012020

cyanoalkoxybiphenyl (nOCB) constituents of E8 LC added to PEO. Roughly, one can deconvolute the XPS C1s core-level spectrum of PEO-E8 composite into the same two components C1s(1) and C1s(2) and in the same manner, as the C1s lineshape of pure PEO. The result is shown in figure 1(d). Comparing figures 1(c) and 1(d), one obtains that the area percentage of C1s(2) peak is decreased from 37 % (pure PEO polymerization processing) to 23 % (PEO polymerization in the presence of 30 wt.% nematic LC E8). This, as well as the clear shift of ca. 0.2 eV to the lower BE observed for the composite sample, suggest an appearance of some type LC–(CH_2CH_2O) molecular complex in the studied PEO-LC blend. The level of concentration of cyanobiphenyls of E8 LC in the composite system ensures their complexation with the ethylene-oxide (CH_2–CH_2O) units in the host polymer.

Further useful information on the structural modification of the host polymer PEO due to the inclusion of E8 LC in the PEO-E8 composite can be obtained by XRD. The XRD records (figure 2) can be used to determine the phase composition and might be indicative of the influence of E8 LC on the crystalline status of PEO. The XRD pattern obtained for pure PEO (figure 2a) demonstrated two distinct crystalline peaks at diffraction angles (in terms of Bragg angle 2θ) of 19.4° and 23.5°. They are attributed to the (120) and (112) crystal planes of the PEO monoclinic crystal structure, as described in the literature. These diffraction peaks of PEO originate from the ordering of polyether side chains and strong intermolecular interaction between PEO chains through the hydrogen bonding [10,12,17,24]. Both characteristic peaks are structural signatures of PEO that exhibit its relatively high degree of crystallinity (the crystallinity is defined as the weight fraction of the crystalline portion of the polymer; as known, the physical properties of polymers are considerably dependent on this parameter). Simultaneously, a broad halo is also present in the XRD pattern (as a hump at the lower angles of XRD) that indicates a partially amorphous phase in the semi-crystalline PEO. As seen from figure 2, this diffraction feature strongly suppresses the weak diffraction peak (doublet) of PEO at 15°.

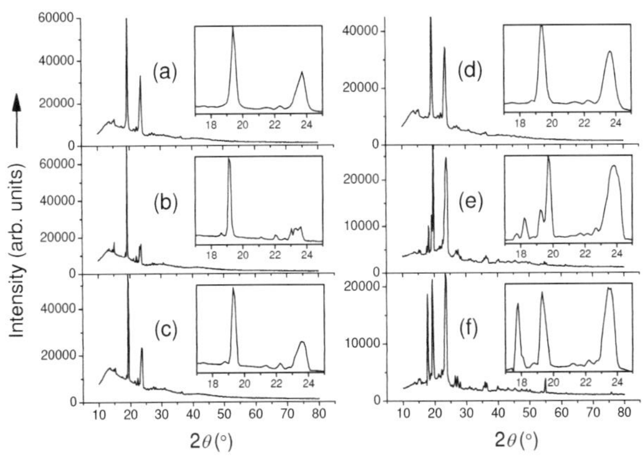

Figure 2. XRD pattern of PEO (a) and PEO-E8 films by weight fraction of the E8 nematic LC (wt.%): 10 (b); 20 (c); 30 (d); 40 (e); 50 (f). The insets show the same data on an expanded scale (Bragg angle 2θ in the range from 17° to 25°).

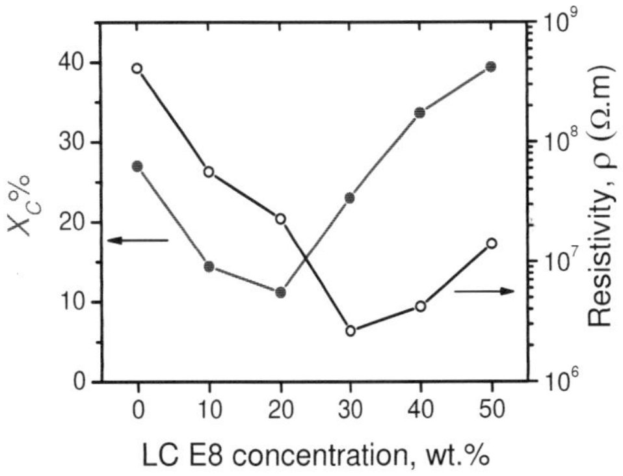

Figure 3. Variation of degree of crystallinity (X_C%) and electrical resistivity (ρ) of the studied PEO-E8 composites at various weight fraction of the LC E8.

The changes in intensity of the two sharp crystalline characteristic peaks of PEO upon the addition of nematic LC E8 should reveal the structural modification of the host polymer PEO, as an indication of the effect from inclusion of E8 LC. In fact, the decrease of their intensity by addition of E8 LC at concentrations 10, 20 and 30 wt.% (figure 2(b-d)) suggests an increase of amorphous content and flexibility in the considered polymer-LC composites, as previously discussed for XRD data for other PEO-based solid-type polymer electrolyte systems with flexibility [17-19]. It is interesting to inspect the change in the XRD pattern of PEO-E8 as depending on E8 LC concentration. XRD study reveals that all the films measured show semicrystalline structure. As seen in figure 2, the mixture of crystalline/semicrystalline phase in which PEO exists is significantly modified upon addition of the LC. Firstly, by increasing percentage of E8 LC (in our case, in the range 10 wt.% – 50 wt.%) gradually occurs a slight transformation from a structure mostly characterized by (120) crystal plane to structure in which the (112) crystal plane of the PEO monoclinic crystal is more pronounced, implying another preferred structural orientation of PEO. Secondly, the addition of E8 LC at higher concentration, e.g., as high as 40 wt.%, leads to appearance of a sharp XRD peak at 2θ near to 18°. This peak arises as a doublet at 2θ = 17.8°/18.25° (figure 2e) and becomes single (2θ = 17.8°), sharp and intense at 50 wt.% of E8 (figure 2f). Its relative intensity increases with increasing E8 weight fraction, whereas the intensity of the PEO diffraction peak at 19.4° declines. The observed additional peak can not be assigned to a crystallographic site of PEO in the pure state and was absent for E8 nematic LCs. For instance, a broad (FWHM ~ 10°) and rather diffuse XRD peak symmetrically centered at 2θ ~ 21° takes place for 5CB LC in its nematic state (the peak for the nematic phase) and reflects the absence of ordering in the system. Also, the additional peak in question can not be associated with a hypothetical LC phase induced by polymer, because usually such induced crystalline phases by nematic LCs exhibit XRD peaks at much lower 2θ values. Therefore, the XRD pattern observed for PEO-E8 composite can be attributed to a crystal modification in PEO. Similarly, additional peak (at 2θ ≈ 16°) has been found by XRD analyses of molecular complexes prepared from PEO with inclusion of ionic liquid [25], where, in general, such a diffraction peak evidences a change in PEO structural geometry due to interaction/complexation with the ionic liquid.

Based on the results from XRD measurements, one can suppose that by PEO:E8 compositions equal to 60:40 wt.% and 50:50 wt.% arises intermolecular crystalline complex promoted from cyanophenyle molecules of E8 LC being in confined volume within the host matrix of the polymer, and anchored, at least to some extent, to segments of PEO chains. According to XRD data in figure 2(e,f), such a structural formation is characterized with an average intercrystallite separation R that is larger than R for undoped PEO ($R \sim 1/\sin\theta$). Most probably such a complex occurs due to electric dipole-dipole intermolecular interactions induced by cyanocompounds in E8 having highly polarized cyanogroups and easily polarizable biphenyl groups. For instance, the longitudinal permanent electric dipole moment of 5CB molecule is ~ 3 D due to the triple –C≡N bond. The dipole moment of PEO molecule mostly due to the charge-polarized group –O– present in PEO is relatively small (~ 1 D, at 25° C), but the PEO ether oxygens are capable of forming hydrogen bonds. Thus, weak hydrogen bonds between the PEO hydroxyl groups and the π-electrons of benzene rings of cyanophenylenes in the E8 LCs are rather possible, because the −OH group of PEO can act as a hydrogen-bond donor, and the −CN group of cyanophenylenes acts as a hydrogen-bond acceptor. A third type of molecular interaction could be also considered – electronic charge-transfer PEO-LCs interactions because the cyano group is strongly electron-withdrawing, and the oxygen atoms in the $[C–C–O]_n$ backbone structure of PEO are electron-donating. The question is about the strength of the contribution of each from these three possible kinds of intrermolecular interactions, which are, actually, LC-matrix surface interactions. In fact, each of them should assist the intermolecular coupling and formation of PEO-LC complex. The dipole-dipole interaction between cyanophenyl core, well known for cyanobiphenyle molecules [26,27], can play a weaker role in the mechanism of formation of PEO-LC complex.

The salvation role of PEO is to ensure multiple contacts between own electron-donating ether oxygens along the highly flexible PEO backbone and electron-withdrawing cyanogroups of the LC dopants, similarly to the case of complexation of PEO with metal (e.g., Na^+) cations in salt-complexed polymer electrolytes [17]. Thus, LC-PEO molecular complex should be formed during the polymerization of PEO, when enough strong chemical interaction between the functional groups of E8 LCs and hydroxyl groups at the ends of PEO chains is realized, enough strong to overcome the coupling effect from the backbone. This happens through the surfaces of both polymer and LC phases. The chains of PEO have no large side-groups, hence can be easily distorted by intermolecular interactions. The chain backbone of PEO is flexible and could adopt other conformations, e.g., as segments of the entire chain, that can result in clusterization into domains [28]. By that, the originally completely helical conformation may be suppressed. In this context, a calculation by means of the Debye-Scherrer formula gives for the average domain size $D \sim 35$ nm corresponding to the XRD peak observed at $2\theta = 17.8°$. Also, clusterization of LC molecules confined in the PEO polymer matrix is rather possible, and thereby the formation of LC clusters attached to PEO backbone. In such clusters (whose dimensions could be close to the above mentioned size $D \sim 35$ nm) the rigid rod-like E8 nematic molecules develop spatial and orientational ordering [29].

The reorganization of the crystalline state of the polymer host has a strong impact on the electro-conducting properties of the studied blends of PEO with the nematic LC E8. The situation is much more complicated, since the polymer-LC molecular interactions are accompanied by a redistribution of electron density in the LC molecules that should be reflected in the change of electro-conductivity and dielectric properties of PEO-E8 composite system. As established by electrical impedance measurements, the electrical resistivity of PEO-E8 decreases by the increase of the relative amount of E8 up to ~ 30 wt.%, and after that it increases. Certainly, this effect can be correlated with the degree of crystallinity (X_C%) of the PEO-E8 composites, as calculated from the XRD patterns in figure 2. X_C% of the studied PEO-E8 composites can be roughly estimated by the ratio of the total area under the crystalline peaks to the total area under the diffractrogram. The X_C% value decreases gradually with increasing concentration of the E8 LC until approximately 20 wt.% and then increases with further increasing E8 concentration (figure 3). The reduction in the degree of crystallinity can be connected with enhanced flexibility of the polymeric chain, which results in the enhancement of segmental motion in the polymer. When the E8 concentration exceeds 20 wt.%, X_C% increases that may be attributed to the reorganization of the polymer matrix.

It is seen from figure 3 that the electrical conductivity of the studied PEO-E8 composites is nearly reciprocal to X_C%. Thus, by dealing with the blend composition (wt/wt %) of the PEO-E8 PE material one can simultaneously achieve a controlled modification of both its structural and electro-conducting properties. By suppression of polymer crystallinity at a proper blend ratio, one can get the most efficient ion transport. It is worthy to note that our analyses of Raman spectra of the studied PEO-E8 composites clearly confirm the variation of X_C% for PEO-E8 composites as depending on the concentration of E8 nematic LC, as well as the conformation of PEO chains (breaking of the intermolecular hydrogen bonds present in the helical configuration of PEO) at higher weight fraction of E8 nematic LC (40 wt.% and more evidently at 50 wt.%).

4. Conclusion

Our claim concerns the preparing of PEO-E8 polymer-LC composites (in form of self-supported, solid-like thin films with a mechanical flexibility) with a good ionic conductivity at room temperature. This was successfully fulfilled by adequate choice of the LC component (the nematic cyanobiphenyl-type LC E8) and its proper concentration. The information obtained from XRD analysis of PEO-E8 composites refers to a transformation of the PEO structure not only to an amorphous state but also to a structural rearrangement and formation of polymer-LC intermolecular complex. The main result of the study reported here is that the inclusion of nematic LC E8 in PEO polymer matrix can lead to a considerable improvement of electrical conductivity of the composed PEO-E8 electrolyte material, as compared to that of the PEO itself. In PEO-E8 composite PE, both components, PEO and nematic LCs, provide effective ion conduction through interactions between PEO oxygen and LC molecules. By that, anchoring effect being induced by interfaces of both PEO and nematic LCs, plays a significant role. On the other side, a negative effect has the conformation of PEO by its molecular complexes with LCs when the LC weight percentage approaches 50 %, that leads to enhanced crystallinity and diminishing ion conductivity. The results reported here will be of interest for engineering of advanced multifunctional polymer-LC flexible composite materials based on interfacial polymer-LC interactions, optimizing the use of LCs as additives to PEs in order to improve their performance aiming their applications in the field of flexible electronics and mechatronics.

Acknowledgments
This work was supported by the European Regional Development Fund within the Operational Programme "Science and Education for Smart Growth 2014 – 2020" under the Project CoE "National center of mechatronics and clean technologies "BG05M2OP001-1.001-0008-C01". The authors gratefully acknowledge also the support by the Ministry of Education and Science of Bulgaria, through the National Science Fund of Bulgaria (under the research Project DFNI-TO2/18).

References

[1] Gray F M 1991 *Solid Polymer Electrolytes: Fundamentals and Technological Applications* (New York: VCH)
[2] Sequeira C and Santos D eds 2010 *Polymer Electrolytes: Fundamentals and Applications* (Cambridge, UK: Woodhead Publ)
[3] Karuppasamy K, Antony R, Alwin S, Balakumar S and Shajan X S 2015 *Mater. Sci. Forum* **807** 41
[4] Ngai K S, Ramesh S, Ramesh K and Juan J C 2016 *Ionics* **22** 1259
[5] Aziz S B, Woo T J, Kadir M F Z and Ahmed 2018 *J. Sci.: Adv. Mater. Devices* **3** 1
[6] Kim S C, Song M, Ryu T I, Lee M J, Jin S H, Gal Y S, Kim H K, Lee G D and Y S Kang 2009 *Macromolec. Chem. Phys.* **210** 1844
[7] Karim M A, Song M, Park J S, Kim Y H, Lee M J, Lee J W, Lee C W, Cho Y R, Gal Y S, Lee J H and Jin S H 2010 *Dye Pigments* **86** 259
[8] Said S M, Zulkifli A Z S, Kamarudin M A, Mainal A, Subramanian B and Mohamed N S 2015 *Europ. Polym. J.* **66** 266
[9] Kamarudin M A, Khan A A, Said S M, Qasim M M and Wilkinson T D 2018 *Liq. Cryst.* **45** 112
[10] Bailey F E, Jr. and Koleske J V 1976 *Poly(Ethylene Oxide)* (New York: Academic Press)
[11] Xue Z, He D and Xie X 2015 *J. Mater. Chem. A* **3** 19218
[12] Arya A and Sharma A L 2017 *J. Phys. D: Appl. Phys.* **50** 443002
[13] Li J, Kamata K, Komura M, Yamada T, Yoshida H and Iyoda T 2007 *Macromolecules* **40** 8125
[14] Stoeva Z, Lu Z, Ingram M D and Imrie C T 2013 *Electrochim. Acta* **93** 279
[15] Tong Y, Chen L, He X and Chen Y 2014 *Electrochim. Acta* **118** 33
[16] Koduru H K, Marinov Y G, Scarpelli F, Hadjichristov G B, Petrov A G, Godbert N and Scaramuzza N 2018 *J. Non-Crystal. Solids* **499** 107
[17] Koduru H K, Marino L, Scarpelli F, Petrov A G, Marinov Y G, Hadjichristov G B, Iliev M.T and Scaramuzza N 2017 *Curr. Appl. Phys.* **17** 1518
[18] Koduru H K, Kondamareddy K K, Iliev M T, Marinov Y G, Hadjichristov G B, Karashanova D and Scaramuzza N 2017 *J. Phys.: Conf. Ser.* **780** 012006
[19] Koduru H K, Scarpelli F, Marinov Y G, Hadjichristov G B, Rafailov P M, Miloushev I K, Petrov A G, Godbert N, Bruno L and Scaramuzza N 2018 *Ionics* https://doi.org/10.1007/s11581-018-2484-8
[20] Russev S C, Tsutsumanova G G, Stefanov I L and Hadjichristov G B 2013 *Vacuum* **94** 19
[21] Martin-Vosshage D and Chowdari B V R 1993 *J. Electrochem. Soc.* **140** 3531
[22] Louette P, Bodino F and Pireaux J J 2005 *Surf. Sci. Spectra* **12** 59
[23] Artyushkova K, Fulghum J E and Reznikov Y 2005 *Mol. Cryst. Liq. Cryst.* **438** 205[1769]
[24] Agrawal R C and Hashmi S A 2007 *Ionics* **13** 295
[25] Chaurasia S K, Singh R K and Chandra S 2011 *Solid State Ionics* **183** 32
[26] Leadbetter A J, Richardson R M and Colling C N 1975 *J. Phys. Colloque C1* **36** 37
[27] Lydon J E and Coacley C J 1975 *J. Phys.* **36** 45
[28] Takahashi Y and Tadokoro H 1973 *Macromolecules* **6** 672
[29] Gwizdała W, Dawid A and Gburski Z 2008 *Solid State Phenom.* **140** 89

IOP Conf. Series: Journal of Physics: Conf. Series **1186** (2019) 012021 doi:10.1088/1742-6596/1186/1/012021

Energy spectra of a spin-$\frac{1}{2}$ XY spin molecule interacting with a single mode field cavity

H Tonchev, A A Donkov, H Chamati

Institute of Solid State Physics, Bulgarian Academy of Sciences,
72 Tzarigradsko Chaussée, 1784 Sofia, Bulgaria

E-mail: htonchev@issp.bas.bg, aadonkov@issp.bas.bg, chamati@issp.bas.bg

Abstract. We report results from our continued investigations on a finite-size chain (molecule) of L spin-$\frac{1}{2}$ spins with XY spin exchange interaction coupled at an arbitrary spin site to a single mode of an electromagnetic field via the Jaynes-Cummings model. Here we study the energy spectra in the case of an arbitrary photon fill number n, as compared to the $n = 1$ limit in our previous work [J. Phys.: Conf. Ser. 682 (2016) 012032] and [J. Phys.: Conf. Ser. 764 (2016) 012017]. By building the appropriate combinations of basis product states involving the photon as well as the up down spins it is possible to simplify the systems that are necessary to diagonalize from $n \times 2^L$ to at most 2^L. This results are in tune with those reported in the recent paper by N. Wu [Phys. Rev. B 97 (2018) 014301], and may prove to be useful for the quantum computing aficionados.

1. Introduction

The Jaynes-Cummings [1–3] model for the interaction of an atomic two level system with a quantized electromagnetic field has an exact solution, see, e.g., [4, 5] for the formulae. Extensions to a field coupled to L 2-level non-interacting systems has also been studied, for example in [6, 7]. In our model consideration the Jaynes-Cummings term is interpreted as a coupling between the field and a spin, of value $\frac{1}{2}$, see [8] for some physical reasoning of such a possibility. Introducing more spins into the spin chain, and switching some kind of a spin-spin interaction between the spins, one can envisage an application of this model to bit manipulation [9].

One can start with an exactly solvable XY model [10], or use a more recent setting [11], and then study the effect of having an additional boson field linearly coupled to the spins, as is the Jaynes-Cummings term. From this point of view, it is useful to obtain some exact formulae for the combined model Hamiltonian. With the help of a transformation to fermion operators [12] this is achieved in [13]. Here, on the other hand, we look into the problem by using the original, spin up and spin down basis states on each site of the chain, "sift" the matrices through the invariant available in the model [14] and then focus also on the corresponding eigenvectors expressed in terms of this states.

Below, in Section 2, we set the Hamiltonian of the model, and consider the resulting more compact form of the matrices to be actually diagonalized, as obtained by the "sifting through" procedure, and show the results for a system of up to 2 spins, concluding with remarks, in Section 3.

2. Bosonic field coupled to a site on a spin chain

The Hamiltonian under consideration is:

$$\widehat{H} \equiv \hat{H}_{JC} + \hat{H}^{XY} = \omega_R \hat{S}^z_k + \omega_f(\hat{a}^\dagger \hat{a} + \tfrac{1}{2}) + G\left(\hat{a}\hat{S}^+_k + \hat{a}^\dagger \hat{S}^-_k\right) + J\sum_i \left(\hat{S}^x_i \hat{S}^x_{i+1} + \hat{S}^y_i \hat{S}^y_{i+1}\right), \quad (1)$$

which is assumed here without further justification to be applicable to a system of spins in a cavity with laser field, so that we have one mode of frequency ω_f of a laser field focused on one of the spins of the spin $\frac{1}{2}$ chain of L sites, (e.g., $L = 5$ in figure 1). The interaction G part is from the Jaynes-Cummings (JC) model for the photon $(\hat{a}, \hat{a}^\dagger)$ interacting with a two level system (spin operator $\hat{S}^i_k$ $(i = x, y, z)$ on site k), the term with J is the XY model of the spin-spin interaction between nearest-neighbour spins, and the index i is enumerating the sites along the spin molecule, $i = 1, \ldots, L-1$ for open ended molecule, and $i = 1, \ldots, L$ for a closed cyclic molecule using the convention $\hat{S}_{L+1} \equiv \hat{S}_1$. Here we focus on the open ended configuration.

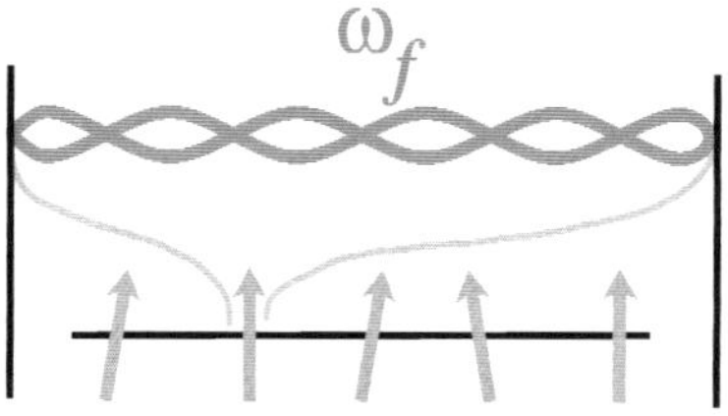

Figure 1. (color online) A view of a spin system in a cavity with an electro-magnetic field of frequency ω_f focused on a particular spin.

The spin basis is chosen as up, down for each site, $|\uparrow_i\rangle$ and $|\downarrow_i\rangle$, while the photon number states are $|n\rangle$, with $n = 0, 1, \cdots$. For example, the action of the Hamiltonian on the state with maximum spin $M_z = L/2$ value we have:

$$\widehat{H}|n, \uparrow_1 \cdots \uparrow_k \cdots \uparrow_L\rangle = \left(\tfrac{1}{2}\omega_R + \omega_f(n + \tfrac{1}{2})\right)|n, \uparrow_1 \cdots \uparrow_k \cdots \uparrow_L\rangle + G\sqrt{n+1}|n+1, \uparrow_1 \cdots \downarrow_k \cdots \uparrow_L\rangle \quad (2)$$

and then for the connected state we get correspondingly:

$$\widehat{H}|n+1, \uparrow_1 \cdots \downarrow_k \cdots \uparrow_L\rangle = \left(\tfrac{1}{2}\omega_R + \omega_f(n + \tfrac{3}{2})\right)|n+1, \uparrow_1 \cdots \downarrow_k \cdots \uparrow_L\rangle + G\sqrt{n+1}|n, \uparrow_1 \cdots \uparrow_k \cdots \uparrow_L\rangle$$
$$+ \tfrac{1}{2}J(|n+1, \uparrow_1 \cdots \downarrow_{k-1}\uparrow_k \cdots \uparrow_L\rangle + |n+1, \uparrow_1 \cdots \uparrow_k\downarrow_{k+1} \cdots \uparrow_L\rangle). \quad (3)$$

A few comments are in order. From these expressions we can deduce than the G part mixes a state with $|n = n_f, M_z = M\rangle$ with states $|n_f + 1, M - 1\rangle$ and $|n_f - 1, M + 1\rangle$ with appropriate factors, and the J part of the Hamiltonian rearranges the spins, without a change in the value of n and M_z, individually, as expected [15, 16], and for their sum as well. This is one way of showing that there is an invariant, given by the value of the sum $n + M_z$, since the mixed vectors belong to the same eigenvalue of this invariant. We will use this below to build the appropriate systems to obtain the eigenvalues. Another conclusion that can be drawn, is that when we truncate the series on n, by assuming a given maximum value n_{max}, this will introduce some spurious zero matrix elements (zero value) of the Hamiltonian acting on the $|n_{max}, M_z >$ state, which can lead to overcounting of the existing dark modes, the modes with $E = 0$, in the spectrum of the Hamiltonian, [17].

Now, looking back into equations (2) and (3), we see that for a given starting value $n = n_f$, by continuing the above procedure of applying the Hamiltonian to the states that appear on the right side, we can build a closed system, which involves states with $0, 1, 2, \cdots, L$ turned spins and correspondingly increased values of n ($n_f+1, n_f+2, \cdots, n_f+L/2$), all the way to the $|n_f+L/2, \downarrow_1 \cdots \downarrow_k \cdots \downarrow_L\rangle$ state. This produces a system of $1+L+(L)(L-1)/2+\cdots 1 = 2^L$ equations (by counting the ways to place 1 downward spin, 2 downward spins, and so on, on L sites). In this most general case with an invariant $n_f+L/2$, $n_f \geq 0$ to obtain the eigenvalues we need to diagonalize a $2^L \times 2^L$ matrix.

We can imagine now a state that initially has 1 turned downward spin somewhere in the chain, $|n = n_f, M_z = L/2-1\rangle$. The above procedure will still lead to a 2^L sized system of equations, with appropriate coefficients, for big enough n_f. On the other hand, if we look at a state $|n=0, \downarrow_1 \cdots \downarrow_k \cdots \downarrow_L\rangle$, it is not possible to simultaneously turn one of the spins up and to decrease n, as n is already at a minimum value. For this particular state, with an invariant value of $-L/2$, we need to diagonalize a 1×1 matrix, as it turns out to be an eigenstate. By doing the above algebra, we find that for all the values of the invariant that are less then $L/2$ we obtain system of equations with size that is less than 2^L. The number of states with an invariant $-L/2+1$ is $1+L$, with an invariant $-L/2+2$ this number is $1+\binom{L}{1}+\binom{L}{2}$, and for the invariant $-L/2+m$ the subspace has the dimension of:

$$Dim_m \equiv 1+\binom{L}{1}+\binom{L}{2}+\cdots+\binom{L}{m}, \quad 0 \leq m \leq L. \tag{4}$$

With this formula we count the states by first assigning $n=m$, with all the spins downward (1 state), then $n=m-1$ and 1 upward spin among $L-1$ downward ones (L number of states), and so on. By extending the definition of Dim_m, so that $Dim_m = 2^L$, for $m \geq L$, we can summarize the above discussion with the conclusion that for an invariant value $-L/2+m$ the corresponding eigenvalues problem requires a diagonalization of a $Dim_m \times Dim_m$ matrix.

As concluding introductory remarks, we mention that the limiting cases, $J=0$ or $G=0$, of the Hamiltonian can be diagonalized analytically. For example, for the Jaynes-Cummings Hamiltonian H_{JC} the corresponding spectrum looks like (compare with, e.g., [5,7]):

$$H_{JC}|n=0,\downarrow\rangle = E_{gnd}|n=0,\downarrow\rangle, \quad H_{JC}(a_\pm|n,\uparrow\rangle + b_\pm|n+1,\downarrow\rangle) = E_\pm(a_\pm|n,\uparrow\rangle + b_\pm|n+1,\downarrow\rangle) \tag{5}$$

where

$$E_{gnd} = -\delta/2, \quad E_\pm = (n+1)\omega_f \pm \Omega_n/2, \quad \Omega_n := \sqrt{\delta^2+4G^2(n+1)}, \quad \delta := (\omega_R - \omega_f), \tag{6}$$

and the normalized coefficients for the eigenvectors of the higher energy states:

$$a_\pm = \sqrt{(\Omega_n \pm \delta)/(2\Omega_n)}, \quad b_\pm = \pm\sqrt{(\Omega_n \mp \delta)/(2\Omega_n)}, \quad a_\pm^2 + b_\pm^2 = 1, \tag{7}$$

and the quantum optics convention is used to denote with δ the detuning and with Ω_n the Rabi frequency.

For the pure H^{XY} part, the energies are given by, e.g., [11]

$$E_J = \frac{2J}{L}\frac{\sin(l\pi/L)}{\sin(\pi/L)}, \quad 0 \leq l \leq L. \tag{8}$$

The explicit form of the matrix elements of the Hamiltonian from equation (1) can be easily obtained, since the spin states correspond to spin 1/2, and it may be just harder to write conveniently the expression, then to derive it. It has the form (compare with [13]), where for convenience we have used the notations $|\uparrow\rangle \equiv |+1/2\rangle, |\downarrow\rangle \equiv |-1/2\rangle$, and $\sigma_i = \pm 1/2$:

$$\widehat{H}|n, \sigma_1 \cdots \sigma_k \cdots \sigma_L\rangle = \sum_{n'+\sum_i(\sigma_I')=n+\sum_i(\sigma_I)\equiv -L/2+m} H_m^{Dim_m \times Dim_m}|n', \sigma_1' \cdots \sigma_k' \cdots \sigma_L'\rangle \tag{9}$$

with the $Dim_m \times Dim_m$ matrix H_m corresponding to the invariant value $-L/2+m$ is given by, (with the extended definition of $Dim_m = 2^L, m \geq L$):

$$\begin{aligned}\langle n', \sigma'_1 \cdots \sigma'_k \cdots \sigma'_L | \widehat{H} | n, \sigma_1\sigma_2 \cdots \sigma_k \cdots \sigma_L \rangle = &\left(\pm\frac{\omega_R}{2}\delta_{\sigma_k, \pm 1/2} + \omega_f(n+\frac{1}{2})\right)\delta_{n',n}\prod_{i=1}^{L}\delta_{\sigma'_i,\sigma_i} \\ &+ G\left(\sqrt{n}\delta_{n',n-1}\delta_{\sigma_k,-1/2} + \sqrt{n+1}\delta_{n',n+1}\delta_{\sigma_k,+1/2}\right)\prod_{i=1,\neq k}^{L}\delta_{\sigma'_i,\sigma_i} \\ &+ \frac{J}{2}\sum_{i=1}^{L-1}\left(\delta_{\sigma_i,-1/2}\delta_{\sigma_{i+1},+1/2} + \delta_{\sigma_i,+1/2}\delta_{\sigma_{i+1},-1/2}\right)\delta_{n',n}\prod_{j=1,\neq i,i+1}^{L}\delta_{\sigma'_j,\sigma_j}.\end{aligned} \tag{10}$$

As an example, lets consider a system with two spins, and the Jaynes-Cummings term being coupled to site 1, this corresponds to $k=1, L=2$ in equation (1). We have two $-1, 0$ special values of the invariant, in the form $-L/2+m$ this is $m=0,1$, with eigensystems of size $Dim_m = 1, 3$ correspondingly, and for the invariant values ≥ 1, or $m \geq 2$ the corresponding matrix is of size $4 = 2^2$. Explicitly, using equations (4) and (9), we have the matrices:

$$H^{1\times 1}_{m=0} = \frac{-\omega_R + \omega_f}{2}, \quad |n=0,-1/2,-1/2\rangle, \tag{11}$$

$$H^{3\times 3}_{m=1} = \begin{pmatrix} \frac{\omega_R+\omega_f}{2} & J/2 & G \\ J/2 & \frac{-\omega_R+\omega_f}{2} & 0 \\ G & 0 & \frac{-\omega_R+3\omega_f}{2} \end{pmatrix} \text{ mixing } \begin{pmatrix} |n=0,1/2,-1/2\rangle \\ |n=0,-1/2,1/2\rangle \\ |n=1,-1/2,-1/2\rangle \end{pmatrix} \tag{12}$$

and

$$H^{4\times 4}_{m=2} = \begin{pmatrix} \frac{\omega_R+\omega_f}{2} & 0 & G & 0 \\ 0 & \frac{\omega_R+3\omega_f}{2} & J/2 & G\sqrt{2} \\ G & J/2 & \frac{-\omega_R+3\omega_f}{2} & 0 \\ 0 & G\sqrt{2} & 0 & \frac{-\omega_R+5\omega_f}{2} \end{pmatrix} \text{ mixing } \begin{pmatrix} |n=0,1/2,1/2\rangle \\ |n=1,1/2,-1/2\rangle \\ |n=1,-1/2,1/2\rangle \\ |n=2,-1/2,-1/2\rangle \end{pmatrix} \tag{13}$$

which can be obtained as a specific case with $n_f = 0$ from the general expression valid for the all the invariant values ≥ 1 ($m \geq 2$) with arbitrary $n = n_f$:

$$H^{4\times 4}_{m\geq 2} = \begin{pmatrix} \frac{\omega_R+(2n_f+1)\omega_f}{2} & 0 & G\sqrt{n_f+1} & 0 \\ 0 & \frac{\omega_R+(2n_f+3)\omega_f}{2} & J/2 & G\sqrt{n_f+2} \\ G\sqrt{n_f+1} & J/2 & \frac{-\omega_R+(2n_f+3)\omega_f}{2} & 0 \\ 0 & G\sqrt{n_f+2} & 0 & \frac{-\omega_R+(2n_f+5)\omega_f}{2} \end{pmatrix} \tag{14}$$

mixing

$$\begin{pmatrix} |n=n_f,1/2,1/2\rangle \\ |n=n_f+1,1/2,-1/2\rangle \\ |n=n_f+1,-1/2,1/2\rangle \\ |n=n_f+2,-1/2,-1/2\rangle \end{pmatrix} \tag{15}$$

With the explicit expressions for the matrices, we can evaluate the spectra by considering the eigenvalues equations $\det(H_m^{Dim_m \times Dim_m} - \lambda \text{ Identity}) = 0$.

IOP Conf. Series: Journal of Physics: Conf. Series **1186** (2019) 012021 doi:10.1088/1742-6596/1186/1/012021

3. Conclusion

In this report we have shown analytical approach to obtain the energy spectra for arbitrary filling number of a one photon mode for Jaynes-Cummings model in a an open spin molecule with XY type spin-spin interaction between the spin-$\frac{1}{2}$ sites.

Acknowledgments

The authors thank I. Boradjiev for pointing us to the reference [13] and N. B. Ivanov, E. Korutcheva, and B. Obreshkov for useful discussions. Support by the Bulgarian Science Foundation grant DN0818/14.12.2016 is gratefully acknowledged.

References

[1] Jaynes E and Cummings F 1963 *Proc. IEEE* **51** 89
[2] Alperin M M, Klubis Y D and Khizhnyak A I 1987 *Introduction to the Physics of Two-Level Systems* (Kiev: Naukova Dumka)
[3] Gerry C and Knight P 2004 *Introductory quantum optics* (Cambridge: University Press)
[4] Scully M O and Zubairy M S 1997 *Quantum Optics* (Cambridge: University Press)
[5] Hur K L, Henriet L, Petrescu A, Plekhanov K, Roux G and Schiró M 2016 *Compt. Rend. Phys.* **17** 808
[6] Tavis M and Cummings F W 1968 *Phys. Rev.* **170** 379
[7] Scharf G 1970 *Helv. Phys. Acta* **43** 806
[8] Boradjiev I *et al* 2018 In preparation
[9] Wang X 2001 *Phys. Rev.* A **64** 012313
[10] Lieb E, Schultz T and Mattis D 1961 *Ann. Phys.* **16** 407
[11] De Pasquale A and Paolo F 2009 *Physical Review A* **80** 032102
[12] Jordan P and Wigner E 1928 *Z. Phys.* **47** 631
[13] Wu N 2018 *Phys. Rev. B* **97** 014301
[14] Juárez-Amaro R, Zúñiga-Segundo A and Moya-Cessa H M 2015 *Appl. Math. Inf. Sci.* **9** 299–303
[15] Dirac P A M 1978 *The Principles of Quantum Mechanics* (Oxford University Press)
[16] Feynman R P 1972 *Statistical Physics* (Benjamin Press)
[17] Tonchev H, Donkov A A and Chamati H 2016 *Jour. Phys.: Conf. Ser.* **682** 012032; *ibid.***764** 012017

Investigation of resistive switching in SiO_2 layers with Si nanocrystals

E Manolov[1], J Paz-Delgadillo[2], V Dzhurkov[1], N Nedev[2], D Nesheva[1], M Curiel-Alvarez[2] and B Valdez-Salas[2]

[1] Institute of Solid State Physics, Bulgarian Academy of Sciences, Tzarigradsko chausse 72 Blvd, 1784 Sofia, Bulgaria
[2] Universidad Autónoma de Baja California, Instituto de Ingeniería, Benito Juárez Blvd., s/n, C.P. 21280, Mexicali, Baja California, México

E-mail: emman@issp.bas.bg

Abstract. MOS structures with two-layer insulator, SiO_2/SiO_x, are tested for switching between high and low resistance state. Part of the structures were annealed at 1000 °C for 60 min in N_2 in order to form silicon nanocrystals (Si NCs) in a SiOx matrix. Cross-sectional transmission electron microscopy revealed that after the high temperature annealing phase separation takes place and nanocrystals with size of ~ 4–5 nm are formed. Direct current-voltage measurements showed that bipolar switching occurs only in the annealed c-Si/SiO_2/SiO_x/Al structures and not in the control samples. A model explaining the resistance change by electric field formation of conductive pathway in the SiO_x–Si NCs layer at negative gate voltages and current driven destruction of the conductive filament at positive voltages was verified by high frequency capacitance-voltage (*C-V*) measurements. From the *C-V* results the dielectric constant of non-annealed SiO_x films with $x = 1.3$ was calculated to be $\varepsilon_{SiOx} = 4.6$.

1. Introduction

Continuously increasing interest in efficient, reliable and economic data storage for diverse applications have attracted much research effort in the field. The well-developed flash memory based on floating-gate silicon MOSFET is approaching its physical limit. Among the technologies proposed to develop the next generation of nonvolatile memories resistive random access memories have received significant attention due to their simple structure, high switching speed, low operating voltage and potential scalability beyond 10 nm feature size [1, 2]. These memories are reversibly switched between a high and low resistance state by direct current (DC) voltage or voltage pulses and the resistance changes are used to store information. Though the development of resistive random access memory has achieved remarkable progress in recent years, the understanding of its conduction mechanisms, dynamic processes and stability issues is still limited.

Resistive switching effect has been observed in many oxides including SiO_x [3], which is particularly interesting compared to other oxides because of its complete CMOS compatibility. SiO_x has been studied as switching oxide in various types of electrode-oxide-electrode and metal-oxide-semiconductor (MOS) structures [4, 5]. Poly-Si and different metals have been used as electrodes [6–8]. For SiO_x a switching mechanisms based on voltage-driven formation and modification of silicon nanocrystals (NCs) in the SiO_x matrix, with SiO_x also serving as source for formation of conductive Si pathway has been proposed [6]. Another model is based on formation of filaments on the edges of columns in sputtered SiO_x [8]. In MOS structures having SiO_2 with Si NCs as gate insulator formation of conductive pathway of aligned nanocrystals has been observed [5].

In this study we present results for the resistive switching properties of thin SiO_x films deposited on thermally grown SiO_2 on n-type Si. The difference in the electron injection properties of the top

(Al/SiO_x) and bottom (crystalline Si/SiO_2) interfaces favors electric field formation of conductive pathways at negative voltages applied to the top contact and current-driven destruction of the filament at positive voltages.

2. Experimental details

Silicon-rich SiO_x layers with initial composition of $x = 1.3$ and thicknesses of 20 and 40 nm were deposited on SiO_2/crystalline silicon (c-Si) substrates maintained at room temperature. The thermal oxide with thickness of ~ 25 nm was grown in dry oxygen atmosphere at 1000 °C on (100) n-type c-Si with $\rho = 4–6\ \Omega$ cm. The silicon wafers were cleaned chemically prior to oxidation applying a standard procedure for the microelectronics industry. Thermal evaporation of SiO at a vacuum of 1×10^{-3} Pa was used for the SiO_x film preparation [9].The deposition rate and film thickness were monitored by a calibrated quartz microbalance. All as-prepared layers were annealed at 250 °C for 30 min in Ar atmosphere to ensure good film stability at ambient conditions. This annealing does not lead to phase separation and Si nanoparticle formation [9]. To form silicon nanocrystals (Si NCs) in SiO_x matrix the samples were furnace annealed at 1000 °C in N_2 for 60 min. Part of the samples without high temperature annealing were used as control samples. After the annealing Al metallization was carried out through a mask and top electrodes (referred from now on as gates) of MOS structures with area of ~ 2×10^{-3} cm^2 were formed. Aluminum electrodes were deposited on the bottom side of the silicon substrates.

The thicknesses of the deposited SiO_x layers were determined ellipsometrically using J.A. Woollam M-2000U instrument and values of approximately 27 and 40 nm were obtained for the thin and thick layers, respectively. For the thermal oxide a value of 24.7 nm was found. The effect of high temperature annealing on the structural modification of SiO_x was studied by cross-sectional Transmission Electron Microscopy (XTEM) using JEOL JEM-2100F microscope. DC current-voltage (*I-V*) and high frequency (1 MHz) capacitance-voltage characteristics were measured by Semiconductor Characterization System Keithley 4200-SCS with preamplifiers. The hold and delay times of all measurements were 2 and 0.5 s, respectively. The polarity of the applied voltage given below refers to the top electrode.

3. Results and Discussion

Figure 1 shows XTEM images of a high temperature annealed c-$Si/SiO_2/SiO_x$ structure with 40 nm thick SiO_x film. In figure 1(a) two regions can be clearly distinguished: an amorphous layer with thickness of ~ 25 nm, the thermal SiO_2 and a second one, which exhibits Si nanocrystals (SiO_x–Si NCs) seen as dark spots. It has been shown that SiO_x films with $x = 1.3$ remain amorphous after 250 °C

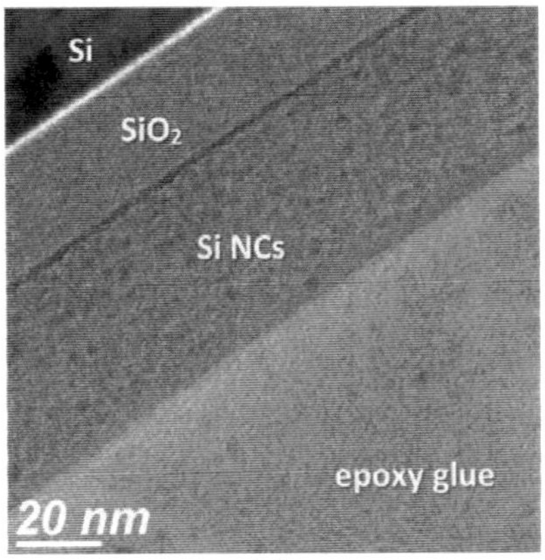

(a)

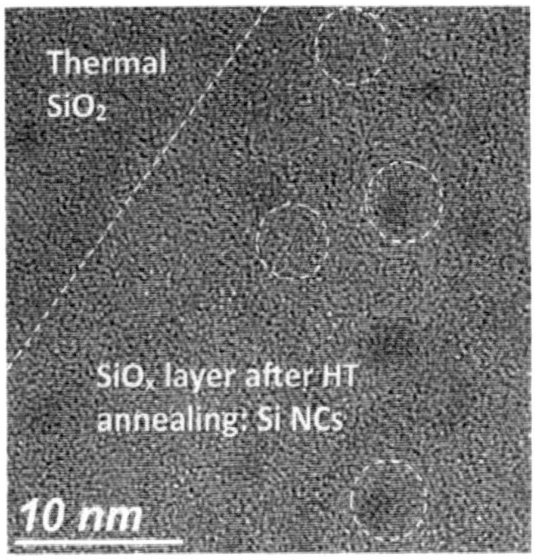

(b)

Figure 1. (a) Cross-sectional micrographs of an annealed c-$Si/SiO_2/SiO_x$ structure with 40 nm SiO_x layer; (b) high resolution XTEM image showing nanocrystals with size of ~ 4–5 nm formed after the high temperature annealing.

annealing in Ar [10]. Therefore, the formation of NCs can be attributed to the high temperature annealing. The nanocrystals are randomly distributed along the film thickness. A thin region free of Si nanocrystals is seen close to the top surface, most likely because the nanocrystals in this region get transformed into SiO_2 due to oxidation after exposure to air. From the high resolution image in figure 1b the size of the NCs was determined to be about 4–5 nm.

Figure 2 shows *I-V* characteristic of a control MOS structure with ~ 27 nm thick SiO_x layer. Large difference was observed between the current flowing through the sample at positive and negative voltages. For example, at gate voltage of $V_g = 30$ V the current was ~ 2.1×10^{-8} A (curve 1), while at $V_g = (-30)$ V it was ~ 1.2×10^{-11} A (curve 2). The current dependence on the voltage polarity can be related to differences in the properties of the injecting interfaces. At positive voltage electrons are injecting from the c-Si in the thermal oxide overcoming the Si/SiO_2 barrier of 3.1 eV and then in the SiO_x. At negative voltage, holes could be injected from the silicon in the SiO_2 and electrons from the top metal electrode in the SiO_x layer. The energy barrier for holes at the Si/SiO_2 interface is much larger (~ 4.7 eV) than the barrier for electrons at the Al/SiO_x interface (less than 3.2 eV, the Al/SiO_2 barrier). Therefore, the main part of the current at negative voltage is due to electrons injected from the top electrode. Most likely, part of these electrons are trapped close to the Al electrode forming a negatively charge region which opposes further electron injection and leads to smaller current values.

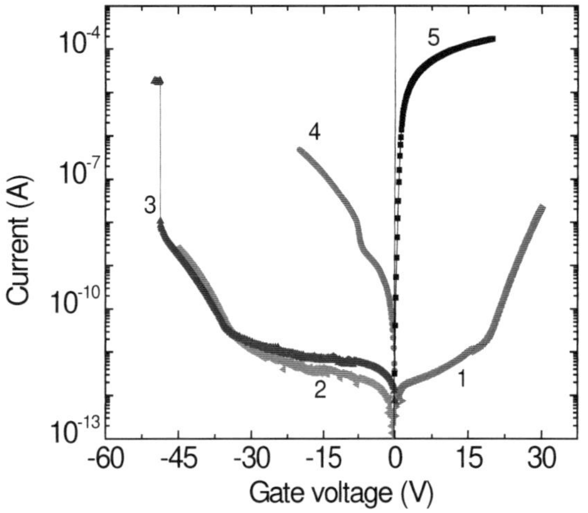

Figure 2. *I-V* characteristic of c-$Si/SiO_2/SiO_x/Al$ structure with 27 nm thick SiO_x layer annealed at 250 °C in Ar.

This assumption was confirmed by high frequency *C-V* measurements on control c-$Si/SiO_2/SiO_x/Al$ structure (figure 3). The curves in Figure 3 were measured with consecutive voltage sweeps between 2 V and a negative voltage V_L in both directions; first from positive to negative voltage and then in the opposite direction. V_L was varied between (-4) and (-14) V with step of (-2) V. The initial curve was measured in the (+2, -3) V range in order to avoid charging of the structure. All curves show hysteresis corresponding to a negative charge trapped in the insulator after a complete voltage cycle. The change of the flat band voltage V_{FB} [11] increases with the increase of the absolute value of V_L (the inset in figure 3). Thus, a sweep to (-14) V results in $\Delta V_{FB} \sim 0.8$ V corresponding to charge density of approximately 4×10^{11} cm^{-2} if assumption is made that all electrons are trapped at the Al/SiO_x interface. No such shifts, corresponding to trapped electrons in the gate insulator were observed when the voltage was varied between (-3) V and positive voltages up to 14 V. Thus, the *C-V* measurements confirmed that electrons are trapped in the SiO_x only when negative voltages are applied to the Al electrode; this could be a possible explanation of the asymmetry of the *I-V* dependencies.

The obtained results indicate that the $Al/SiO_x/SiO_2$/c-Si structure favors electric field switching from high to low resistance state by formation of conductive pathway at negative gate voltages because large electric fields ($E_{ox} \sim 8$ MV cm^{-1} at $V_g \sim (-45)$ V) can be achieved at relatively small currents, ~ 10^{-10} A (figure 2, curves 2, 3). On the other hand, current driven destruction of the conductive filament may be expected at positive voltages because large currents flow at smaller

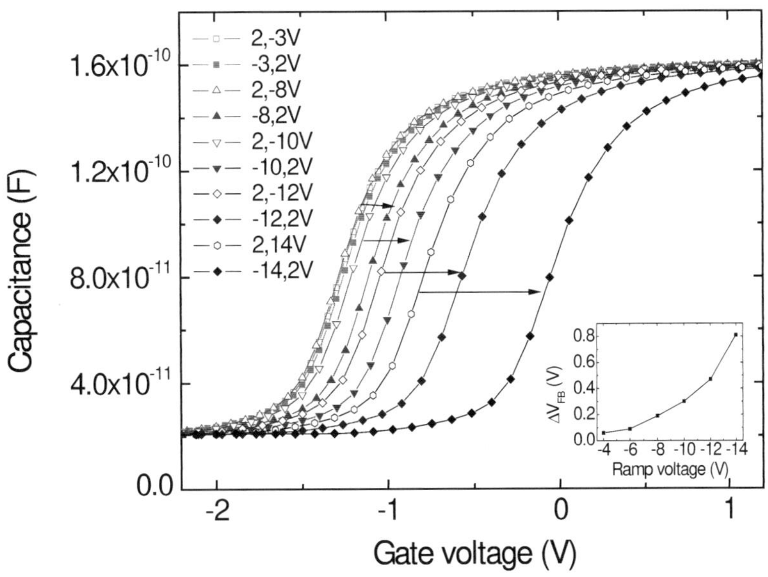

Figure 3. *C-V* characteristics of c-Si/SiO_2/SiO_x/Al structure with 27 nm thick SiO_x layer annealed at 250 °C in Ar. The inset shows ΔV_{FB} as a function of the negative voltage V_L in the ramp scans (+2,V_L) V.

electric fields, i.e. ~ 10^{-8} A at E_{ox} ~ 5.8 MV cm^{-1} corresponding to V_g ~ 30 V. However, no high-low resistance switching of the control sample was observed for negative voltages in the range between 0 and approximately (-50) V ($E_{ox} \geq 10$ MV cm^{-1}) when destructive breakdown occurs (figure 2, curve 3). No switching was also observed at positive voltages. The current through the gate insulator after destructive breakdown shows similar dependence on the voltage polarity (figure 2, curves 4, 5) as before breakdown, faster current increase in the positive direction.

Thickness of 25 nm was calculated for the thermal oxide from the accumulation capacitance of c-Si/SiO_2/Al structure, a value in good agreement with the ellipsometry result. Using the measured capacitance in accumulation of the c-Si/SiO_2/SiO_x/Al structures (161 pF) and the thicknesses of the SiO_x (27 nm) and SiO_2 (25 nm) the dielectric constant of the SiO_x layer was calculated to be $\varepsilon_{SiOx} = 4.6$.

In contrast to the control samples the structures annealed at 1000 °C showed resistive switching (figure 4). A strong current increase of more than 3 orders of magnitude was obtained at negative voltages between (-20) and (-25) V (curves 1, 4). After switching the structures maintain the low resistance state at voltages varying between (0, -30) V range (curve 2). Transitions from low to high resistance state were observed with ramps in the positive direction at voltages between 25 and 30 V (curves 3, 5). After switching, *I-V* curves very close to the initial ones (in both high and low resistive states) were measured. Similar result for resistive switching only in SiO_x with Si nanocrystals and not in control samples was reported in [5].

Interestingly, after several switching cycles some structures remained in high resistance state when the voltage was varied in the negative direction until breakdown was reached (curve 6). It may be assumed that in these capacitors the defect density and other structural modifications caused by the high temperature annealing, which facilitate the formation of conductive pathways is very low. After several cycles the defects are probably completely eliminated and no further high-low resistance transition is possible.

From the obtained results it may be concluded that the studied structures show bipolar switching. None of the measured capacitors was switched from high to low resistance state at positive voltage and from low to high resistance state at negative voltage.

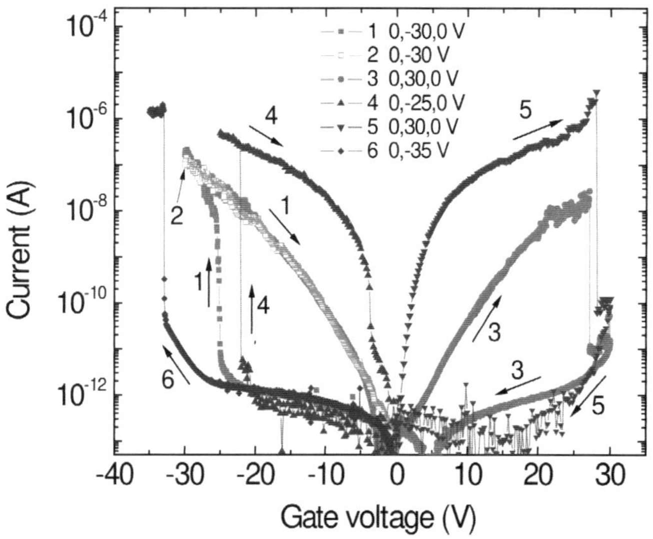

Figure 4. *I-V* characteristic of c-Si/SiO$_2$/SiO$_x$/Al structure with 27 nm thick SiO$_x$ layer annealed at 1000 °C in N$_2$.

4. Conclusion

Resistive bipolar switching was obtained in annealed c-Si/SiO$_2$/SiO$_x$/Al structures, which contain Si nanocrystals in the SiO$_x$ layer. The transition from high to low resistance state was accomplished by electric field formation of conductive pathway in the SiO$_x$-Si NCs layer at negative gate voltages, while current driven destruction of the conductive filament was observed at positive voltages. No resistive switching was found in the control c-Si/SiO$_2$/SiO$_x$/Al structures. From *C-V* measurements at 1 MHz the dielectric constant of non-annealed SiO$_x$ (x = 1.3) films was calculated to be $\varepsilon_{SiOx} = 4.6$.

Further optimization of the proposed structure for memory applications may be achieved by the following modifications: first, decrease of the SiO$_x$ and thermal oxide thicknesses, which will result in lower switching voltages. Second, increase of the defect density in the SiO$_x$, which could facilitate the formation of conductive pathways in the annealed structures and eventually lead to resistive switching in non-annealed (control) c-Si/SiO$_2$/SiO$_x$/Al structures.

Acknowledgements

J. Paz-Delgadillo gratefully acknowledges the fellowship from Consejo Nacional de Ciencia y Tecnología (CONACYT), Mexico.

References

[1] Pan F, Chen Ch, Wang Zhi-shun, Yang Yu-chao, Yang J and Zeng F 2010 *Prog. Nat. Sci-Mater.* **20** 1-15
[2] Wang Ch, Wu H, Gao B, Zhang T, Yang Y and Qian H 2018 *Microelectron. Eng.* **187–188** 121-33
[3] Hong X-L, Loy D J-J, Dananjaya P A, Tan F, Ng Ch-M and Lew W-S 2018 *J. Mater. Sci.* 53 8720-46
[4] Mehonic A, Cueff S, Wojdak M, Hudziak S, Jambois O, Labbe Ch, Garrido B, Rizk R and Kenyon A J 2012 *J. Appl. Phys.* **111** 074507
[5] Xia G, Ma Zh, Jiang X, Yang H, Xu J, Xu L, Li W, Chen K and Feng D 2012 *J. Non-Cryst. Solids* **358** 2348-52
[6] Yao J, Sun Zh, Zhong L, Natelson D and Tour J M 2010 *Nano Lett.* **10** 4105-10
[7] Yao J, Zhong L, Natelson D and Tour J M 2011 *J. Am. Chem. Soc.* **133** 941
[8] Mehonic A, Munde M S, Ng W H, Buckwell M, Montesi L, Bosman M, Shluger A L, Kenyon A J 2017 *Microelectron. Eng.* **178** 98-103
[9] Nesheva D, Bineva I, Levi Z, Aneva Z, Merdzhanova T and Pivin J C 2003 *Vacuum* **68** 1

[10] Curiel M A, Nedev N, Nesheva D, Soares J, Haasch R, Sardela M, Valdez B, Sankaran B, Manolov E, Bineva I and Petrov I 2010 *Mat. Sci. Eng.* B **174** 132-136

[11] Nicollian E H and Brews J R 2002 *MOS* (*Metal Oxide Semiconductor*) *Physics and Technology*, (U.S.A.: Wiley-Interscience)

IOP Conf. Series: Journal of Physics: Conf. Series **1186** (2019) 012023 doi:10.1088/1742-6596/1186/1/012023

Room temperature sensitivity of ZnSe nanolayers to ethanol vapours

V Dzhurkov, Z Levi, D Nesheva and T Hristova-Vasileva

Institute of Solid State Physics, Bulgarian Academy of Sciences, Tzarigradsko chausse 72 Blvd, 1784 Sofia, Bulgaria

E-mail: valeri.dzhurkov@gmail.com

Abstract: Layers of ZnSe (50, 110 and 160 nm thick) are deposited at room substrate temperature by thermal evaporation of ZnSe powder in vacuum. As-deposited and relaxed layers are investigated by spectroscopic ellipsometry to get information about their optical properties, porosity and surface roughness. The obtained data indicate that all layers are nanocrystalline and the lattice relaxation causes a porosity decrease and surface roughness increase. Ethanol sensing experiments are carried out at room temperature in air and results show an increase of the layers sensitivity with decreasing thickness while the relaxation causes sensitivity decrease. The sensitivity behavior is discussed considering the changes in the films porosity and the ethanol related decrease of the potential barriers at the nanocrystallite interfaces.

1. Introduction

The gas detection is related mainly with surface processes and therefore thin films are very suitable for gas sensors. Semiconductor thin films are widely used for resistive gas sensing, especially metal oxides, which have high sensitivity, stability and relative short response times at high working temperatures (200 – 300 °C). However at room temperature their sensing properties worsen significantly and during the last few decades much attention has been given up to the problem of room temperature sensing [1]. Nanostructured semiconductor films, in particular metal chalcogenides [2], are considered as promising materials for sensitive element in gas sensors due to their high surface-to-volume ratio and high porosity. Good gas sensitivity at room temperature has been reported for nanocrystalline thin films from CdS and CdSe [3], as well as in ZnO nanostructured thin films [4, 5].

ZnSe has been widely used in fabrication of various optoelectronic devices [6] such as blue-green laser diodes, tunable mid-IR laser diodes for remote sensing, white-light LEDs, continuous wave ZnSe-based laser diodes and UV photodetectors. However, there are almost no reports on ZnSe as a gas sensor probably because of its decomposition and oxidation at temperatures higher than 200 °C. It can be appropriate for detection of gases at room temperature or temperatures below 200 °C if a good sensitivity is achieved.

In this study ZnSe thin layers have been prepared by thermal evaporation of ZnSe powder in high vacuum. The effect of the microstructure and porosity of the layers on the room temperature sensitivity to ethanol vapors have been investigated.

2. Experimental

Layers of ZnSe having thicknesses of 50, 110 and 160 nm were deposited on Corning 7059 glass substrates kept at 25 °C. The layers were produced by physical evaporation of powdered ZnSe (Merck, Suprapure) at a residual pressure of 3×10^{-4} Pa from a tantalum crucible located at the bottom of a cylindrical screen (not intentionally heated). The top of the screen is close to the substrates; thus evaporation in a quasi-closed volume was carried out. The deposition rate ($V_d = 0.5$ nm s^{-1}) and layer thickness were controlled by a preliminary calibrated quartz microbalance system. The films were grown in a step-by-step manner i.e. during the film deposition the substrates were rotated at a rate of 8 turns min^{-1} spending only 1/12 part of the turn time over the ZnSe crucible. At each step a small portion of ZnSe vapors condense on the substrates with a time interval between the steps and the films were formed from a large number of such portions. It has been shown that this deposition approach results in nanocrystal size reduction in CdSe layes and this increases gas sensitivity of those films [3].

Ellipsometic measurements were carried out by an automatic M2000D spectroscopic ellipsometer (J.A. Woollam, Germany). Angles of incidence from 50° to 65° by 5° were used in the wavelength range of 400–1000 nm and the obtained data were further fitted using Woollam's CompleteEASE program. A model based on the Bruggeman effective medium approximation was applied on a modeled structure which consists of glass substrate and a porous ZnSe layer. The surface roughness was also taken into account. It has been modeled using the standard procedure of an effective medium composition with 50% voids and 50% layer material.

The layer's response to ethanol vapors at room temperature was studied by measuring the changes of the dc dark current through the layers at room temperature using planar Ohmic-like indium contacts on the top surface of the films. The measurements were carried out under a closed glass test chamber, under atmospheric conditions, having a volume of 1500 cm^3 in which vapors were produced by evaporating 100-500 μL of liquid ethanol (99.7%). The amount of the liquid ethanol is used below as measure of the ethanol vapor concentration. During the measurement the vapors were homogeneously distributed inside the chamber by a propeller. A current increase was observed upon film exposure to ethanol vapors as the maximum current value I_{max} was achieved 3 min after starting ethanol evaporation. For all samples concentration dependences were measured starting with the minimum concentration and then the ethanol concentration was gradually increased. The ethanol vapors were removed by opening the glass chamber and exposing the sample to atmospheric air.

3. Results and discussion

The porosity and surface roughness of sensing films are important characteristics which strongly influence their chemical sensitivity. In order to get information about ZnSe layers' porosity and roughness they were characterized by spectroscopic ellipsometry (SE). ZnSe is a direct band gap semiconductor and from the optical absorption data obtained by SE one can get information about the films crystallinity – are they predominantly crystalline or amorphous. Figure 1(a) shows the absorption spectrum of a 50 nm thick film presented in coordinates corresponding to direct allowed electron transitions between the conduction and valence bands ($\alpha E \sim (E-E_g)^{1/2}$, E_g - optical band gap, E – light energy). Such transitions are typical for crystalline materials with a direct band gap. Figure 1(b) presents the same spectrum in coordinates corresponding to indirect allowed transitions ($\alpha E \sim (E-E_g)^2$), typical for amorphous materials. It is seen from the figure that the $\alpha(E)$ dependence shows good linearity in the coordinates corresponding to direct electronic transitions typical for crystalline ZnSe. The results for the other two layer thicknesses are very similar and one can conclude that the films with all thicknesses are predominantly crystalline. The optical band gap determined from the intersection of the straight line with the x-axis is 2.64 eV for all thicknesses. It is in good agreement with the value we obtained for 1 μm thick microcrystalline ZnSe films (2.67 eV) [7]. Normally the thickness reduction results in a decrease of the crystallite size and if it is less than ~10 nm a band gap increase can be observed due to carriers' confinement. No such effect is observed even for the 50 nm film which implies that the average crystallite size in the films with all thicknesses is larger than 10 nm which is in agreement with the atomic force microscopy (AFM) results showing that the grain size in ZnSe films having thickness in the range 30–100 nm is 25–30 nm [8, 9].

IOP Conf. Series: Journal of Physics: Conf. Series **1186** (2019) 012023 doi:10.1088/1742-6596/1186/1/012023

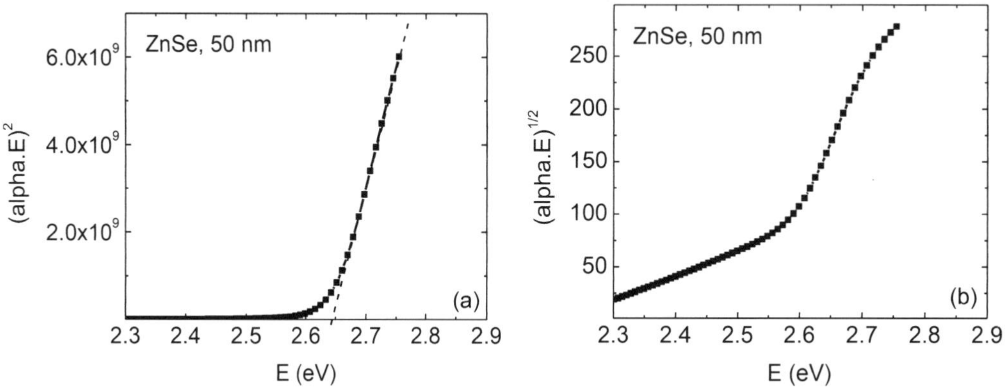

Figure 1. Absorption spectrum of 50 nm thick films of ZnSe presented in coordinates corresponding to allowed direct (a) and indirect (b) electronic transitions between valence and conduction bands.

It has been reported [10–13] that, independently of the deposition method, the structure of ZnSe layers with $d \leq 100$ nm is a mixture of nanocrystals embedded in an amorphous matrix. Our previous investigations on ZnSe layers (prepared at same other conditions and deposition rate of 1.5 nm sec^{-1}) with thicknesses between 30 nm and 100 nm, have shown that the layers with all thicknesses contain two phases, amorphous and crystalline, and the portion of the crystalline phase decreases with decreasing thickness [9]. The predominantly crystalline structure of the 50 nm films here studied can be due to considerably lower deposition rate (0.5 nm s^{-1}) than that applied for the deposition of the previously studied films. It seems that the smaller the material portions at each step of the step-by-step deposition process, the higher the film crystallinity. This result is interesting and could be used for preparation of ZnSe films with different ratio of crystalline-to-amorphous phases. Thus one could affect the films' conductivity, gas sensitivity etc.

ZnSe layers having thicknesses less than 500 nm are highly strained and strain relaxation with time has been observed [11]. Our previous AFM investigations have shown that lattice relaxation takes place in the ZnSe layers during 2-month storage at 25 °C in air and this strongly affects the films surface [9]. In this study information for the surface roughness has been obtained from the results of SE measurements performed on as-deposited and relaxed films with thickness of 50 nm. The surface roughness of as-deposited layer is very small ($\leq$ 0.5 nm) while the roughness of relaxed films (kept for four months at 25 °C in air) is around 2.6 nm. This observation is in very good agreement with the previous AFM results for 50 nm thick ZnSe film which showed that the surface of as-deposited films is homogeneous and rather smooth (root mean square roughness $R_q = 0.44$ nm). After relaxation for 2 months the AFM R_q increased by ~ 5 times and pits of various sizes were observed. A third set of measurements, performed four months after the second ones, did not show any further structural changes. Values of around 7.6 nm and 6.2 nm have been obtained from the SE data for the roughness of relaxed films with thickness of 160 nm and 110 nm, respectively. The roughness values obtained for the relaxed films are in good agreement with those reported by other authors for ZnSe films prepared by thermal evaporation in vacuum [9, 14–16].

The volume fraction of pores in all relaxed ZnSe films, determined by applying the Brugemann's effective medium approach, is in the range of 0.15–0.19 which is in good agreement with the value of about 0.17 obtained previously for relaxed ZnSe films [8]. However the volume fraction of pores in as-deposited 50 nm films is much higher, around 0.3. Taking into account the surface roughness (AFM and SE) results for 50 nm thick films one can assume that the pores in as-deposited films are disposed in the layers' volume, while the lattice relaxation results in appearance of pits on the layers surface but reduction of the pore fraction in the volume.

The sensitivity of ZnSe films to ethanol vapors was measured on both as deposited and relaxed films at identical conditions. The ratio $S = (I_{max}-I_0) / I_0$ is used as measure of the films ethanol sensitivity *S*, in which I_0 denotes the current in the reference gas (air) and I_{max} is the maximum current achieved upon exposure to ethanol vapors. Figure 2 shows concentration dependences of the sensitivity of as-deposited and relaxed films with three different thicknesses. Distinct thickness dependence is seen from the figure. The chemical sensitivity of the films increases with decreasing thickness. The as-deposited films with thickness of 50 nm and 110 nm show a linear increase with the increase of the concentration of ethanol vapors while after the lattice relaxation only the 50 nm thick film shows such linearity. The linear response is an important parameter for practical application of sensing materials. A comparison of the *S*-values in figure 2(a) and (b) shows that the sensitivity of each as-deposited film is higher than that measured after lattice strain relaxation. Taking into account that the relaxation causes porosity decrease and roughness increase one can conclude that porosity has stronger effect on the layer sensitivity than surface roughness.

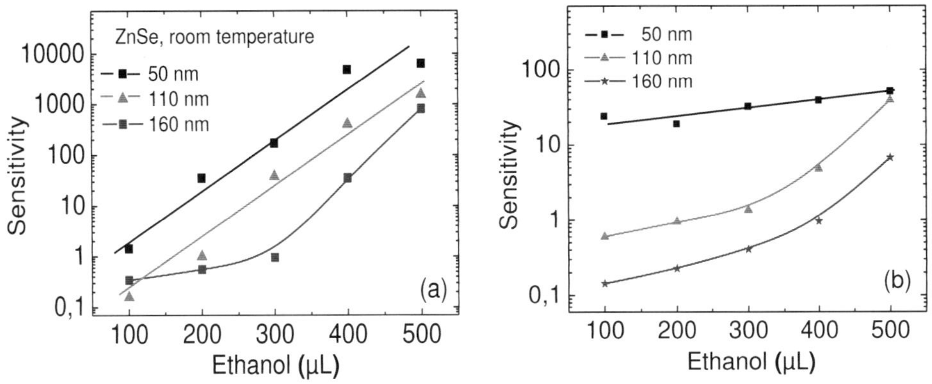

Figure 2. Room temperature sensitivity to ethanol of as-deposited (a) and relaxed (b) ZnSe layers with different thicknesses denoted in the figure.

Dark current in air and the films' response I_{max} of the 50 nm thick as-deposited and relaxed ZnSe layer to ethanol vapors are shown in figure 3 for illustration. The results for the layers with other two thicknesses are very similar. It is seen from the figure that both I_0 and I_{max} obtained for the relaxed layer are lower than the corresponding values in as-deposited one.

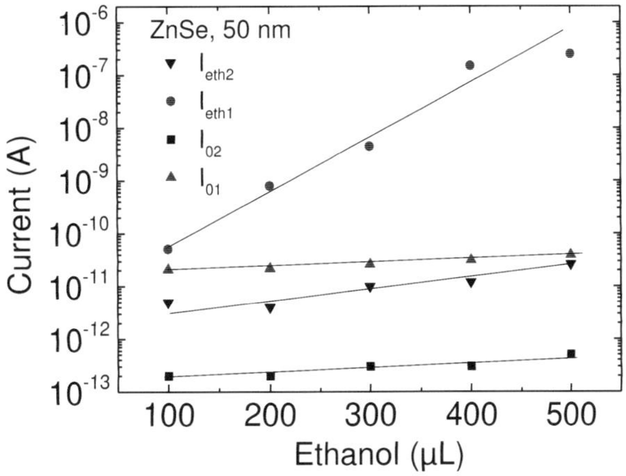

Figure 3. Dark current in air (I_0) and the maximum value of the current measured upon exposure to ethanol vapors of various concentration obtained for a 50 nm thick ZnSe layer at room temperature. 1 - as-deposited layer, 2 - relaxed layer.

According to a common model based on the fact that polycrystalline films consist of a large number of grains, contacting at their boundaries [17, 18], the electrical behaviour of such films is governed by the formation of double Schottky potential barriers at the interface of adjacent grains. These barriers are caused by charge trapping at the interface. In n-type semiconductors it is considered that electrons are most frequently trapped in states which originate from adsorbed oxygen atoms and/or molecules. The height of the barriers determines the conductance of the film; the higher the barrier, the lower the current. When exposed to a chemically reducing gas, like ethanol, interaction between gas molecules and the oxygen occurs, which results in removal of oxygen from the grain interfaces, the barrier height decreases and the dark current increases.

Adopting this model one can explain the observed thickness dependence of the layers sensitivity and the relaxation effect in the following way. Among the parameters affecting the sensitivity of the layers is the penetration of the analyte molecules/atoms into the layers depth; the higher the depth, the higher the sensitivity. The penetration is influenced by the mobility of the gas molecules i.e. of the measurement temperature, as well as by the existence of pores and interfaces which facilitate the movement of molecules. Our sensing experiments were carried out at room temperature and one can expect a low mobility of the ethanol molecules and a limited penetration depth. AFM results have shown that the grain size in all ZnSe films with thickness in the range 30–100 nm is 25–30 nm, the SE results for the relaxed layers show similar porosity of the layers. Hence one can assume that the ethanol penetration depth i.e. the layer part in which the barrier height is reduced and the electrical conductivity is increased is approximately equal in all layers. This means that the relative part with an increased conductivity is smaller in the thicker layers, which explains the sensitivity increase with decreasing layer thickness. The higher sensitivity of the as-deposited layers can be rated to the higher porosity while the lower conductivity of the relaxed films is probably due to continued oxygen adsorption and electron trapping by the adsorbed molecules.

4. Conclusions

First results on the effect of microstructure and porosity of as-deposited layers from ZnSe with thickness 50, 110 and 160 nm on their room temperature sensitivity to ethanol vapors have been presented. The effect of the crystal lattice strain relaxation was also explored. It has been found that all as-deposited and relaxed films are nanocrystalline. The porosity of the layers in as-deposited state is high (~30%) and significantly decreases upon relaxation (down to 15-19%), while the surface roughness of the relaxed films is higher than that of the as-deposited ones. The ethanol sensing experiments have shown that the layers sensitivity increases with decreasing thickness which has been related to low mobility and relative small penetration depth of the ethanol molecules at room temperature. The relaxation causes a sensitivity decrease and this result has indicated that the layer porosity has stronger effect on the layer sensitivity than the surface roughness. The 50 nm thick films have demonstrated good sensitivity and linear response in both as-deposited and relaxed states which makes them perspective for practical detection of high concentrations of ethanol vapors at room temperature.

Acknowledgements

This work was partially supported by the European Regional Development Fund within the OP "Science and Education for Smart Growth 2014 - 2020", project № BG05M2OP001-1.001-0008-C01References. A poster presentation at 20ISCMP, Varna, 2018 was supported by the Ministry of Education and Science, Bulgarian National Science Fund (project number DPMNF 01/7–23.07.2018).

References

[1] Deshpande N and Gosavi S 2015 *Metal Chalcogenides Thin Films-Fundamentals and its Sensor Applications* (LAP Lambert Academic Publishing)
[2] Gaiardo A, Fabbri B, Guidi V, Bellutti P, Giberti A, Gherardi S, Vanzetti L, Malagù C and Zonta G 2016 *Sensors* **16** 296

IOP Conf. Series: Journal of Physics: Conf. Series **1186** (2019) 012023 doi:10.1088/1742-6596/1186/1/012023

[3] Nesheva D, Aneva Z, Reynolds S, Main C and Fitzgerald A G 2007 *J. Optoel. Adv. Mat.* **8** 2120 and references therein
[4] Lee Yi-Mu, Huang Ch M, Chen H W and Yang H W 2013 *Sens. Actuators* A **189** 307
[5] Nesheva D, Dzhurkov V, Stambolova I, Blaskov V, Bineva I, Moreno J M C, Preda S, Gartner M, Hristova-Vasileva T and Shipochka M 2018 *Mater. Chem. Phys.* **209**, 165
[6] Pathak R K and Bais S 2014 *Int. J. Adv. Eng. Technol.* **7** 1300
[7] Nesheva D, Aneva Z, Levi Z, Vuchkov N, Temelkov K and Bineva I 2007 *Nanoscience & Nanotecnology* **7**, 83
[8] Šcepanovic M, Grujic-Brojcin M, Nesheva D, Levi Z, Bineva I and Popovic Z V 2009 *Acta Phys. Pol.* A **116** 708
[9] Nesheva D, Scepanovic M, Levi Z, Askrabic S, Aneva Z, Petrova A and Popovic Z V 2009 *J. Optoel. Adv. Mat.* **11** 1351
[10] Imran M, Saleem A, Khan N, Khurram A and Mehmood N 2018 *Thin Solid Films* **648** 31
[11] Lee H Y, Kim S J, Kim J W and Chung H B 2003 *Thin Solid Films* **441** 214
[12] Guziewicz E, Godlewski M, Kopalko K, Lusakowska E, Dynowska E, Guziewicz M, Godlewski M and Phillips M 2004 *Thin Solid Films* **446** 172
[13] Kale R B, Lokhande C D, Mane R S and Han Sung-Hwan 2006 *Appl. Surf. Sci.* **252** 5768
[14] Mittal V, Sessions N P, Wilkinson J S and Murugan G S 2017 *Opt. Mater. Express* **7 (3)** 712
[15] Raghu P, Naveen C S, Sanjeev G and Mahesh H M 2013 *Int. J. Eng. Res. Technol.* **2** 3568
[16] Shaaban E R 2013 *J. Alloys Compd.* **563** 274
[17] Srivastava R, Lal P, Dwivedi R and Srivastatva S 1994 *Sens. Actuators* B **21** 213
[18] Timmer B, Olthuis W and van den Berg A 2005 *Sens. Actuators* B **107** 666

Zeolites from fly ash embedded in a thin niobium oxide matrix for optical and sensing applications

K Lazarova[1], S Boycheva[2], M Vasileva[1], D Zgureva[3], B Georgieva[1] and T Babeva[1]

[1]Institute of Optical Materials and Technologies ''Acad. J. Malinowski'', Bulgarian Academy of Sciences, Acad. G. Bonchev str., bl. 109, 1113 Sofia, Bulgaria.
[2]Technical University of Sofia, Department of Thermal and Nuclear Power Engineering, 8 St. Kl. Ohridsky Blvd., 1000 Sofia, Bulgaria.
[3]Technical University of Sofia, College of Energy and Electronics, 8 St. Kl. Ohridsky Blvd., 1000 Sofia, Bulgaria.

E-mail: babeva@iomt.bas.bg

Abstract. Thin films consisting of a sol-gel Nb_2O_5 matrix doped with fly ash zeolites in different volume fractions were deposited by the method of spin-coating. Zeolites of Na-X phase were prepared by long-term alkaline atmospheric conversion of coal ash collected from the electrostatic precipitators in TPP "AES Galabovo" supplied by lignite coal from the "Maritza East" basin using two different molarities of alkaline activator (NaOH). The surface morphology and structure of zeolite powders and films were studied by scanning electron microscopy and X-ray diffraction, respectively. Optical constants (refractive index and extinction coefficient) along with thickness of the films were calculated using a previously developed procedure comprising two steps of nonlinear minimization of the discrepancies between the measured and calculated reflectance spectra of films deposited on silicon substrates. The liquid adsorption and vapor sensing ability of fly ash zeolites were tested by measuring the reflectance spectra of the films prior to and after exposure to probe acetone molecules. The application of zeolites from fly ash for optical sensing is demonstrated and discussed.

1. Introduction

Fly ash (FA) is an industrial waste material produced by the combustion of coal in coal fired power stations when used as a fuel source. It can cause many environmental problems if discharged to the air. It is also harmful for people's health because it contains a variety of heavy metals and radionuclides carried by small particles, typical with diameters of less than 2.5 microns, that can easily be inhaled or digested through drinking water. Thus, efficient utilization of fly ash that diminishes its disposal is one important issue for environmental protection. FA has been studied for many potential applications such as the synthesis of geopolymers, the manufacture of glass and ceramics, soil amelioration agent, for use as low-cost catalysts, and for adsorbents for air and water pollutants [1,2]. In addition, due to the high silica and alumina content, fly ash can be converted into zeolite [3-6]. The synthesized fly ash zeolites (FAZ) are preferable when adsorption issues are concerned because they are more crystalline, possess mixed micro-mesoporous structures that facilitate mass transport phenomena, and have more voids and channels on their surfaces [7]. Zeolites synthesized from fly ashes have already been implemented for adsorption of NO_x, SO_2, CO and other harmful air pollutants [8], removal of fluoride from drinking water [7] and heavy metal cations such as Cu(II), Pb(II) and Zn(II) from wastewater [9,10], and as an anticancer agent for alternate or complementary therapy [11].

In most cases, the synthesized zeolites are used in powder form that sometimes is inconvenient from the viewpoint of reutilizing. Therefore, in the present paper we study fly ash zeolites deposited in thin film form which according to our knowledge is done for the first time. In order to deposit thin films with good optical quality that is essential for optical sensing application, we incorporate FAZ particles in a niobium oxide matrix thus overcoming to a great extend the scattering. We have already demonstrated that the sensing properties of oxide/zeolite composites substantially improve as compared to an undoped matrix benefitting additionally from tunable optical properties [12].

The goal of this paper is to prepare composite thin films consisting of Nb_2O_5 matrix doped with synthesized FAZ in different volume fractions and to study their optical and sensing properties.

2. Experimental part

Fly ash (FA) used as a starting material for the synthesis of zeolites was collected from the electrostatic precipitators in one of the biggest Thermal Power Plants in Bulgaria TPP "AES Galabovo", which is supplied by domestic lignite coal from the "Maritza East" basin. Chemical and phase composition of the raw FA studied by a combination of classical and instrumental analytical methods was summarized in [6].

Zeolite samples FAZ1 and FAZ2 were prepared by long-term alkaline atmospheric conversion of fly ash at molarities of 1.25 M and 1.50 M of the alkaline activator NaOH, respectively. The ratios of the alkaline activator to coal fly ash were 0.5 and 0.6 for FAZ1 and FAZ2, correspondingly. The obtained fly ash zeolites were characterized with respect to their phase composition by X-ray diffraction using Bruker D8 Advance diffractometer with CuKα-radiation and a Ni filter, while their morphology was investigated by scanning electron microscopy using a MIRA 3 TESCAN microscope.

Thin composite films comprising a Nb_2O_5 matrix with embedded FAZ particles were deposited on silicon substrates by the method of spin-coating and annealed at 320 °C. Niobia matrix, prepared by the sol-gel method [13] was mixed with 5 wt% dispersion of FAZ particles in ethanol in a volume concentration of 1, 2.5 and 5 %. The refractive index, extinction coefficient and thickness of FAZ-Nb_2O_5 composites were calculated from reflectance measurements (Cary 5E, Varian) using previously developed two-stage curve fitting [13]. A liquid sensing experiments were performed by measuring reflectance spectra prior to and after exposure to liquid acetone at one and the same spot using an optical profiler (Zeta-20, Zeta Instruments) with a built-in spectrometer. For the vapor sensing experiments the samples were put in a measuring cell and their reflectance spectra are collected in air, argon and acetone vapors atmospheres [14].

3. Results and discussion

It was found [6] that the raw FA contains about 47 wt. % SiO_2 and 23 wt % Al_2O_3, which means that the aluminosilicate part predominates in its composition and makes it suitable for the synthesis of zeolites. The molar ratio of SiO_2/Al_2O_3 is approximately 3.4 making this FA a suitable starting material for the synthesis of high-silica zeolites of the faujasite type [4]. Experimental X-ray diffractograms of the studied FAZ are plotted in figure 1 in comparison with a pattern of a faujasite type (FAU) zeolite regarded as a referent one. Obviously, both studied samples refer to the faujasite phase as their main diffraction lines coincide with those of the referent zeolite. It may be expected that the supplementary crystalline phases that are resistant to alkaline treatment be transferred from the raw FA into the reaction products. Therefore, the reflexes of quartz (Q), magnetite (M) (Fe_3O_4), hematite (H) (α-Fe_2O_3) and anorthite (A) ($CaAl_2Si_2O_8$) are also observed in figure 1.

Figure 2 presents SEM pictures of FAZ1 and FAZ2 in powder form. Both zeolites are composed of agglomerates with a size of 1-2 μm and have an octahedral shape typical of the FAU crystalline phase. Although the two zeolites are very similar, they have different behaviour when added in Nb sol. From the SEM pictures of the surface of thin Nb_2O_5 films doped with FAZ1 and FAZ2 presented in figure 3 it is seen that FAZ1 is buried in the oxide matrix and is mostly covered with it, while the FAZ2 particles are mainly found on the surface and are not veiled by the oxide. Different optical and sensing properties could be expected due to the different morphology observed.

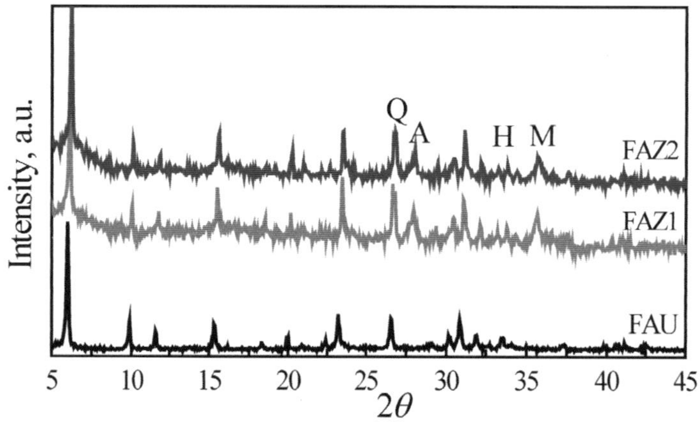

Figure 1. X-ray diffractograms of FAZ1, FAZ2 and a referent zeolite of faujasite type (FAU): Q-quartz; A-anorthite; H-hematite; M-magnetite.

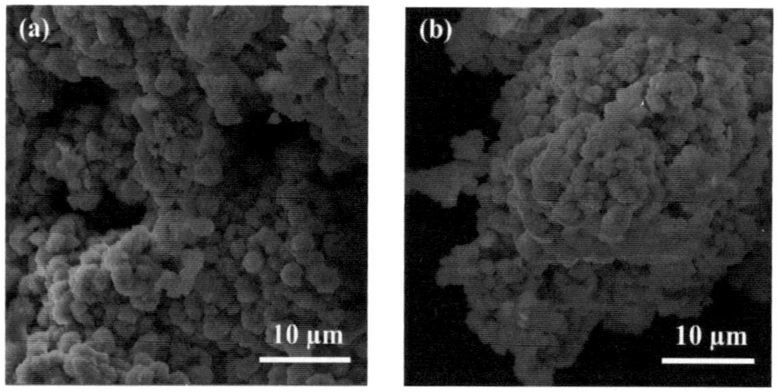

Figure 2. SEM pictures of fly ash zeolites powders FAZ1 (a) and FAZ2 (b)

The next step of our investigation concerns determination of the optical properties of deposited composites comprising niobium matrix and fly ash zeolites. Refractive index of Nb_2O_5 films with FAZ1 and FAZ2 particles with different doping levels are calculated from their reflectance spectra and are presented in figure 4(a). It is seen that the refractive index (n) of the films decreases with the addition of the FAZ particles mostly pronounced in the case of FAZ2 incorporation. Generally there is a substantial drop in refractive index of the doped thin film samples as compared to the undoped ones: the values of n decrease from 2.14 to 1.65 / 1.61, 1.52 / 1.51 and 1.45 / 1.43 for FAZ1 / FAZ2 with concentrations of 1.0, 2.5 and 5.0 %, respectively. Simultaneously there is a slight increase in extinction coefficient from 0.018 for the undoped sample to 0.020 for the heaviest doped one that could be associated with slightly increased scattering due to the rougher surface.

The thickness of the studied thin films as a function of FAZ concentration is presented in figure 4(b). It is seen that the addition of FAZ particles results in an increase in films thickness mostly pronounced for FAZ2 zeolites.

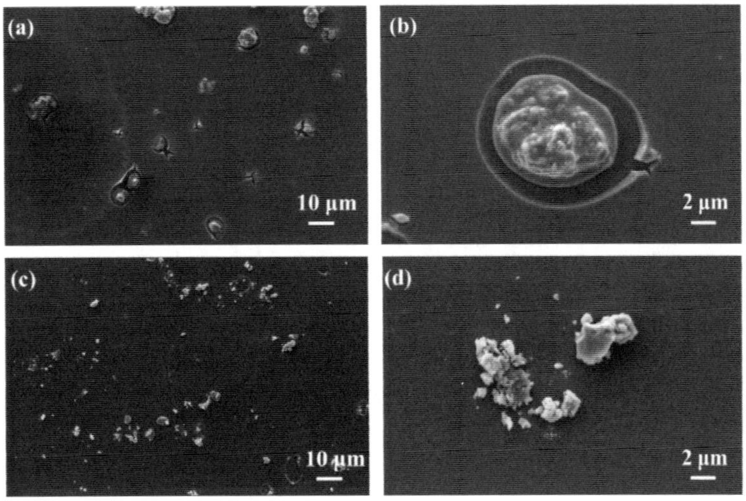

Figure 3. SEM pictures of thin films comprising sol-gel Nb_2O_5 matrix with embedded fly ash zeolites FAZ1 (a,b) and FAZ2 (c,d) at different magnifications.

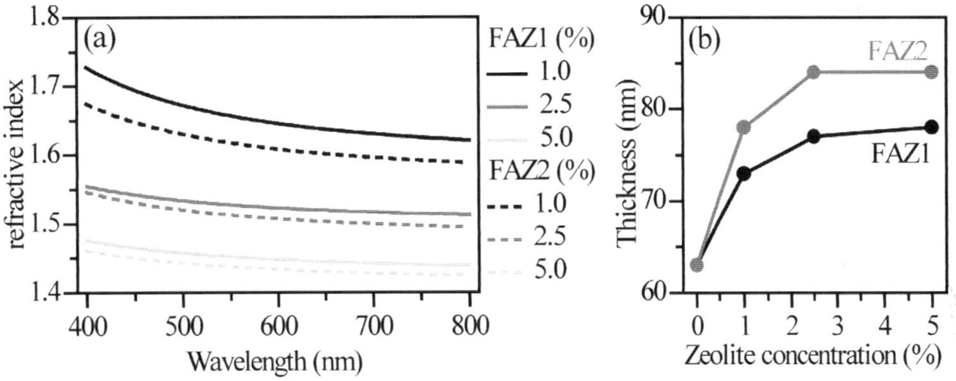

Figure 4. Dispersion curves of refractive index (a) and dependence of thickness on zeolite concentration (b) of thin Nb_2O_5 films embedded with fly ash zeolites FAZ1 and FAZ2.

Considering both that the weight amount of FAZ1 and FAZ2 is the same and their shapes and sizes are very similar we could assume that the incorporation of FAZ particles is accompanied with the introduction of additional free volume (air) in the samples which volume fraction is different for the two FAZ-doped samples thus leading to different levels of swelling of Nb_2O_5 matrix. Further, the stronger swelling in the case of FAZ2 indicates more free volume in these films. This assumption is confirmed by the smaller refractive index values of films doped with FAZ2. It is seen from figure 4(a) that for all the studied concentrations the incorporation of FAZ2 particles leads to lower values of the refractive index as compared to FAZ1 particles. Therefore the FAZ2 composites are less dense as compared to the FAZ1 ones and have more free volume. Thus considering the presence of porosity we assumed that the immersion of samples in liquid analyte will lead to analyte penetration in the empty pores and a subsequent increase in the effective refractive index and drop in the measured reflectance of the sample. To verify this we measure reflectance spectra for Nb_2O_5 thin films doped with FAZ2 particles before and after immersion in liquid acetone for 5 min and calculate the acetone induced change in reflectance (figure 5).

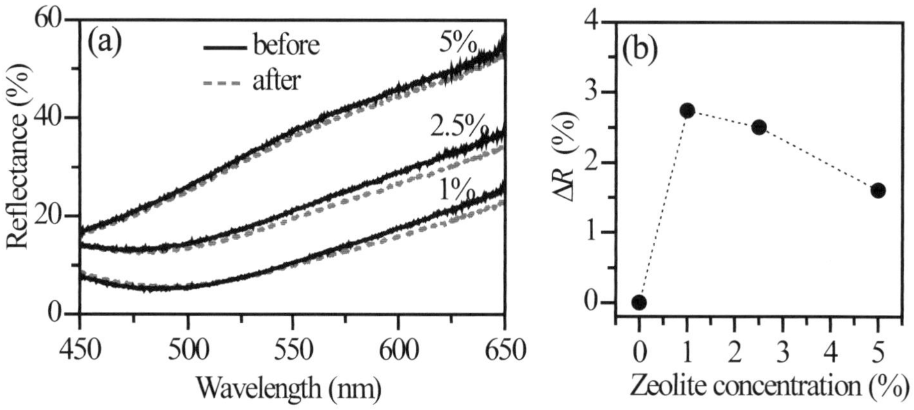

Figure 5. Measured reflectance before and after immersion in liquid acetone (a) and liquid acetone induced change of reflectance at wavelength of 650 nm (b) for thin Nb_2O_5 films doped with FAZ2 particles in denoted concentrations.

A decrease in reflectance after immersion is seen in all samples that confirms the expected increase in effective refractive index due to replacement of air in the pores with acetone with a higher refractive index. Interestingly, with an increasing in the zeolite concentration above 2.5 % a decrease of the reflectance change is observed (figure 5b) although the free volume is the highest for the heaviest doped sample. A possible reason could be the different ratio of open to closed porosity with a predominantly closed type for 5% doped sample.

The final step of our investigation concerns vapor adsorption of the oxide-zeolite composites. Accordingly, the reflectance spectra of films is measured prior to and after acetone vapor exposure and vapor-induced relative changes in reflectance ($\Delta R/R$) are presented in figure 6. It should be noted that all adsorption measurements are performed at room temperature. The vapor-induced changes are similar for both types of FAZ particles and are slightly influenced by the concentration. As in the case of liquid acetone, a small decrease in the response for the 5% FAZ2 doped films is observed.

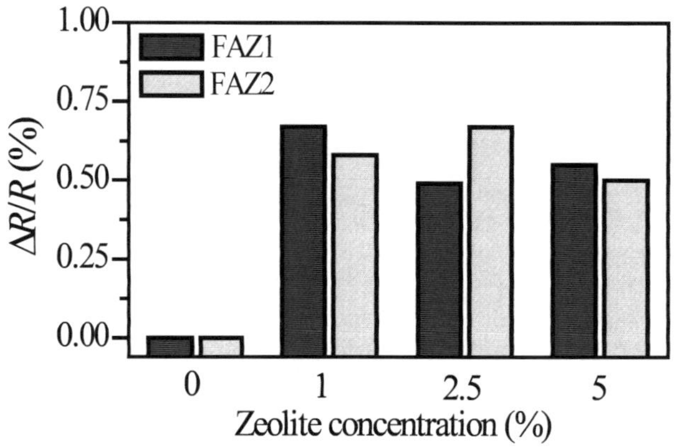

Figure 6. Zeolite concentration dependence of relative change in reflectance induced by exposure to acetone vapors of FAZ1 and FAZ2 zeolites embedded in thin Nb_2O_5 films annealed. Measurements are conducted at room temperature.

In our previous investigation [14] we demonstrated an enhancement of the sensitivity of acetone sensing by Bragg stacks comprising Nb_2O_5 and zeolite thin films after annealing at 450 °C. However, in the present case, for Nb_2O_5 / FAZ composite, annealing at 450 °C does not lead to an improvement. Indeed there is a slight drop in relative reflectance change (not shown here). This could be due to the contraction of the oxide matrix and a decrease in the total pore volume.

4. Conclusions

The successful deposition of composite thin films comprising a Nb_2O_5 matrix and fly ash Na-X zeolite in concentrations of up to 5% is demonstrated. Long-term alkaline atmospheric conversion of fly ash at two different molarities (1.25M and 1.5M) of the alkaline activator (NaOH) is applied for Na-X zeolite synthesis. A decrease of refractive index and increase of thickness of the composite films relative to the undoped matrix are obtained in both cases associated with porosity generation in the films and mostly pronounced for the zeolite sample synthesized at higher molarity. The presence of porosity is verified and confirmed by reflectance measurements of the films before and after exposure to acetone in liquid and vapour physical states. Changes in reflectance are detected in both cases but they are stronger for liquid acetone. The vapor-induced changes are similar for both types of fly ash zeolites and are slightly influenced by concentration, while for liquid acetone a reduction of response is observed for heavily doped sample (5%).

Acknowledgments

The financial support of Bulgarian National Science Fund (BNSF) under the project DN 17/18 (12.12.2017) is highly appreciated.

References

[1] Blissett R and Rowson N 2012 *Fuel* **97** 1
[2] Ge J, Yoon S and Choi N 2018 *Applied Sciences* **8** 1116
[3] Holler H and Wirsching U 1985 *Fortschr. Miner* **63** 21
[4] Jha V, Nagae M, Matsuda M and Miyake M 2009 *J. Environ. Manage* **90** 2507
[5] Querol X, Moreno N, Umaa J, Alastuey A, Hernández E, López-Soler A and Plana E 2002 *Int. J. Coal Geol.* **50** 413
[6] Boycheva S, Zgureva D, Václavíková M, Kalvachev Yu, Lazarova H and Popova M 2018, *J. Haz. Mat.* in press
[7] Panda L, Kar B and Dash S 2018 *AIP Conference Proceedings* **1953** 080003
[8] Siddharth S, Maiti S, Raj H, Bisht R, Minocha A, Panigrahi S, Alexander S, Sameer and Singh M 2018 *Materials Research Express* **5** 085507
[9] Hui K, Chao C and Kot S 2005 *J. Hazard Mater.* **127** 89
[10] Qiu W and Zheng Y 2009 *Chem. Eng. J.* **145** 483
[11] Subhapriya S and Gomathipriya P 2018 *Materials Research Express* **5** 085401
[12] Lazarova K, Awala H, Fallah J El, Vassileva M, Mintova S and Babeva T 2017, *Bulg. Chem. Commun.* **49** 88
[13] Lazarova K, Vasileva M, Marinov G and Babeva T 2014 *Optics & Laser Technology* **58** 114
[14] Lazarova K, Awala H, Thomas S, Vasileva M, Mintova S and Babeva T 2014 *Sensors* **14** 12207

IOP Conf. Series: Journal of Physics: Conf. Series **1186** (2019) 012025 doi:10.1088/1742-6596/1186/1/012025

Leakage currents in Al_2O_3/HfO_2 multilayer high-*k* stacks and their modification by post-deposition annealing steps

D Spassov[1], A Paskaleva[1], T A Krajewski[2], E Guziewicz[2] and Tz Ivanov[1]

[1]Institute of Solid State Physics, Bulgarian Academy of Sciences, Tzarigradsko Chaussee 72, Sofia 1734, Bulgaria
[2]Institute of Physics, Polish Academy of Sciences, Al. Lotników 32/46, 02-668 Warsaw, Poland

E-mail: d.spassov@issp.bas.bg

Abstract. Leakage currents of Al_2O_3/HfO_2 multilayer stacks deposited by atomic layer deposition in the low voltage region ($0 \div \pm 4$V) are investigated in dependence on Al_2O_3 sublayers thickness and the post-deposition annealing in two ambients (O_2, N_2,). The stacks with 5cy of Al_2O_3 show highest current, compared to structures with lower (2cy) and higher (10cy) Al_2O_3 content. Oxygen treatment does not affect significantly the leakage current at low voltages, while annealing in nitrogen definitely increases it. The current follows power law dependence, V^n, with n varying in the interval of about 0.6 to ~1. The application of ±12V pulses shift the leakage curves towards higher or lower current values depending on the polarity of the pulse. Leakage curves after the pulse stress are described either with V^n or $(V-V_t)^n$ relations depending on both pulse and measurement bias polarity. The results are found consistent with a model similar to the collective transport in array of small metal dots model in which the current is limited by the available sites for charge carriers to move on.

1. Introduction

The importance of non-volatile semiconductor memory devices for the portable electronics and computer systems has been increased dramatically in the recent years. The demands for larger amount of information storage, higher transfer rates as well as price per bit reduction fuels the further scaling down of these devices. The majority of the existing non-volatile semiconductor memories (flash memories) are based on the floating gate cell technology being in production since the early 1980s [1]. The future scaling of this technology, however, becomes problematic as it faces several significant challenges [2]. Among the various potential replacements of floating gate technology, the charge trapping memory (CTM) is clearly gaining momentum recently promising to become main-stream in the near future. CTM storage cell shares the same operational principle as floating-gate one, except for the fact that in floating gate cell the charge is stored in a special poly-Si floating gate whereas in CTM charges are kept directly in the gate dielectric, in discrete traps distributed in the dielectric band-gap. CTM was introduced in 1967 and used Si_3N_4 as a CT layer [3] but the real progress of CTM is associated with the implementation of high-*k* dielectrics. The high-*k* materials have focused researchers' attention in the field of microelectronics since mid-1990s. The numerous studies demonstrated the trap-rich nature of these dielectrics, inferring that they are particularly suitable for the CTM applications in the nano-size memories. In fact, future generations of CTM could be realized only by implementation of high-*k* dielectrics in them. HfO_2 is one of the most intensively studied high-*k* materials, and it has been already implemented in some advanced devices. Therefore, the possibility for its employment in other kind of microelectronics devices seems very attractive. It has been reported that HfO_2 possesses the necessary potential for CTM applications. It has been also suggested [4] that incorporation of Al into HfO_2 matrix can further improve its parameters from CTM point of view. Recently we have investigated the charge trapping in atomic layer deposited multilayered HfO_2/Al_2O_3 stacks, demonstrating the feasibility for employment of the stacks as CT medium, and the possibility of tailoring the CT characteristics by variation of Al_2O_3 content and thermal treatments. [5,6]. In this work we focus on leakage currents through these stacks and their dependence on the voltage pulses charging the capacitors.

2. Experimental procedure

Multilayered HfO_2/Al_2O_3 charge trapping stacks were deposited on p-type (100) Si wafers with resistivity 6–8 Ωcm. The substrates received standard RCA chemical cleaning prior to the deposition. The dielectric stacks were prepared by atomic layer deposition (ALD) in Savannah-100 ALD system at a temperature of 135°C. The precursors were tetrakis(dimethylamido)hafnium (TDMAH) for HfO_2 sublayers and trimethylaluminum (TMA) for Al_2O_3 ones. The multilayered stacks were composed of equivalent bilayer blocks of HfO_2 and Al_2O_3 sublayers with constant thickness; the total stack was obtained by 5 repetitions of the bilayer block. The thickness of HfO_2 sublayers was the same for all stacks – 20 ALD cycles, while the Al_2O_3 sublayer thickness was varied. The leakage currents of three types of stacks with Al_2O_3 sublayers thickness of 2, 5 and 10 cy (achieving this way variable Al_2O_3 content in the stack) were investigated. The samples are designated as 5x(20:10), 5x(20:5) and 5x(20:2) reflecting the number of bilayer blocks (5) and the number of HfO_2 and Al_2O_3 sublayer deposition cycles, respectively. The stack thicknesses, *d* were determined ellipsometrically and were as follows 19 nm for 5x(20:10), 16.5 nm for 5x(20:5) and 15 nm for 5x(20:2) stack. After the deposition part of the samples received rapid thermal annealing (RTA) in O_2 or N_2 at 800 °C for 1 min. The measurements were carried out on MOS capacitor structures with evaporated Al top (square, 2.25×10^{-4} cm^2, patterned by photolithography) and bottom electrodes. The currents were measured at room temperature with Keithley 236 SMU; in all cases the bias was applied to the Al top (gate) electrode while the Si substrate was at zero potential.

3. Results and discussion

The study of the effect of Al_2O_3 content and the post deposition annealing (PDA) on the *I-V* characteristics is predominantly concentrated to the low voltage region (up to ± 4V applied bias, *V*). (The maximum *V* corresponds to electric field in stack of ~ 2.1 to 2.4 MV/cm in dependence on *d*.) The reason behind is that at higher applied voltages the inevitable charge trapping in the stacks is influencing the *I-V* characteristics making it difficult to assess the impact of the Al_2O_3 content and PDA. Additionally, although the stacks sustain applying of quite high biases for times of ~ 1s [5], in the course of *I-V* measurements the total time under elevated *V* is largely increased leading to wear-out and breakdown.

Figure 1 shows the obtained current densities *J* of the as-deposited samples. The *I-V* characteristics are almost symmetrical in respect to zero bias up to ~ 3 V. At higher positive *V*, the slope of the curves for stacks with lower alumina amount increases implying a change of the conduction mechanism. As seen stacks with the highest Al_2O_3 content exhibit the lowest *J*. The decrease of alumina concentration from10 cy to 5 cy in the bilayer block increases *J* about 3 times; further reduction of Al_2O_3 (to 2 cy), however, results in a drop of *J* to levels close to the values of samples with 10 cy Al_2O_3. The observed behavior suggests that introducing a small Al amount into HfO_2 matrix suppresses some of the defects present there, causing a current reduction. With the increase of Al, new defect sites tend to appear in the stack and *J* increases. When Al_2O_3 reach a certain critical value, the introduced Al_2O_3 starts to behave rather as a separate layer than a dopant. As the band gap of Al_2O_3 is larger than HfO_2 one, *J* through the stack is diminished. The conjecture that the impact of 10 cy Al_2O_3 is akin to the effect of well-defined Al_2O_3 layer can be deduced from the lack of a transition point in *J-V* curves at $V>0$. The electrons injected from Si substrate in this case experience not only an interfacial SiO_2 layer which thickness

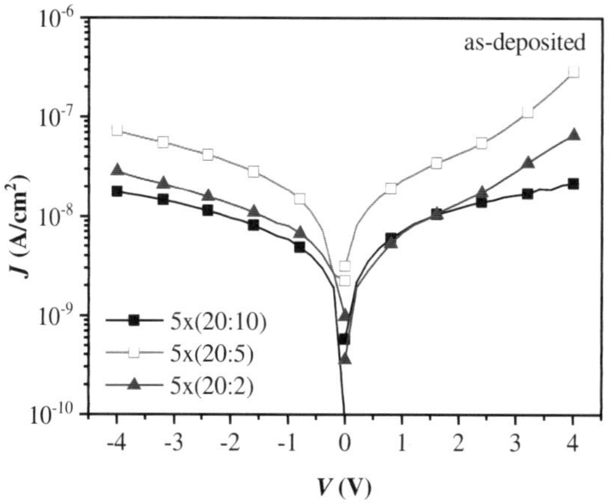

Figure 1. Leakage current characteristics of as-deposited stacks.

and quality is the same for all structures but also the first Al_2O_3 sublayer.

In figure 2 the effect of PDA in N_2 or O_2 on the *J-V* characteristics of the three kinds of stacks is shown. It is seen that O_2 annealing decreases the leakage currents at both bias polarities. The reduction is generally small at low voltages. It is more significant at positive biases higher than about 2.5 V for the samples with smaller Al_2O_3 content (2 and 5 cy). PDA in N_2 increases leakage currents in all stacks. While this increase is small for 5 cy Al_2O_3 stack, for the stack with 2 and 10 cy Al_2O_3 the leakage current is one order of magnitude or even more (depending on *V*) higher compared to their as-deposited counterparts. Consequently, the leakage currents in N_2 treated samples is lower in the 5x(20:5) stack, unlike the as-deposited ones. It should be also noted that both PDA change the shape of *I-V* curves as well. For 5x(20:5) and 5x(20:2) stacks steeper part of *J-V* curves at V>2V is not present after PDA in O_2; it is shifted to higher positive biases. The shape of leakage curves of 5x(20:10) stacks is not amended by O_2 treatment. The effect of N_2 on *I-V* curves shape is just the opposite. For 5x(20:10) samples at $|V|\gtrsim$ 2V slope of the curves substantially increases. Similar transformation of *I-V* curves are detected for the other stacks as well, but only at negative biases. Investigation on the memory effect of the stacks by capacitance-voltage (*C-V*) measurements [5] suggested that RTA in O_2 significantly increases hole and electron trapping compared to the as-grown stacks while nitrogen annealing suppresses both trappings especially for the stacks with smaller Al_2O_3 content. It was also found that O_2 annealing increases the initial positive oxide charge in the structures while RTA in N_2 introduces negative initial charge. The leakage data obtained here also supports the assumption of the distinctly different effect of the annealing ambient on stacks properties. Moreover they suggest that worsening of the memory windows in N_2

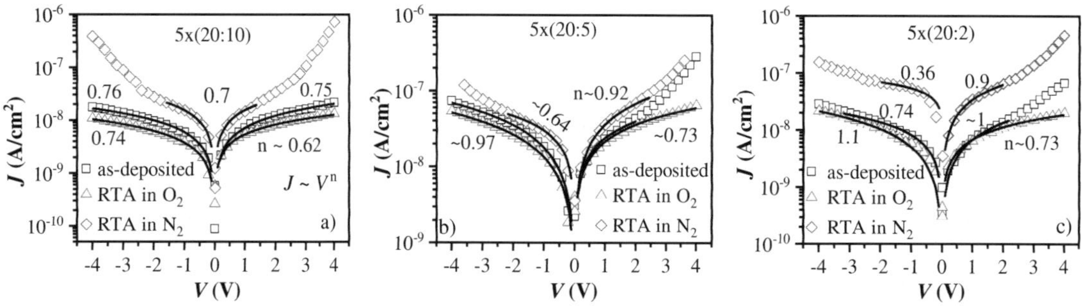

Figure 2. Effect of PDA on the leakage currents of stacks with different Al_2O_3 content: a) stack with 10 cy Al_2O_3; b) 5cy Al_2O_3 and c) 2 cy Al_2O_3.Symbols represent experimental data and solid lines fit to J~V^n..

treated stacks could be due to higher leakage in them. The current reduction after O_2 annealing in high-*k* films is usually ascribed to a partial removal of oxygen vacancies [7,8], on the other hand N_2 treatment was found detrimental for *J* [8,9]. In case of Al_2O_3 RTA in N_2 was shown to increase also the density of the interface states through creation of oxygen deficient defects [8] or even AlSi formation [9]. In the present study, the nature of the traps involved in the CT memory effect [5] are not known, but there are some indications that they are related rather to introduction of Al [4,5] than to O vacancies in HfO_2 or Al_2O_3. So the leakage data may be explained either by an annealing/introduction of oxygen deficient centers and even generation of new defects upon O_2/N_2 annealing, respectively or by some modification of Al-related ones. In the former case the defects involved in the low voltage conduction differ from those associated with the CT memory effect, and in the latter, one could assume that the modification of the Al-related defect makes it shallower energetically. Based on the *C-V* measurements of air annealed stacks [5] showing intermediate results, we speculate that chemical reaction involving N is highly probable. More detailed investigations, however, are needed to clarify this issue.

The almost symmetrical *J-V* curves (apart from the case of 5x(20:5) and 5x(20:2) under $+V$ where change of the slope is found) suggest that conduction in the investigated voltage range is rather bulk limited one. The *J-V* curves are fitted well with $J \sim V^n$ dependence where n is in the range of about 0.6÷1 except for the case of 5x(20:2) stacks after RTA in N_2 (figure 2). The data could be interpreted as a manifestation of Ohmic conduction (thermally activated hopping conduction [10] where the sublinearity arises from the fact that the voltage drop across the dielectric is affected by existing charges there and differs from the applied bias. Alternatively, the conduction might be described with collective charge transport in arrays of metal quantum dots [11,12]. According to this mechanism injected electrons are transported along tunneling paths made from the traps in the stacks, and J is expressed as $J \sim (V-V_t)^n$ (with n=1 or 5/3 for 1D and 2D conduction paths; V_t is a threshold voltage, characteristic of Coulomb blockage of transport). The obtained V_t=0 implies negligible charge trapping during the measurements [12]. Similarly the electric field distortion in the dielectric could lead to n slightly lower than 1. As pointed earlier, the change of the slope of the *J-V* curves for as-grown 5x(20:5) and 5x(20:2) at $V \gtrsim 2$V and for all N_2 treated samples is indicative for a change in the dominant conduction mechanism. The estimation of these voltage regions shows linear behavior of both ln(J) and ln(J/V) vs. $\sqrt{V}$ suggesting Schottky or Poole-Frenkel conduction, respectively. Taking into account the relative symmetry of the curves in respect to 0V, the conduction is most probably via Poole-Frenkel mechanism.

Next we will consider the response of the leakage characteristics to voltage pulses (V_p) with amplitude of ±12V and a duration of 1s. The results are illustrated in figure 3 for one of the stacks (5x(20:10)) and the overall trends can be described as follows. After a positive pulse the current at positive applied V diminishes and at $-V$ increases. The negative voltage pulse has an opposite effect - J for $-V$ range decreases and for $+V$ region the current increases. More generally, J corresponding to the negative or positive branch of *J-V* curve decreases after stress if V_p has the same polarity. And vice versa, J of the

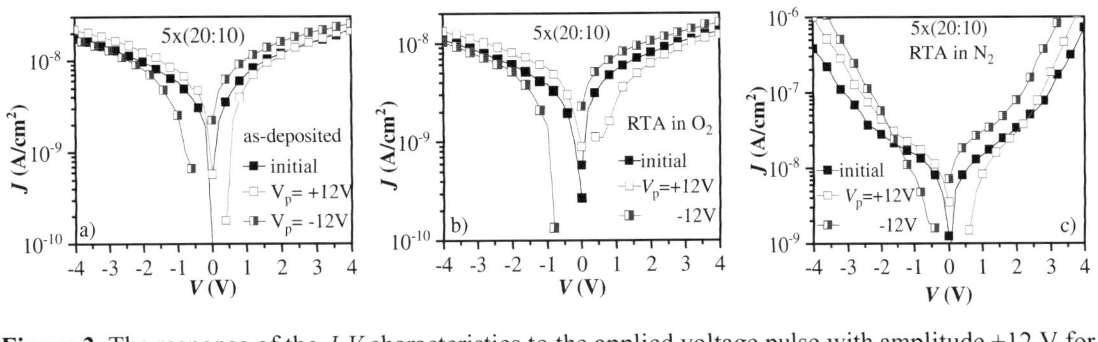

Figure 3. The response of the *J-V* characteristics to the applied voltage pulse with amplitude ±12 V for 1s.

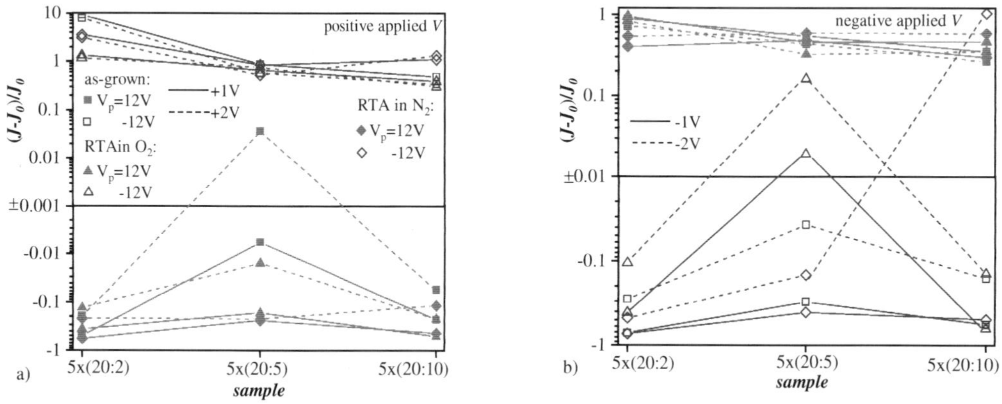

Figure 4. The relative change of the leakage current values after applying ±12 V pulses at applied biases of: a) 1 (solid line) and 2 V (dash); b) -1 (solid line) and -2V (dash).

corresponding branch of J-V curve increases if the polarity of V_p is opposite. (Under $+V_p$ and positive applied bias for J-V measurements electrons are injected from the Si substrate, and under $-V_p$ and negative V at J-V recording electrons are injected from Al-gate electrode). The variation of J after the pulse stress is more pronounced at lower biases. Figure 4 summarizes the obtained relative changes of J at ±1 and ±2 V after the stress for all the samples. The current reduction for 5x(20:2) and 5x(20:10) stacks is almost the same and without clear dependence on PDA (the maximum obtained decrease of J is ~1.5 times for both V polarities, and it diminishes with the increase of V). As is seen the relative reduction of J of 5x(20:5) stacks for both V_p polarities is smaller as compared to other stacks, and for applied bias of 2V J becomes even higher than the initial one. Only after RTA in N_2 this stack behaves similarly to the other stacks. The increase of J in the positive bias J-V branch after applying a negative pulse (figure 4a) is larger than the J increase for the negative biases after $+V_p$ stress (figure 4b). In the former case increase of J up to an order of magnitude is found for as-grown 5x(20:2) stacks, and in the latter the maximum increase is ~2x again for the 5x(20:2) stack. A clear dependence of the increase of J on the PDA as well as the Al_2O_3 content of the stacks is difficult to be traced. Nevertheless, the stacks with lower Al_2O_3 content tend to show a higher increase of J especially for the as-grown samples. For these structures PDA diminishes the current increase after pulse stress. We also note that the response of J-V curves at $|V|\gtrsim 2$V to the voltage pulses of the thickest stack (5x(20:10)) treated in N_2, figure 3c diverge from the case of as-grown and O_2 annealed samples. We will also mention that changes in J upon pulsing are reversible and sustainable switching between the states with higher and lower conduction can be obtained by cycling the pulse polarity.

In an effort to clarify the effect of voltage pulsing on the leakage characteristics of the stacks and evaluate the possible conduction mechanisms, the J-V curves after stress were plotted in log-log scale. Figure 5 depicts the results for 5x(20:2) stack. The J-V curves at positive biases after -12V pulse along with those at negative V after +12V pulse are well described by straight lines with slopes n in range of 0.5÷0.9, implying that $J \sim V^n$. The current at positive V after +12V and J at $-V$ after -12V stress, however, are fitted with $J \sim (V-V_t)^n$, where n are in the interval of 0.65÷1.6, and $|V_t|$ varies between 0.2 and 0.8V. A clear dependence of the values of n on the pulse stress is not observed, but for the as-grown and annealed in O_2 stacks the stress tends to reduce slightly n. For the structures treated in N_2 an increase of n for the J curves at positive V after +12V pulse and at negative V after -12V pulse is found ($n \gtrsim 1$ and for the 5x(20:10) stack are ~1.6). The samples processed in N_2 demonstrate also lower V_t. The obtained transition of $J \sim V^n$ for the pre-stressed J-V curves towards $J \sim (V-V_t)^n$ suggests that the conduction through the stacks is governed by a mechanism similar to the collective charge transport in arrays of metal quantum dots in which the role of quantum dots is taken by electron traps [11,12]. When +12V pulse is applied the injected form the substrate electrons are trapped in the stack reducing the available sites

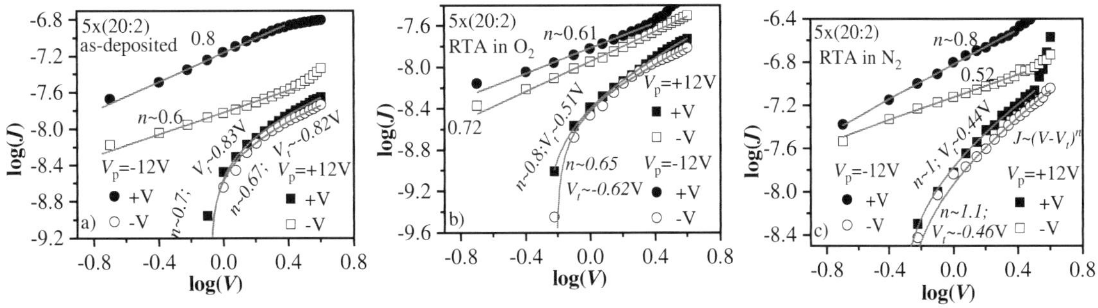

Figure 5. log-log J-V plots of 5x(20:2) stacks after pulse voltage stress; symbols – experimental data, solid lines – fits to $J{\sim}V^n$ and $J{\sim}(V-V_t)^n$. a) as-grown, b) RTA in O_2 and c) RTA in N_2.

in the conductive paths thus reducing J at positive biases and shifting the J-V curves by V_t. The electron trapping, however, is likely to be confined in the area near the injecting electrode, as evidenced by almost the same current values under positive applied biases after $+V_p$ and at negative voltage after $-V_p$ (figure 5). The observed increase of J at negative biases after $+V_p$, however, suggests that during the stress electron detrapping near the opposite gate electrode occurs and the current measured at $-V$ increases and V_t becomes 0. Similarly, at -12V stress the electrons are injected from the gate and are trapped near gate electrode and electrons at sites near the Si are detrapped, hence J under $+V$ is increases and J at $-V$ decreases and the J-V curve is shifted to more negative V. The higher increase of J at positive biases after -12V compared to the current increase at negative biases after +12V stress means that during the -12V pulse a hole substrate injection also takes place thus enhancing the detrapping process. The values of n imply 1D path, but in case of PDA in N_2 conduction through 2D paths ($n{\sim}5/3$) after pulse stress could be assumed reflecting the higher density of the defects through which conduction takes place. The smaller values of V_t for these stacks (compared to as-grown and O_2 annealed ones) is in an agreement with the reduced electron trapping detected from C-V investigations [5].

4. Conclusion

Leakage currents in multilayered Al_2O_3/HfO_2 stacks in the low voltage region are influenced by the stack composition as well as by the post deposition treatments. The results imply that depending on the amount of Al_2O_3 in the stack a reduction or increase of the leakage current levels can be obtained, i.e there is an optimum Al_2O_3 content (thickness of Al_2O_3 sublayers) providing minimum leakage currents. Additional improvement of the leakage currents can be achieved by PDA in oxidizing environment. The treatment in nitrogen, however, deteriorates the leakage currents. The origin of the effect of Al content and PDA can be sought in an elimination of existing and an introduction of new types of electronic traps predominantly in HfO_2. A switching between a state with higher and lower conductivity for a given bias polarity can be obtained by applying a voltage pulses with a corresponding polarity and amplitude. The observed currents are governed by a bulk limited conduction mechanism. A good agreement with the model of collective transport in an array of metal quantum dots is found, suggesting that conduction through the structures could be controlled by charging pulses.

Acknowledgement
The work is conducted under a joint research agreement between Bulgarian and Polish Academy of Sciences, and supported by the European Regional Development Fund within the Operational Programme "Science and Education for Smart Growth 2014 - 2020" under the Project CoE "National center of mechatronics and clean technologies BG05M2OP001-1.001-0008-C01.

References

[1] Masuoka F, Asano M. Iwahashi H, Komuro T and Tanaka S 1984 IEDM'84 464
[2] Zhao C, Zhao C Z, Taylor S and Chalker P R 2014 *Materials* **7** 5117-45
[3] Brewer J and Gill M 2008 *Nonvolatile Memory Technologies with Emphasis on Flash: A Comprehensive Guide to Understanding and Using NVSM Devices* (Piscataway/USA: IEEE Press)
[4] Paskaleva A, Rommel M, Hutzler A, Spassov D and Bauer A J 2015 *ACS Appl. Mat. Interf.* **7** 17032
[5] Spassov D, Paskaleva A, Krajewski T A, Guziewicz E, Luka G and Ivanov Tz 2018 *physica status solidi* a in press DOI: 10.1002/pssa.201700854
[6] Spassov D, Paskaleva A, Guziewicz E, Luka G, Krajewski T A, Kopalko K, Wierzbicka A and Blagoev B 2016 *J. Phys.: Conf, Ser.* **764** 012016
[7] Paskaleva A, Atanassova E, Lemberg M and Bauer A J 2006 *Defects in High-k Gate Dielectric Stacks: Nano-Electronic Semiconductor Devices* ed Gussev E (Dordetch: Springer) p 411
[8] Winzer A, Szab N, Wachowiak A, Jordan P M, Heitmann J and Mikolajick T 2015 *J Vac. Sci. Technol. B* **33** 01A106
[9] Papanasam E and Kailath B J 2018 *Microelectron. Intern.* **35** 65-73
[10] Paskaleva A, Spassov D and Danković D 2017 *Facta Universitatis Series: Electronics and Energetics* **30** 511-48.
[11] Zhu W, Chen T P, Liu Y, Yang M, Zhang S, Zhang W L and Fung S 2009 *IEEE Trans. Electron Dev.* **56** 2060
[12] Romero H E and Drndic M 2005 *Phys. Rev. Lett.* **95** 156801

In-depth evolution of tellurium films deposited by Frequency Assisted Thermal Evaporation in Vacuum (FATEV)

T Hristova-Vasileva[1,2*], I Bineva[1], R Todorov[2], A Dinescu[3] and C Romanitan[3]

[1] Institute of Solid State Physics, Bulgarian Academy of Sciences, 72 Tsarigradsko Chaussee Blvd., 1784 Sofia, Bulgaria
[2] Institute of Optical Materials and Technologies, Bulgarian Academy of Sciences, 109 Acad. G. Bontchev Str., P.O. Box 95, 1113 Sofia, Bulgaria
[3] National Institute for R&D in Microtechnologies – IMT Bucharest, 126 Erou Iancu Nicolae Str., 077190 Bucharest, Romania

E-mail: teddie@issp.bas.bg

Abstract. In order to enlighten the in-depth organization of thin tellurium films, deposited by the frequency assisted thermal evaporation in vacuum (FATEV) approach, morphological, structural and spectroscopic investigations were performed. Spectroscopic ellipsometry (SE) and cross-section scanning electron microscopy (SEM) showed a change of the growth mechanism upon application of higher vibrations` input frequencies during the films deposition. The films, deposited by conventional thermal evaporation in vacuum with no vibrations applied, as well as these, deposited by FATEV with application of 50 Hz vibrations, were characterized by initial densification followed by 2D nanoparticles growth when a certain threshold thickness was reached. On the other hand, the high-frequency vibrations of 4 and 10 kHz preconditioned growth of tellurium nanoribbons oriented towards the z-axis from the very beginning of the film formation. The topography changes, observed by atomic force microscopy (AFM) and scanning electron microscopy, showed highly porous surfaces (with high root mean square surface roughness) formed by distinct nanoblades for the films, deposited at low input frequencies, while the films, deposited under the impact of vibrations in the kilohertz range, were much more ordered, and hence their surface was significantly smoother. The structural parameters of the samples were investigated by X-ray diffraction (XRD) analysis.

1. Introduction

The frequency assisted thermal evaporation in vacuum (FATEV) is a new approach for physical vapor deposition (PVD) of thin films. It is based on application of mechanical vibrations with audio frequencies on the substrate during condensation of an evaporated material. The technological specifics and the range of FATEV application, determined until now, are widely described in our previous works [1, 2]. As the method is still under evaluation only the changes of the deposits' surface have been studied. It has been found that the application of vibrations with frequencies from the lower part of the audio spectrum leads to formation of rough undulated surfaces with period of about 300 nm, while the excitement with kilohertz frequency causes smoothening of the films surface [1]. Both effects can be useful in practice for production of active layers in sensorics, optics, microelectronics, etc.

During our investigations of tellurium films with nominal thickness of 100 nm, deposited by FATEV at input frequency of 4 kHz, an interesting fact was observed - together with the significant smoothening of the surface (a drop of the root mean square surface roughness Sq from 34 nm (for a film, deposited by conventional PVD) to 14 nm (for a sample, deposited by FATEV with 4 kHz impact) a formation of an ordered columnar structure from the bottom to the top of the film was observed [1].

The aim of the present work is to study the in-depth changes in FATEV deposited tellurium films with nominal thickness of 60 nm, caused by vibrations with input frequencies of 0, 50, 4000 and 10000 Hz. In addition, we will present new data about the thickness effect on the surface morphology of FATEV deposited Te films, which would be useful for widening the knowledge about the possibilities of the FATEV method.

2. Experimental procedures

Series of tellurium (Merck, 5N) films with nominal thickness of 60 nm were deposited using FATEV approach [1, 2] with deposition rate of ~ 0.5 nm/s during the impact of mechanical vibrations with input frequencies of 0, 50, 4000, and 10000 Hz on Si (100) substrates. The deposition rate and the film thickness during the deposition were controlled using a quartz-crystal microbalance device Miki MSV 1841/A.

SEM observations have been performed using an e-Line EBL equipment in SEM mode (Raith GmbH., Germany), working with an accelerating voltage of 10 kV. The as-deposited Te films were observed on top and cross-section with magnification of 50-, 100- and 200-thousand times.

The optical properties of the films deposited on silicon substrates were examined by UV-Visible-Near IR phase modulated spectroscopic ellipsometric platform UVISEL2 (HORIBA JobinYvon) in the spectral range of 1.5 - 6 eV at 70 ° incident angle.

AFM measurements have been carried out on a scanning probe microscope Multimode V (Bruker, Santa Barbara, CA). The images have been taken in tapping mode, as each sample has been investigated in multiple points of the surface. The measurements in scales of 1 and 3 μm have been performed with scanning rates in the interval of 0.5-2.0 Hz and with an image resolution of 512 lines per scan direction. Al-coated silicon cantilevers TAP150-Al-G (Budget Sensors Innovative Solutions Bulgaria Ltd., Bulgaria) with a nominal resonant frequency of ~150 kHz and spring constant of 5 N/m have been used. The radius of the cantilever's tip is smaller than 10 nm. The root mean square roughness (Sq) of the samples was determined in the scale of 3 μm. The images have been just flattened before the analysis, made with the SPIPtm 6.1.0 program.

The structure of the films was evaluated through grazing incidence X-ray diffraction (GIXRD) using a Rigaku SmartLab X-ray thin film diffraction system (Rigaku Corporation, Japan) at CuK$\alpha_{1,2}$ irradiation (λ=1.5418 Å) and an angle of incidence of 0.5 °. The XRD patterns were taken in the Bragg angle (2θ) range from 10 to 60 ° with a step of 0.01 °.

3. Results and discussion

The XRD patterns of the films are presented in figure 1. The phase analysis of the latter shows presence of a hexagonal tellurium phase, indexed in the JSCPD database under the powder diffraction file (PDF) 101-1098. A change of the main peak from [101], as described in the literature, to [100] is observed, which most probably is caused by the influence of the Si (100) substrate orientation.

The results from the structural analysis of the samples are presented in Table 1. The lattice constants are similar to these described in PDF 101-1098 (a=b=4.454 Å, c=5.924 Å; $\alpha=\beta$=90 °, γ=120 °). The average crystallites size D and the micro-strain ε of the samples, revealed by the

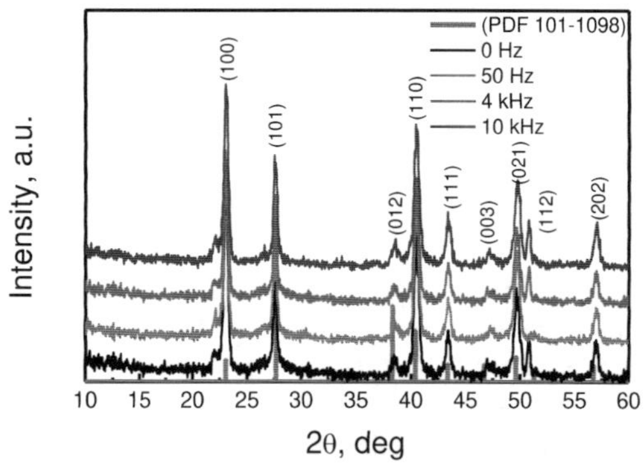

Figure 1. XRD spectra of as-deposited films and a reference pattern of Tellurium (PDF 101-1098).

Williamson-Hall method [3], are also presented in Table 1.

Table 1. Crystal lattice constants (distances a, b, c; angles – α, β, γ), position of the main XRD peak (2θ), full width at the main peak half maximum (FWHM), average crystallites size (D), micro-strain (ε), and root mean square surface roughness (Sq) of as-deposited Te films.

Frequency, Hz	a, Å	b, Å	c, Å	α=β, °	γ, °	2θ, deg	FWHM, deg	D, nm	ε	Sq, nm
0	4.457	4.457	5.908	90	120	22.990	0.448	17.72	0.11	7.25
50	4.457	4.457	5.919	90	120	22.997	0.424	16.28	0	9.77
4000	4.453	4.453	5.914	90	120	23.011	0.430	16.92	0.23	3.39
10000	4.459	4.459	5.922	90	120	23.002	0.45(2)	16.61	0.12	4.22

The morphological and topological investigations of the surface showed that the films are homogeneous with randomly distributed nanoparticles peaking, whose oragnization increases with the increase of the frequency of the vibrations applied during deposition. SEM microphotographs visualizing this are presented in figure 2. The surface roughness of the films, determined by AFM and presented in Table 1, follows the pattern, observed for the thicker Te films [1], and decreases with the increase of the vibrations frequency. The cracks, observed on the surface of the sample, deposited at input frequency of 4000 Hz (figure 2c) are most probably due to an increased stress in the films, as also shows the micro-strain value from the XRD evaluations of the same sample (Table 1).

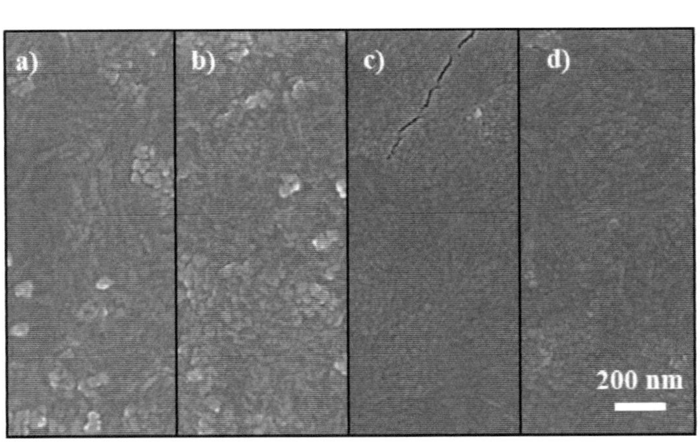

Figure 2. Top view SEM microphotographs of tellurium films deposited at impact of vibrations with input frequencies of a) 0 Hz; b) 50 Hz; c) 4 kHz; d) 10 kHz. Magnification – 50 000ˣ.

Cross-section SEM microphotographs of the films are presented in figure 3. A significant difference in the growth mechanism of the films can be seen, depending on the frequency of the vibrations applied. More particularly, the films deposited by conventional method (0 Hz) and at low frequency of 50 Hz are bulk, homogeneous and without distinct nanoparticles forming the layer, while the films deposited using frequencies in the kilohertz range are consisted of singular nanoribbons growing predominantly towards the z-axis from the very beginning of the film formation. The average widths of the nanoribbons, estimated by averaging of the visible formations from the morphological investigations in both cross-section and top at magnification of 200 000ˣ, are ~ 30 nm for the samples deposited under the impact of 4 kHz vibrations, and ~ 20 nm for those, deposited using 10 kHz input frequency.

In order to reveal a quantitative presentation of these observations, spectroscopic ellipsometry evaluations were performed. In our calculations the samples were modelled either as: (*i*) an isotropic layer with a top overlayer with variable ratio between the bulk and the void parts caused by surface roughness, called Model 1 in the text below, or (*ii*) a single porous layer (Model 2). The optical constants of the tellurium films were described by the Tauc-Lorentz dispersion model [4]. Due to the irregular shape of the pores in the thin films (see figure 3), the Bruggeman approximation [5] was applied in the present work for calculation of the void fraction. The validity of the applied models was determined by a calculation of a common mean square error function (χ^2), which accounts for the discrepancies between the measured (*"meas"* indices) and simulated (*"calc"* indices) data for ψ and Δ for each photon energy value $h\nu$ [4]:

$$\chi^2 = \frac{1}{2N-P-1}\sum_{i=1}^{N}\{[\Psi_{calc}(h\nu) - \Psi_{meas}(h\nu)]^2 + [\Delta_{calc}(h\nu) - \Delta_{meas}(h\nu)]^2\} \quad (1)$$

where N and P are the total number of data points and the number of fitted parameters, respectively.

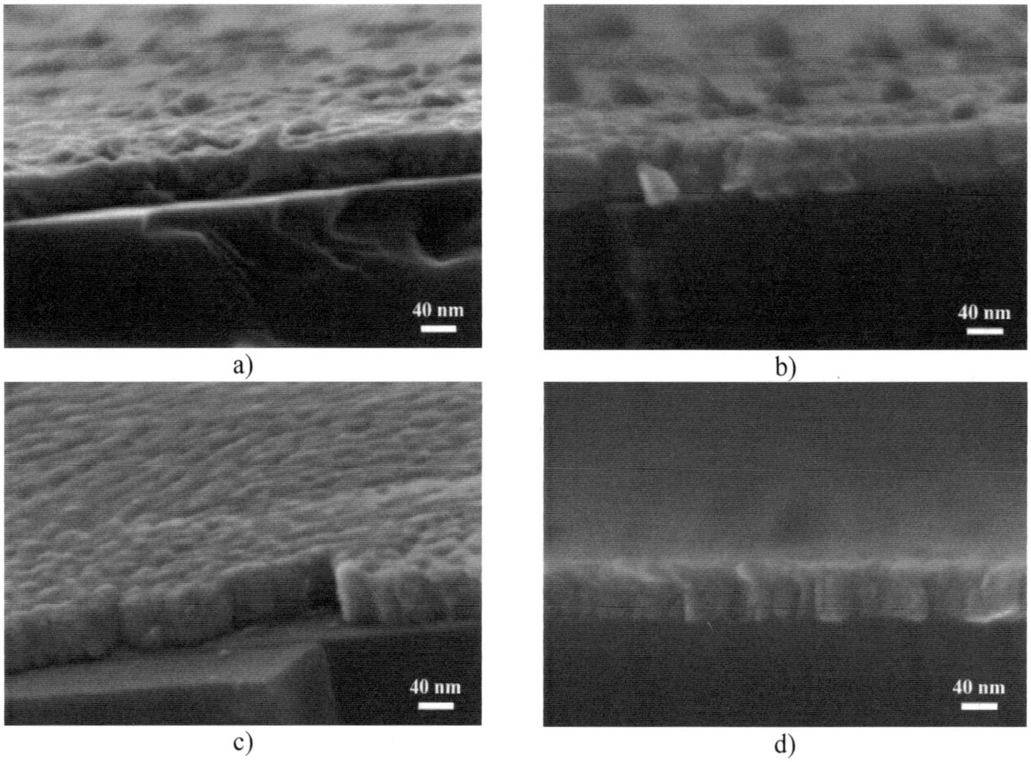

Figure 3. Cross-section SEM microphotographs of Te films deposited at impact of vibrations with input frequencies of a) 0 Hz; b) 50 Hz; c) 4 kHz; d) 10 kHz. Magnification – 200 000^{x}.

The calculations showed that the best fits for the films, deposited by conventional method and by FATEV at input frequency of 50 Hz, were obtained when Model 1 was applied (figure 4a), while these, deposited at high frequency impact, were described better by a single porous layer (Model 2, figure 4b).

The results from these evaluations, as well as the thickness of the films (determined from cross-section SEM observations), and the root mean square surface roughness Sq (determined by AFM, and shown again here for convenience), are presented in Table 2.

Table 2. Results from the SE investigations (applicable model, layer's thickness, bulk/void ratio), average thickness (determined by cross-section SEM microphotographs), and root mean square surface roughness Sq of Te thin films.

Frequency, Hz	Model	d, nm (bulk layer)	d, nm (porous layer)	Te/void ratio, %	d (SEM), nm	Sq, nm
0	1	45.88 ± 4.16	17.89 ± 3.42	50/50	56.94	7.25
50	1	59.90 ± 1.80	17.20 ± 4.30	44/56	60.94	9.77
4000	2	-	69.73 ± 2.72	86/14	63.52	3.39
10000	2	-	68.21 ± 2.69	84/16	66.47	4.22

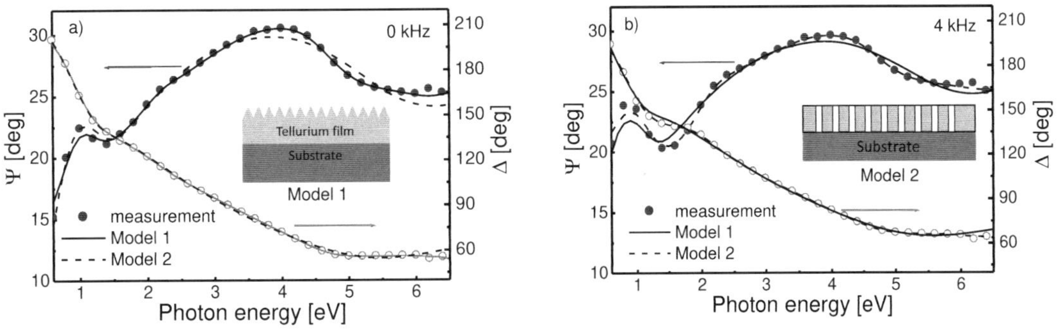

Figure 4. Experimental spectra (symbols) and theoretically calculated values of ψ and Δ for Models 1 and 2 (solid and dashed lines, respectively), for thin Te films deposited by conventional method (a) and at input frequency of 4 kHz (b).

The thicknesses and roughnesses determined by the spectroscopy and microscopy methods correspond well when one takes into account that the whole thickness containing pores is defined in the ellipsometry evaluations, while during the AFM and SEM investigations the thickness and the roughness are determined at an average level.

Looking at the porosity estimations it can be concluded that the thin films can be divided in two groups – (1) bulks with highly porous surface layer, and (2) porous layers through their whole volume with average voids content of ~ 15 %.

The observations of the first group are expected and confirm the effects observed in the thicker Te films, investigated previously by us [1]. On the other hand, the presence of determinable pores in the columnar-like tellurium films, deposited using a frequency impact of 4 and 10 kHz, is a new fact, which can be discussed and analyzed.

It has been stated before [1, 2] that the mechanical vibrations with input frequencies in the audio range cause activation of the substrate surface and formation of undulated surfaces with approximate period of 300 nm when frequencies of 50 and 150 Hz were applied. The change of the preferential growth direction, the presence of distinct nanoparticles forming columnar structure and the presence of empty voids between them from the very beginning of the films deposition, all observed in the present study, give us grounds to assume that the width of the nano-formations is limited by the corresponding resonance period of the substrate, caused by the outer mechanical excitement applied. Taking into account the widths of the nanoribbons of ~ 30 nm and ~ 20 nm respectively for 4 kHz and 10 kHz excitement, one can assume that these input frequencies cause resonance period with approximately the same length on the substrate.

The development of structures with opened nanopores is of constant interest for practical usage starting from sensors and nanosieves (because of the possibility for selective transmission and/or registration of atoms and molecules with defined dimensions), even to optical devices as waveguides and photonic crystals. The observed presence of straightened nanowires with total amount of 15 % pores between them in the present study gives grounds for further investigations on the possibilities for practical usage of both: application of FATEV deposition to other crystalline materials, and possibilities of the FATEV deposited tellurium films. However, it has to be mentioned that the deviation from the straight line observed in both figures 2c,d and 3c,d is most probably due to beginning of influence of the self-resonance of the tellurium nanoribbons, caused by the continuous excitement of the substrate. Probably, this disadvantage could be fixed by decreasing the amplitude of the vibrations during deposition. But, in principle, since these distortions are more distinct for the film, deposited at input frequency of 10 kHz, it can be said that the better sample for purposes which require continuous porosity is that, deposited under 4 kHz frequency impact.

IOP Conf. Series: Journal of Physics: Conf. Series **1186** (2019) 012026 doi:10.1088/1742-6596/1186/1/012026

4. Conclusions

Tellurium films with nominal thicknesses of 60 nm were deposited by frequency assisted thermal deposition in vacuum (FATEV) using input frequencies of the vibrational impact of 0, 50, 4000 and 10000 Hz. It has been found through XRD phase analysis that all deposits correspond to a hexagonal tellurium. The structural analysis of the films revealed that the crystalline lattice parameters of the tellurium phase correspond to these described in the literature with a micro-strain varying from 0 up to 0.2 (for the films, deposited at 4 kHz vibrational impact). The average crystallites size was found to be around 17 nm for all samples. The morphological observations on the samples' surface confirm the FATEV effect determined previously for thicker films - the oragnization of the particles forming the films increases with the increase of the frequency of the vibrations applied during deposition with a peak of the rms surface roughness at input frequency of 50 Hz.

During the observations in cross-section, a change of the growth mechanism of the films was observed when higher frequencies of 4 and 10 kHz were applied. While the samples deposited without excitement impact and under vibrations with a frequency of 50 Hz were consisted of bulk homogeneous bottom layer and a very well defined roughness layer on top, these deposited using frequencies in the kilohertz range were consisted of singular nanoribbons growing predominantly towards the z-axis from the very beginning of the film formation. A quantitative presentation of these observations was offered through spectroscopic ellipsometry evaluations using two models describing the layers' organization and a Bruggeman effective medium approximation. The microscopy observations were confirmed by the ellipsometry estimations – the samples are divided in two groups – (1) bulks with highly porous surface layer, and (2) bottom-to-top porous layers with average voids content of ~ 15 %. The latter observations give grounds to assume that the excitement of the substrate with input frequencies in the kilohertz range causes a width limitation of the nanoformations defined by the corresponding resonance period of the substrate.

References

[1] Hristova-Vasileva T, Bineva I, Dinescu A, Arsova D and Nesheva D 2016 *Surf. Coat. Technol.* **307** 542
[2] Hristova-Vasileva T, Bineva I, Dinescu A, Danila M and Arsova D 2017 *J. Phys.: Conf. Series* **794** 012015
[3] Birkholz M 2005 *Thin film analysis by X-ray scattering* (Wiley-VCH, Weinheim)
[4] Jellison Jr G E and Modine F A 1996 *Appl. Phys. Lett.* **69** 371
[5] Bruggeman D A G 1935 *Ann. Phys.* **24** 636

Spin – coating of SnO_2 thin films

M Ganchev[1], A Katerski[2], S Stankova[1], J S Eensalu[2], P Terziyska[3], R Gergova[1], G Popkirov[1], P Vitanov[1]

[1] Central Laboratory of Solar Energy and New Energy Sources, Bulgarian Academy of Sciences, 72 Tzarigradsko Chaussee blvd, 1784 Sofia, Bulgaria
[2] Laboratory of Thin Film Chemical Technologies, Department of Materials and Environmental Technology, Tallinn University of Technology, Ehitajate tee 5, 19086 Tallinn, Estonia
[3] Solid State Physics Institute, Bulgarian Academy of Sciences, 72 Tzarigradsko Chaussee blvd, 1784 Sofia, Bulgaria

E-mail: mganchev@gmail.com

Abstract. The functional characteristics of thin film planar solar cells are very sensitive to the presence of pores, voids or holes. Thus deposition of compact, dense and uniform layers is essential in that technology. These issues are especially important in organometalic perovskite solar cells where the active layers are in the nano metric scale. In this emerging technology the reduction of cost-efficiency ratio is a serious barrier for commercialization. In this work we have investigated different parameters of the process of deposition of SnO_2 thin films by spin coatings which influence its compactness and impenetrability. Influences of the thermal treatments on the properties of the films were investigated by UV – vis spectrometry, X-ray diffractometry (XRD), Scanning Electron Microscopy (SEM), Atomic Force Microscopy AFM), X-Ray Photo-electron spectroscopy (XPS) and Spectral Elipsometry (SE). Finally the coverage ability of the SnO_2 films was checked by Cyclic Voltammetry (CV) in aqueous redox solution.

1. Introduction

Organometallic perovskite photovoltaics rival chalcopyrites and silicon based devices and appears to be most promising technology for planar thin film solar cells on flexible substrates. Currently, the record efficiency for perovskite solar cells jumped over 23% [1, 2] and there is still some gain to achieve by further optimization of materials properties [2], processing or interface engineering [3]. The most popular electron selective layer (ESL) in these devices is the meso-porous TiO_2 owing to its good photoelectric properties [4]. However, the electron mobility of TiO_2 is far below the mobility of commonly used hole – transport layers (HTL) resulting in accumulation of charge at the interface and decrease of efficiency [5]. Recently, a more prospective candidate to replace TiO_2 was presented, SnO_2 having higher electron mobility and better match to the energy levels of organometalic perovskite, yielding solar cells with efficiencies of up to 21 % [6, 7].

In the present work we report the application of the spin coating method for deposition of dense, voids – free SnO_2 thin films and investigation of its properties in order to apply it as electron selective layer in organometallic perovskite solar cells (PSC).

2. Experimental

Solutions of $SnCl_4.5H_2O$ (Acros organics) in concentrations 0.05 or 0.2 M in Isopropyl alcohol – (IPA Macron®) were used for spin coating of the films onto quartz substrates, polished monocrystalline Silicon wafers or glass substrates, covered by conductive $TiO_2/Pt/TiO_2$ sandwich layer. Working volume was 40 μl, dispensed onto the rotating sample with a size of 15mm x 20 mm for 30 s at rotation speed 3500 min^{-1}. The reaction of formation of the desired films is:

$$SnCl_4.5H_2O \rightarrow SnO_2 + 4HCl\uparrow + 3H_2O\uparrow \quad (1)$$

XPS analyses were performed on a Kratos AXIS Supra spectrometer with an un-monochromatic Al X-ray source under the vacuum better than 10^{-8} Pa at 90 degree take-off angle. All the spectra have been charge-corrected by taking C1s peak (Binding Energy (BE) = 285.0 eV) as reference. Film thickness, spectral distribution of the refractive index and the extinction coefficient were measured on an automatic spectroscopic ellipsometer - M2000D, Woollam (Santa Barbara, CA). Diffractometer Rigacu Ultima IV with Cu Kα radiation (λ = 1.5418 Å) at 40 kV accelerating voltage was used in steps 0.01° for 2 s in an angular range 10 to 80° 2θ. Atomic Force Microscopy (AFM) was performed by "Easyscan 2" (Nanosurf®) TAP 190 – Al G. Optical spectroscopy was carried out on UV – VIS – NIR spectrometer JASCO – V 670 in the range 190 - 1500 nm. Cyclic voltammetry (CV) analysis was done in a three electrode configuration in dark. The SnO_2 working electrode was positioned in the back side of the cell squeezed against round window (diameter 4.5 mm) to define equal and constant active surface. Pair of platinum wires were used as electrodes - counter and reference (redox), respectively. The redox electrolyte was 1mM $K_3Fe(CN)_6$+ 1mM $K_4Fe(CN)_6$ in aqueous 0.5 M KCl, intensively purged with nitrogen to remove dissolved oxygen. The potentials are referred versus redox potential of the solution, estimated as E_{redox} = 0.280 V (vs AgCl/3M KCl).

3. Results and discussion

3.1. Deposition

The as deposited films were dried at 100 °C for 1 hour and afterwards have been annealed at temperatures 120 °C, 140 °C, 160 °C and 180 °C. To form tick layers the deposition was repeated several times followed in every case of drying and annealing in related temperature. Thickness measurements of the layers were carried out by ellipsometric measurements. Figure 1 presents the final thickness of the layers depending on the number of spin coated films. In the limits of the error it is seen that every succeeding film adds about 5 nm to the common structure.

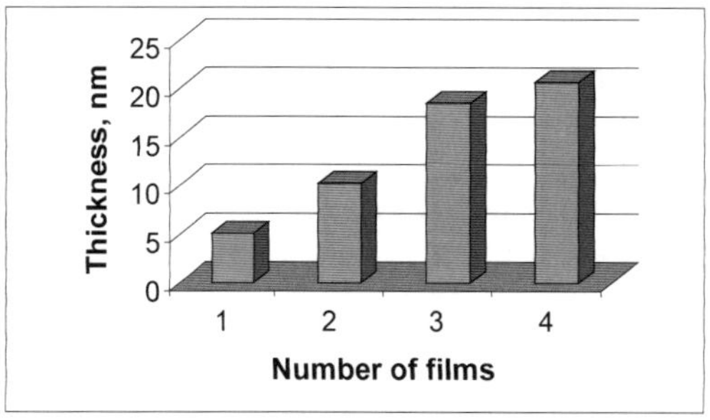

Figure 1. Dependence of layers' thickness on number of deposited films, annealed at 140 ° C for 1 hour.

3.1.1 XPS analysis

XPS analysis data of an SnO_2 film deposited from solution of 0.05 M Sn^{4+} in IPA and annealed at

140 °C in air are presented on figure 2. XPS survey scan in figure 2a represent detected emission signals from tin, oxygen, carbon and trace amounts of chlorine and platinum respectively from the surface of the SnO_2 sample. The presence of minor quantity of Pt obviously originates from the substrate.

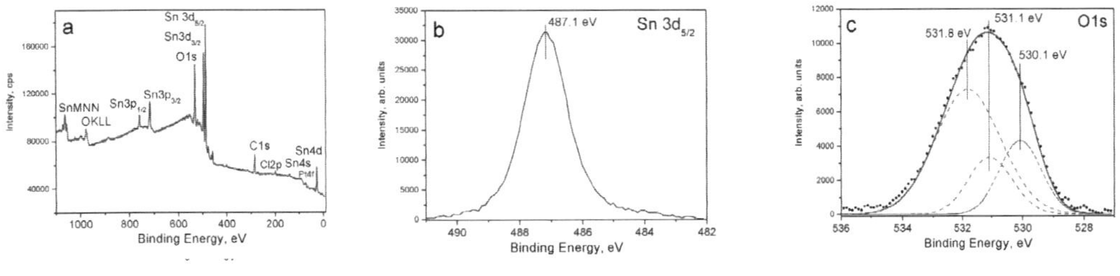

Figure 2. XPS analysis of SnO_2 thin film, annealed at 140 °C: a – survey; b – Sn $3d_{5/2}$ core level; c – O 1s core level.

Figure 2b shows the Sn $3d_{5/2}$ core level peak which is symmetric and centered at Binding Energy (BE) 487.1 eV corresponding to Sn^{4+} ionic state bounded by oxygen as a SnO_2 compound in the investigated samples [8, 9, 10]. Figure 2c represents the O1s core level spectrum which is asymmetric. The spectrum was deconvoluted using Lorentzian-Gaussian (pseudo-Voigt function) fitting analysis. In the O1s core level spectrum, the peak at BE of 530.1 eV is ascribed to Me-O [11, 12], the peak at BE of 531.1 eV is ascribed to the oxygen deficient region i.e. oxygen vacancies (Vo) [11, 12] and the peak at BE of 531.8 eV is ascribed to the surface hydroxyl groups, which are connected to the surface contaminations [11].

3.2. Structural properties

Figure 3 shows the XRD patterns of SnO_2 films with different thickness, deposited on polished Si monocrystalline substrates and annealed at 120 °C for 1 hour.

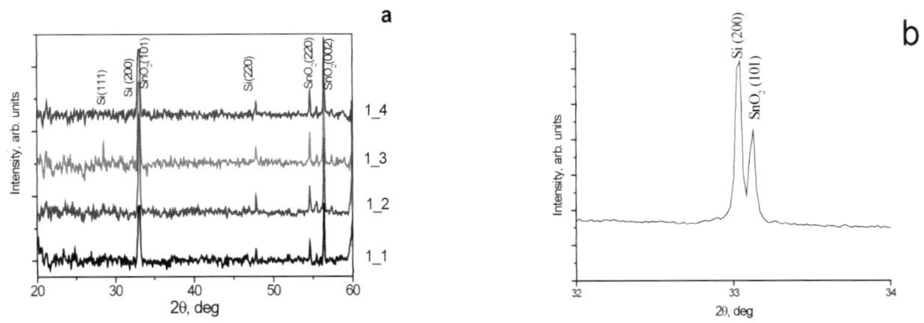

(a) (b)

Figure 3. a - XRD patterns of thin films SnO_2, annealed at temperature 120 °C with different thickness: 1 – 5 nm; 2 – 10 nm; 3 – 15 nm; 4 – 20 nm. b - Pattern between 32 ÷ 34 degrees 2θ for distinction of reflexes of Si (200) and SnO_2 (101).

Besides the reflexes of the substrate Si (111), Si (200) and Si (220) peaks indexed as planes (002), (220) for tetragonal SnO_2 are seen [13]. Detailed analysis of the patterns revealed twinning of the reflex at 33° angle 2θ. First of them, at 33.04°, could be attributed to the substrate Si (200) [14,15,16] and the second one, at 33.12° should be attributed to the (101) plane reflex from SnO_2. The results are close to the observations reported in [17] for similar SnO_2 thin films.

3.3. Optical properties

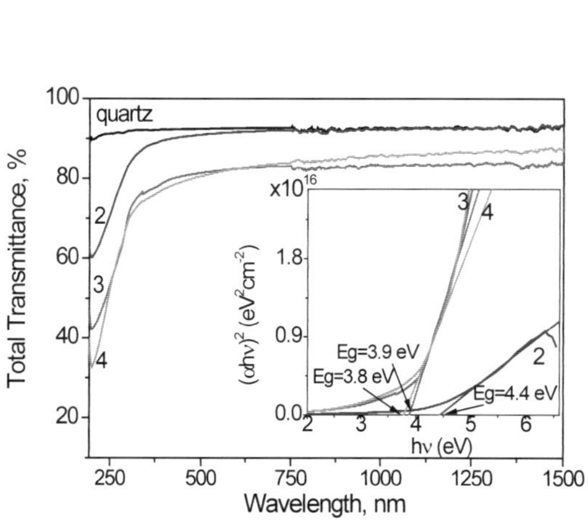

Figure 4. Total transmittance spectra of bare quartz substrate and 2, 3 and 4 SnO_2 thin films sequence deposited onto the same quartz substrate and annealed at 180 °C. The inset shows corresponding bandgap determination on a Tauc plot.

Figure 4 presents total transmittance spectra from bare quartz substrate and quartz substrate covered with 2, 3 and 4 SnO_2 thin films annealed at 180 °C. We deposited SnO_2 thin film onto quartz substrate due to needs of transparency in the SnO_2 absorption edge region. Presented deposited SnO_2 thin film demonstrated a high optical transparency, with an average transmittance around 90 % in the visible range (400–800 nm), which indicates their applicability in transparent electronics. The absence of interference fringes on the spectrum of the film shows that it is extremely thin. The corresponding optical direct bandgap energy of all deposited SnO_2 thin film was calculated after taken into consideration the reflectivity, absorption coefficient, and film thickness applying expression (2) on a Tauc plot:

$$(\alpha h\nu) = A(h\nu - Eg)1/n \quad (2)$$

,where hν is the photon energy, A is a constant called critical absorption, n = 0.5 for direct transition and Eg is the optical band gap. The values of Eg for deposited SnO_2 films were obtained by plotting (αhν)2 against the photon energy (hν) and extrapolating the linear part of the curve to zero on the energy axis as shown in the inset of figure 4. The band gaps of SnO_2 films annealed at 180 °C were determined to be from 3.8 to 4.4 eV which is in agreement with previously reported results [17]. Gradually increasing of the optical band gap is attributable to the quantum size effect, as reported previously in several different systems [18, 19].

3.4. Spectral elipsometry

Figure 5 shows the spectral distribution of refractive index **n** and extinction coefficient **k** obtained by spectral elipsometry analysis of spin coated SnO_2 thin films on glass substrates.

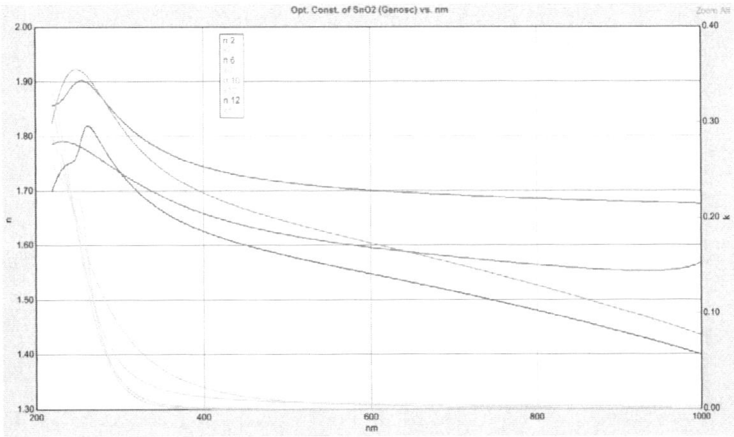

Figure 5. Dispersion Refractive index (n) and extinction coefficient (k) of SnO_2 thin films, deposited by spin coating on TiO_2/Pt/TiO_2 covered glass substrates.

As can be seen for curves of extinction – Figure 5 the region for strong absorption from the curves of extinction coefficient is before 300 nm which is in agreement with the data for optical absorbance in figure 4.

3.5. Atomic force microscopy

Figure 6 presents the AFM image of an SnO_2 thin film sample, deposited with solution 0.05 M $SnCl_4.5H_2O$ in IPA and annealed at 160 °C for 1 hour. Measured root mean square (RMS) of the surface roughness is 3.9 nm. There are no voids or cracks on the layers and these values characterize the spin coated SnO_2 thin films as flat and uniform.

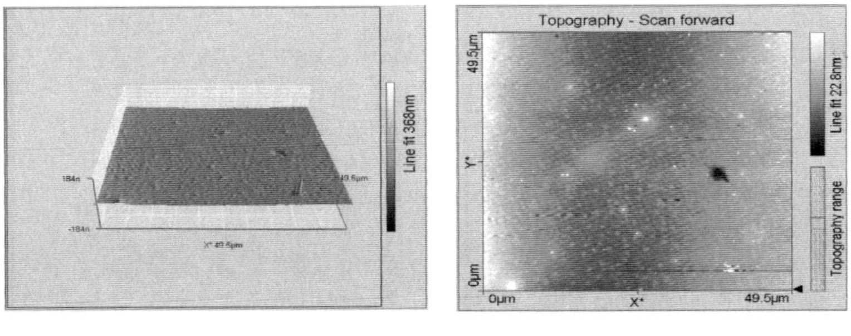

Figure 6. Atomic force microscopy images of SnO_2 layers, deposited from solution 0.05 M $SnCl_4.5 H_2O$ in IPA and annealed at 160 °C for 1 hour.

3.6. Electrical properties

Thin compact SnO_2 layers are used in perovskite solar cells to extract electrons from the absorber and prevent recombination with the holes. Characterization of these properties of the spin coated SnO_2 thin films was evaluated with Cyclic Voltammetry (CV) measurements in model redox solution of 1mM $K_4Fe(CN)_6$ + 1mM $K_3Fe(CN)_6$ in 0.5 M KCl, with redox potential E^{redox} = 0.2803 V (vs AgCl/3M KCl) electrolyte, deaerated by bubbling N_2. Figure 7 presents the CV curves for redox solution, measured by Pt electrode, background current, measured on conductive substrate electrode – **a**, CV curve of an SnO_2 layer deposited by 0.2 M solution and dried at 100 °C – **b** and CV curve of the SnO_2 layer, dried and annealed at 140 °C - **c**.

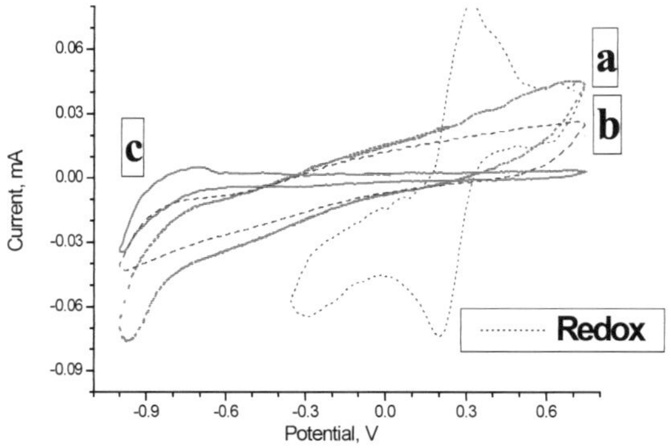

Figure 7. Cyclic voltammograms of samples SnO_2 annealed at 140 oC for 1 hour in 1mM $K_4Fe(CN)_6$ + 1mM $K_3Fe(CN)_6$ in 0.5 M KCl solution; a – current on the substrate; b – current on the film SnO_2 dried at 100 °C for 1 hour; c – current on the film SnO_2 dried and annealed at 140°C for 1 hour.

As can be seen the current for the redox couple $[Fe(CN)_6]^{4-}$ / $[Fe(CN)_6]^{3-}$ on the metal Pt electrode is highest in the interval near redox potential. The current amplitude on the conductive substrate layer $TiO_2/Pt/TiO_2$ - **a** decreases in wide potentials range, between -1000 mV ÷ 750 mV (vs. E_{redox}). The same electrode, covered by spin coated SnO_2 film, dried at 100 °C for 1 hour shows lower current in the same conditions, whereas for sample **c** – thin film SnO_2 annealed at 140 °C current amplitude decreased sharply in order of magnitude.

4. Conclusions

By spin coating, we deposited flat and uniform void- and crack-free SnO_2 films (Eg 3.8-4.4 eV depending on the thickness) with RMS square mean roughness 3.9 nm after annealing at 140 – 160 °C according to AFM. XPS analysis has shown characteristic signals for Sn^{4+} and oxygen in the form of O^{2-}. The XRD patterns presented a tetragonal fine structure of SnO_2. For determination of the phase of tetragonal SnO_2 was used reflex near 33 ° 2θ, having in mind the complexity for XRD analysis of reflexes from thin films on mono - crystalline silicon in this vicinity of angles 2θ [14]. In our case the reflex is well defined away of those for (200) plane of Si followed of the pair of reflexes from (220) and (002) planes of SnO_2. The electron selective ability of SnO_2 thin films were assessed by Cycling Voltammetry measurements and were shown that layers deposited from 0.2 M $SnCl_4.5H_2O$ in IPA and annealed at 140 – 160 °C presented noticeable current far away in the cathodic area and negligible anodic current.

Acknowledgements

Research is sponsored by the Bulgarian Scientific Found under Contract No. T02-12, Bilateral program between Bulgarian Academy of Sciences and Estonian Academy of Sciences – 2017 - 2020 and Program for Young Scientists in Bulgarian Academy of Sciences - 2017, DFNP-17-128, the Estonian Ministry of Education and Research project IUT19-4 and the European Regional Development Fund project "Center of Excellence" TK141.

References:

[1] NREL efficiency chart. https://www.nrel.gov/pv/assets/images/efficiency-chart-20180716.jpg (2018)
[2] D.Yang.R.Yang, K.Wang, C.Wu, X.Zhu, J.Feng, X.Ren, G.Fang, S.Priya, S.(F.) Liu, Nat.Comm. (2018) 9:2335; DOI 10.1038/s41467-018-05760-x/

IOP Conf. Series: Journal of Physics: Conf. Series **1186** (2019) 012027 doi:10.1088/1742-6596/1186/1/012027

[3] K.T.Cho, S.Paek, G.Granchini, R.Roldan Carmona, P.Gao, Y.H.Lee, M.K.Nazeeruddin, Energy Environ.Sci., 2017, DOI:10.1039/C6EE03182J

[4] T.Sing,T. Miyasaka, Adv.Energy.Mater.8 1700677 (2018)

[5] W. Tress et al. Energy Environ.Sci. 8, 995 – 1004 (2015)

[6] J.P.Correa Baena, L.Steier, W.Tress, M.Saliba, S.Neutzner, T.Matsui, F.Giordano, T.J.Jacobsson, A.R.Srimath Kandada , S.M.Zakeeruddin, A.Petrozza, A.Abate, M.K.Nazeeruddin, M.Gratzel, A.Hagfeldt, Energy Environ.Sci., 2015, 8, 2928 – 2934

[7] E.H.Anaraki, A.Kermanpur, L.Steier, K.Domanski, T.Matsui, W.Tress, M.Saliba, A.Abate, M.Gratzel, A.Hagfeldt, J.P.Correa Baena, Energy Environ.Sci., 2016, DOI:10.1039/c6ee02390h

[8] J.-M.Themlin, M.Chtaib, L.Henrard, P.Lambin, J.Darville, J.-M. Gilles, Phys.Rev. B. V 46, 4, 15.07.1992-II

[9] M.A.Stranic, A.Moskwa, Surf.Sci.Spec. 2 (1993); doi:10.1116/1.1247723

[10] Dilip L. Kamble, Namdev S. Harale, Vithoba L. Patil, Pramod S. Patil, Laxman D. Kadam, Journal of Analytical and Applied Pyrolysis 127, 38–46 (2017)

[11] I. Gromyko, M. Krunks, T. Dedova, A. Katerski, D. Klauson, I. Oja Acik, Applied Surface Science 405, 521–528 (2017)

[12] Bathula Babu, I. Neelakanta Reddy, Kisoo Yoo, Dongseob Kim, Jaesool Shim, Materials Letters 221, 211–215 (2018)

[13] ICDD PDF - JCPDS file card № 41-1445

[14] P.Zaumseil, J.Appl.Cryst.(2015), 48, 528 – 532

[15] A.R.Chaudhuri, A.Fissel, H.J.Osten, Appl.Phys.Lett. 104, 012906 (2014)

[16] A.R.Chaudhuri, A.Fissel, H.J.Osten, Appl.Phys.Lett. 106, 149903 (2015)

[17] G.Sanon, R.Rup, A.A.Mansingh, Thin Solid Films, 190 (1990) 287 – 301

[18] S.F. Rus a, T.Z. Ward, A. Herklotz, Thin Solid Films 615 (2016) 103–106

[19] Mi Zhang, Maji Xu, Mingkai Li, Qingfeng Zhang, Yinmei Lu, Jingwen Chen, Ming Li, Jiangnan Dai, Changqing Chen, Yunbin He, Applied Surface Science 423 (2017) 611–618

IOP Conf. Series: Journal of Physics: Conf. Series **1186** (2019) 012028 doi:10.1088/1742-6596/1186/1/012028

Experimental investigations on ultrashort laser ablation for micro and nanomachining of materials

Katarzyna Garasz[1], Marek Kocik[1], Todor Petrov[2]

[1]The Szewalski Institute of Fluid-Flow Machinery, Polish Academy of Sciences, Fiszera 14, 80-231 Gdańsk, Poland

[2]Georgi Nadjakov Institute of Solid State Physics, Bulgarian Academy of Sciences, Tzarigradsko Chaussee Blvd.72, 1784 Sofia, Bulgaria

kgarasz@imp.gda.pl

Abstract. With the advantage of ultrashort pulse width and ultrahigh peak power, subpicosecond lasers can provide extremely precise machining, and therefore has been widely applied in micro and nanoscale manufacturing. We present the results of the experimental parametric study on efficiency, accuracy and quality of subpicosecond laser micromachining of materials, that have a potential for application in innovative devices and technologies. The laser micromachining process was performed with a Yb:KYW subpicosecond laser and a Ti:S femtosecond laser. The study contains a comparison between the two methods, and results analysis using advanced metrology techniques, such as 3D microscopy.

1. Introduction

The so-called ultrafast or ultrashort pulse lasers provide unique material processing possibilities. It is due to pulse duration, which is shorter, than the conduction time of the machined material. That offers an opportunity for a cold ablation, which usually appears with pulses shorter than 10 ps [1].

Ultrashort pulse durations became available since the development of mode-locked lasers and the chirped pulse amplification (CPA) technique [2,3]. Pulse energies in the range of milljoules became easily obtained via CPA. Since then, a numerous groups studied the subject of ultrafast ablation processes analytically and numerically. Considering the energy transfer from electrons to the lattice occurs via carrier-phonon scattering on a timescale estimated from several hundred femtoseconds to a few picoseconds, the physics of the ablation process in a subpicosecond regime differs from the ablation caused by longer laser pulses [4].

According to Liu et al. [5], who first studied the ablation dynamics with pulse width varying from 10 ns to 100 fs, the ablation threshold is lowering with a decreasing pulse duration. The low ablation threshold with ultrashort lasers translates into near zero heat affected zone (HAZ), minimal burring and debris, very high accuracy and the ability to produce small, precise features in target materials. However, it must be carefully considered, if the micromachining efficiency and quality always improves with the shortening of the laser pulse. Femtosecond (10^{-15}) pulse lasers targeting development of new technology, appeared in research labs years ago. The ability to process almost any material with almost no side effects and no post processing tools, enabled innovation in the area of micro and nanotechnology. Still, these complex and difficult to calibrate systems, are considered unstable and not reliable enough for fully commercial use. The efficiency of the micromachining process depends, on the one hand, on the laser radiation parameters, such as pulse duration, wavelength, pulse energy, repetition rate or irradiation time, and on the other hand on the target material properties.

In this article, we consider two categories of ultrafast lasers:

- A lab-designed subpicosecond Yb:KYW laser (L1), that emits 650 fs pulses.
- A commercial-type femtosecond Ti:S laser (L2), that emits 35 fs pulses.

We have undertaken the problem of subpicosecond laser interactions with matter, under variable laser pulse parameters and considering various materials (metals and plastics). In laser irradiated material samples, a various features, such as crater profiles, ablated volume, local changes in crystallography and chemistry, surface modifications are related to various dynamical mechanisms and the ablation threshold and ablation rates can be readily obtained from the analysis of the final state of material [6]. Therefore, a detailed analysis of the irradiated materials were performed, using various microscopy techniques.

2. Experimental setup

The laser micromachining of materials was carried out using two abovementioned lasers. One, available at the Centre for plasma and laser engineering, at the Institute of Fluid-Flow Machinery PAS (Poland) and another, available at the Laboratory of metal vapor lasers at the Institute of Solid State Physics BAS (Bulgaria). Figure 1 presents the micromachining experimental setups for both lasers.

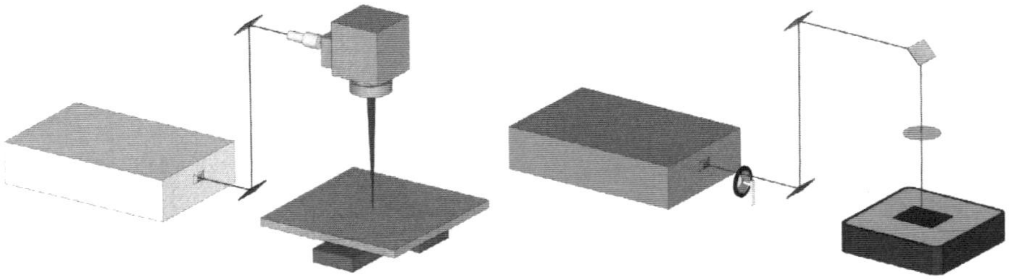

Figure 1. The Experimental setup for L1 (left) and L2 (right) ultrafast laser micromachining.

The subpicosecond Yb:KYW laser offers 100-400 kHz repetition rates and power level up to 15 W, as long as the possibility of generating three different wavelengths of UV, VIS, IR range, of which 1030 nm and 345 nm were used as stable for the the micromachining processes. The femtosecond TiS laser provides a 1 kHz beam with a tunable wavelength from 266 nm to 800 nm. Three different wavelenghts from these range were used. The summary of micromachining parameters is presented in Table 1.

From the laser output, the beam is passing through the optical collimator and focusing system, containing of mirrors and lenses adequate for the applied laser wavelength. The concept of micromachining implies, that the output laser beam, after passing through the optical collimator, will be directed into the positioning system. To move and focus the UV laser beam on the surface of the workpiece, an optical scanner with focusing lens was used for L1 setup. The scanner is equipped with two galvanometric mirrors that deflect the laser beam, making it possible to move according to a given pattern in the XY plane. The samples are put on the z-axis adjustable table, directly beneath the scanning area, on the exact level of the laser beam focus [6].

Table 1. The micromachining parameters.

	Yb:KYW (L1)	Ti:S (L2)
Wavelength	1030/345 nm	266/532/800 nm
Repetition rate	100 – 400 kHz	1 kHz
Output power	< 15 W	< 15 W
Pulse width	650 fs	35 fs
Scanning speed	0,7 mm/s – 13 mm/s	7 mm/s – 100 mm/s

For other wavelengths, a galvo-scanner could not be applied, due to optical elements limitations. Instead, two high-precision linear motors were provided to move the sample in the XY plane and a separate focusing optics, dedicated for the given wavelength range, is used. The L2 setup is using the

Aerotech planar table for the same purpose. Varying the laser pulse energy during the experiment is realized differently in each setup. In the L1 case, the laser output power is dependend on the current set to the amplifier. In the L2 setup, a variable ND filter is used to adjust the pulse energy.

3. Results

The section presents visual and numerical results of the performed experiments. Using L1 and L2 lasers working in a configuration as in Figure 1, micromachining of materials was performed with variable wavelength, repetition rate, pulse energy and scanning velocity. During the process, in each case the target material was scribed with approx. 1 cm long, straight line. Stainless steel, aluminum, copper, nickel, zinc, brass, nylon and polypropylene were used as target materials. The Nikon Eclipse microscope was used to measure a post-machining crater dimension and HAZ range on the surface of each sample. A Hirox KH8700 3D microscope was applied to visually judge the surface of the irradiated material and examine a cross-section profile, i.e. measure the depth of the crater.

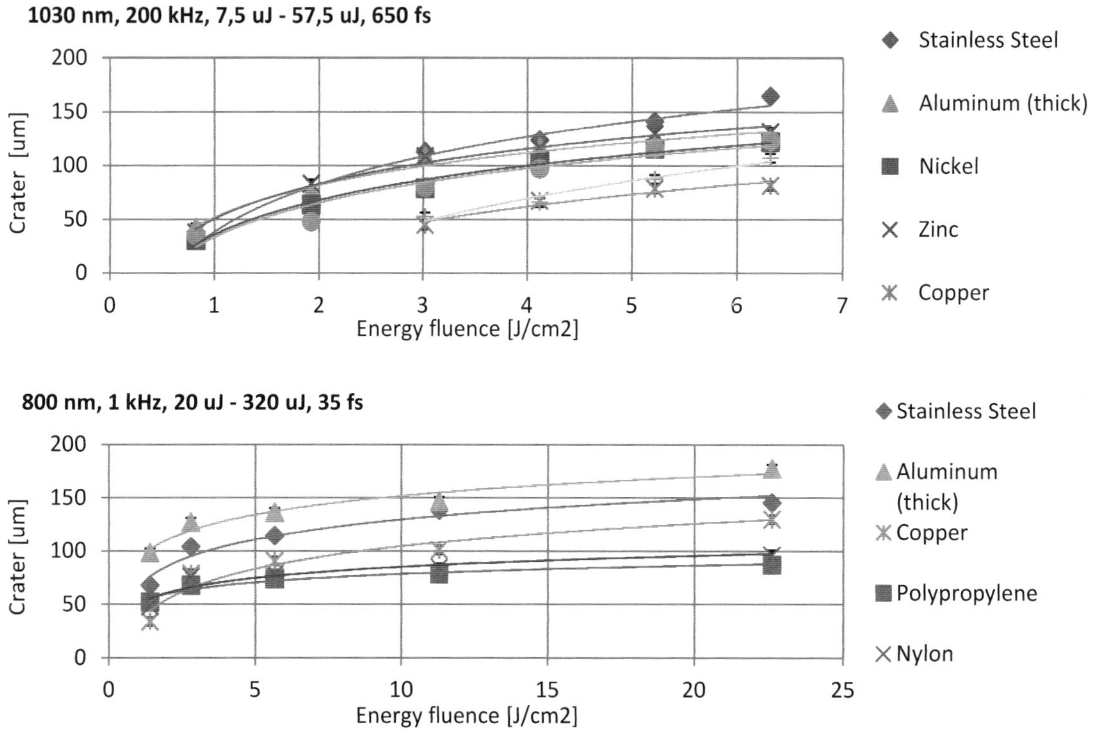

Figure 2. Numerical results of L1 and L2 laser micromaching of materials with variable output power.

The results revealed very good efficiency and accuracy of the micromachining process. Shortening the duration of the laser pulse to the range of tens of femtoseconds results in near zero HAZ, which translates into the high quality of the cut and no unwanted effects in the surrounding area (i.e. cracks, deformations, debris). However, the heat transfer in the subpicosecond (hundreds of fs) range is also very insignificant and does not affect the crater itself, resulting in a clean and precise cut.

The crater diameter is increasing with the increase of laser fluence and a decrease of pulse repetition rate. Decreasing of the repetition rate, however, will affect the depth of the crater, whereas changing the output power with a repetition rate remaining constant, makes much smaller difference in the crater depth. Both lasers offer wavelength choices of IR, VIS, UV. Certain wavelengths work best for specific materials. One can select the wavelength based on a particular feature size required, the smallest focus spot size achievable is directly related to the wavelength. The use and argument for different wavelengths matched to different materials is less obvious for shorter pulses. The normal material absorption dependency on wavelength would no longer apply, due to the multiphoton absorption domination.

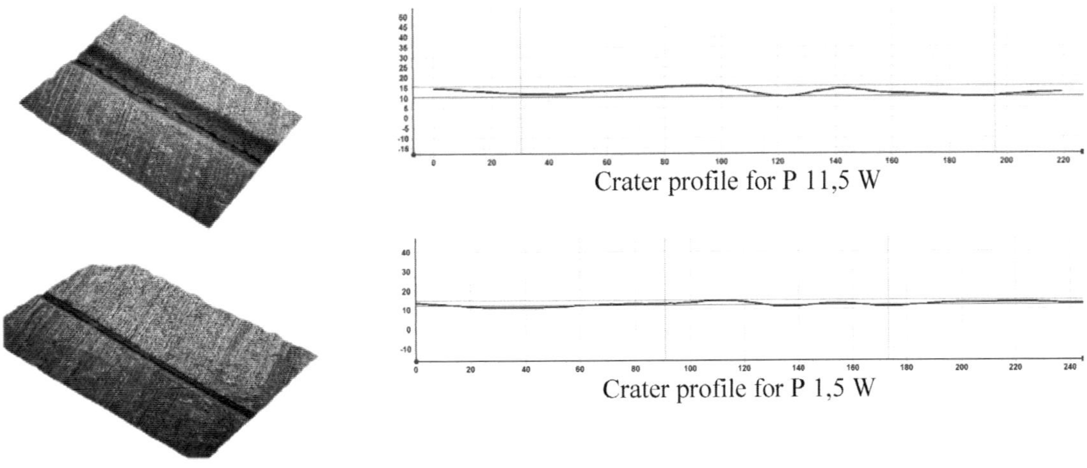

Figure 3. 3D analysis results of L1 laser micromaching of aluminum with variable output power.

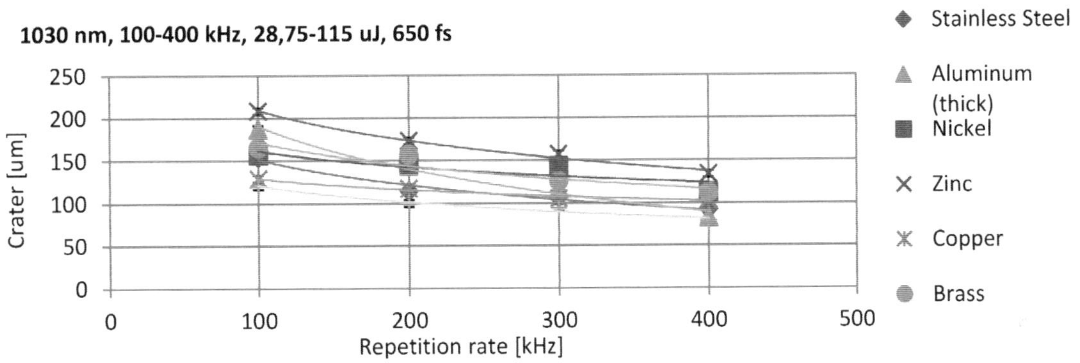

Figure 4. Numerical results of L1 laser micromaching of materials with variable pulse repetition rate.

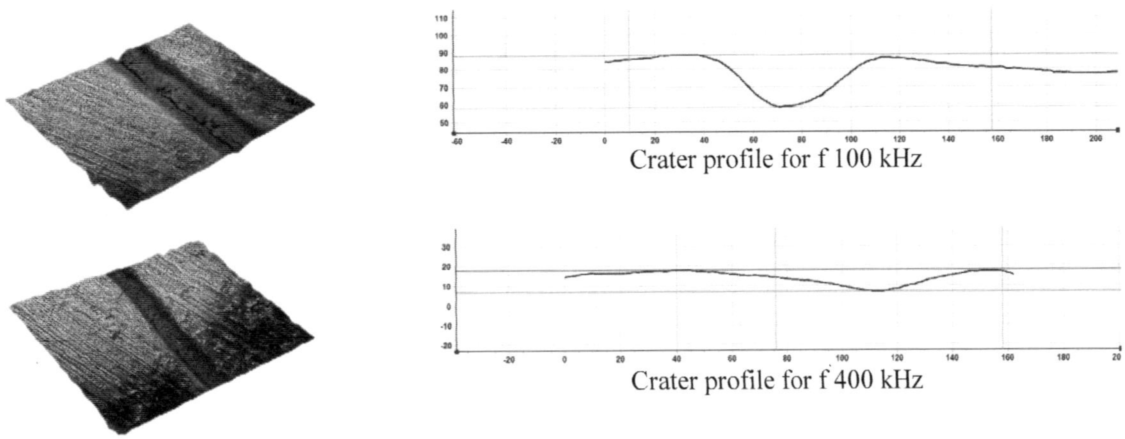

Figure 5. 3D analysis results of L1 laser micromaching of aluminum with variable repetition rate.

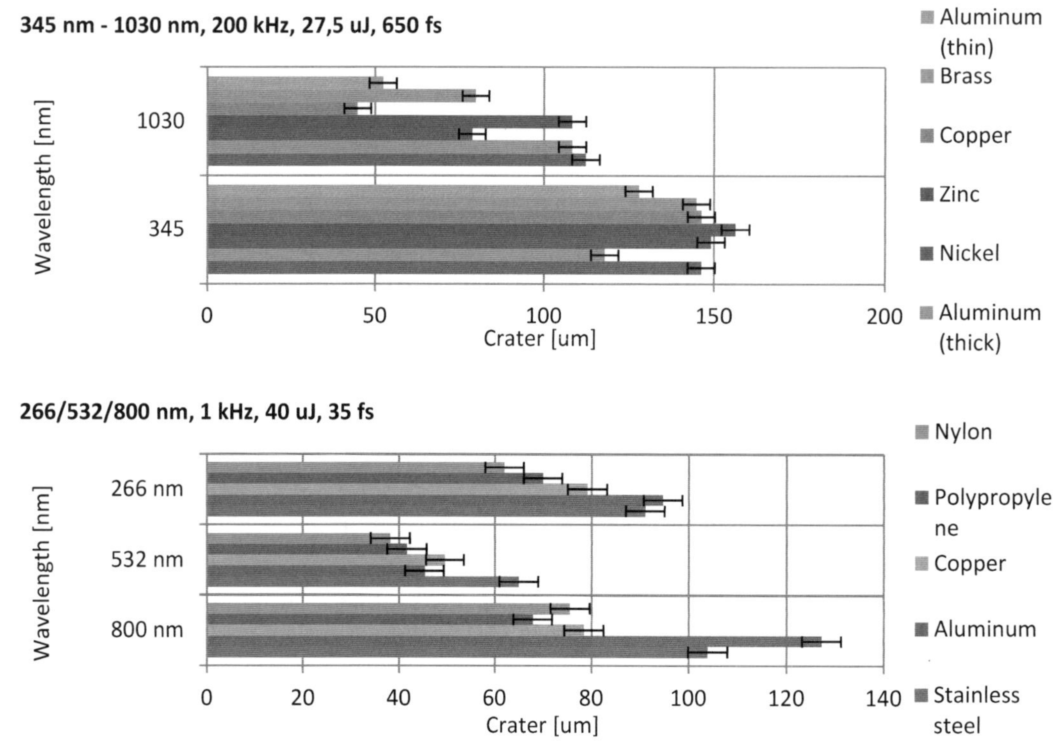

Figure 6. The Experimental setup for L1 (left) and L2 (right) ultrafast laser micromachining.

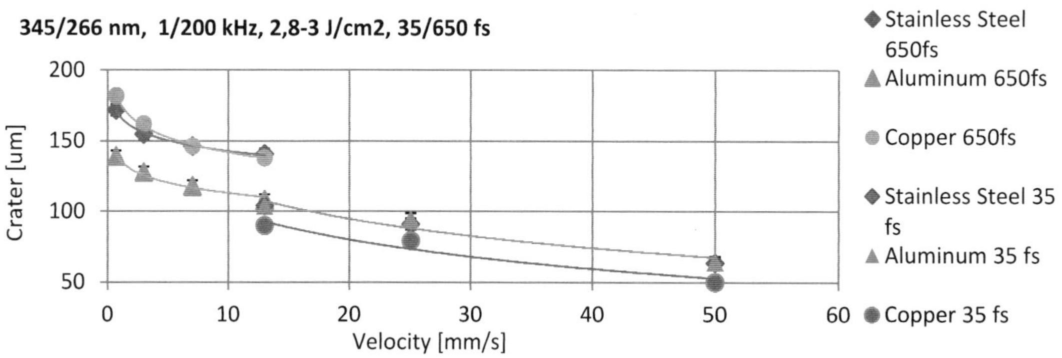

Figure 7. Numerical results of L1 and L2 laser micromaching of materials with variable velocity.

The crater diameter increases with decreasing scanning speed, but the depth of the crater is not significantly affected. Higher micromachining velocity means less time for laser-matter interaction and heat transfer. An optimal velocity must be determined for each material, depending on its parameters. It is especially visible with copper, which has a specific heat of higher value than other materials examined. For shorter pulses, faster micromachining rates were applied. The material samples seem to be less affected by scanning speed in the femtosecond regime.

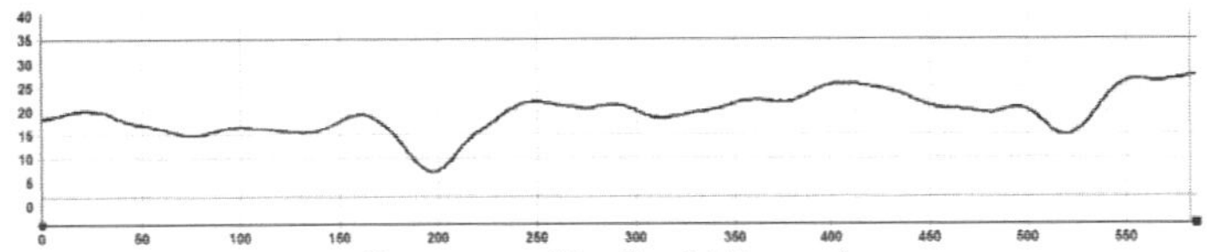

Crater profile for V 7 mm/s

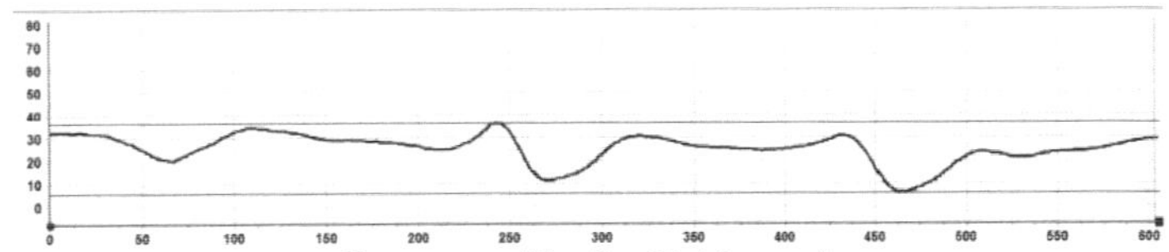

Crater profile for V 13 mm/s

Figure 8. 3D analysis results of L1 laser micromaching of aluminum with variable velocity.

4. Conclusions

Longer (subpicosecond) laser pulses gave more predictable results. The laser fluence – crater size characteristics are more coherent in this case. However, the 650 fs micromachining reflects attributes of both, femtosecond and picosecond regime.

The experimental and theoretical investigations so far, lead to some improvement in the physical understanding of the ultrafast laser ablation phenomena. The dynamics of the ablation process can be roughly divided into several stages: energy absorption, energy transfer to the lattice and subsequent material removal. The first step of the ablation process is deposition of energy into the material. The primary absorption mechanism involves excitation of electrons from the valance to the conduction band and free carrier absorption. The interband excitation can occur through multiphoton and avalanche ionization, with high enough laser intensity. Nonlinear absorption is very important in femtosecond interaction due to the high intensity of the incoming radiation [7].

During the laser-mater interaction all of the processes occur simultaneously and it is difficult to estimate the contribution of each one. Due to the complexity of the process, it is also difficult to calculate or measure the effective penetration depth of the radiation. Therefore, the pulses of subpicosecond regime should be considered "good enough" for micro and nanomachining of materials.

References

[1] Weingarten K 2009 *Laser Technik Journal* **6**, 3, 51-54
[2] Strickland D, Mourou G 1985 The Dawn of Ultrafast Science and Technology at the University of Rochester Opt. Commun. **56,** 219, 272-279
[3] Mourou G 1997 Appl. Phys. B **65,** 205-211
[4] Chichkov B, Momma C, Nolte S, Van Alvensleben F, Tunnerman A 1996 *Applied Physics A* **63,** 109-115
[5] Liu X, Du D, Gourou G 1997 IEEE 1. *Quantum Electronics* **33**, 1706, 6234-6240
[6] Garasz K, Tański M, Kocik M, Mizeraczyk J 2015 3rd International Conference on Photonics, Optics and Laser Technology PHOTOPTICS - Doctoral Consortium, 3-9
[7] Sokolowski-Tinten K, Bialkowski L, Doing M, Cavalieri A, Von der Linde D, Oparin A, Mayer-ter-Vehn J, Anisimov S 1998 *Physical Review Letters* **81**, 224-227

IOP Conf. Series: Journal of Physics: Conf. Series **1186** (2019) 012029 doi:10.1088/1742-6596/1186/1/012029

Femtosecond-laser-assisted LIPSS generation on chitosan/hydroxyapatite thin layers for biomedical applications

I Bliznakova[1], A Daskalova[1], A Zhelyazkova[1], L Angelova[1], A Trifonov[2], I Buchvarov[2]

[1]Institute of Electronics, Bulgarian Academy of Sciences
72, Tsarigradsko Chaussee blvd., 1784 Sofia, Bulgaria
[2]Faculty of Physics, St. Kliment Ohridski University of Sofia
5, J. Bourchier blvd., 1164 Sofia, Bulgaria

e-mail: irbliznakova@abv.bg

Abstract. In this work, chitosan/hydroxyapatite (Ch/HAp) thin composites were synthesized and examined before and after femtosecond laser irradiation in order to investigate the substrate response and surface changes occurring during the process. The zones of interaction formed after treatment by fs laser pulses were analysed by scanning electron microscopy (SEM) and Fourier transform infrared spectroscopy (FTIR). The results revealed that the laser processing of Ch/HAp thin films leads to generation of well-defined laser-induced periodic surface structures (LIPSS) for a laser fluence $F = 2.07\ \mathrm{J/cm^2}$ at 800 nm wavelength with pulse durations in the range of 130 fs. It is demonstrated experimentally that the femtosecond interaction regime results in the formation of ripples beneficial for the functionalization of biopolymer/ceramic samples for further tissue engineering applications. In the transmittance spectra, profile variations for the non-treated and laser-processed surfaces were noticed depending on the number of pulses applied.

1. Introduction

Traumas, fractures, surgery, infections, tumours, congenital abnormalities, progressive disease of bone and bone structure, are only some of the causes of injury to and deformations and loss of bone tissue [1], which is why reconstruction of bone tissue defects is a major challenge facing orthopaedists in their efforts to restore the normal functioning of the human body [2]. The most commonly clinically-used bone grafts are autologous and allogeneic bone tissue materials, which, apart from being biocompatible, non-immunogenic and carry no risk of disease transmission, have their drawbacks, such as increased morbidity, infection in the donor bone, a second surgery in the donor location, limited supply of nutrition and allograft rejection, which can cause neurosensory deficit [3]. As an alternative, bone graft materials, such as metals, polymers and ceramics are being applied to fill the bone defect and provide mechanical and structural support.

The essential elements needed for bone reconstruction in tissue engineering are cells, scaffolds, and growth factors. In recent years, temporary "platforms" of various materials have been extensively studied for seeding different types of cell cultures and improving cell adhesion, proliferation and/or differentiation. These scaffolds act as support structures for cell attachment and growth into bone

[1]irbliznakova@abv.bg

tissues and, therefore, must have adequate mechanical properties. The clinical success of the scaffolds for bone regeneration could be ascribed to their combining inorganic and organic biomaterials.

Due to this fact, research the field of bone tissue regeneration has been focused on hybrid biomaterials [4]. Such temporary "platforms" contain composites of hydroxyapatite, chitosan, collagen, stimulant molecules, etc. In order to avoid a reduction in the bioactivity and biocompatibility and to enhance the process of cell adhesion and growth, it has been found appropriate to combine more than one element, such as a biopolymer and ceramics. The combination of hydroxyapatite (HAp) with chitosan for preparation of matrix substitutes has attracted interest in the field of tissue engineering as it combines the advantages of both materials and minimizes the drawbacks of using them separately – for example, the linear unbranched structure of chitosan and its high molecular weight makes it an excellent viscosity-enhancing agent in acidic environments, which gives scaffolds osteogenic and biodegradation properties [5, 6]; on the other hand hydroxyapatite is characterized by a high level of biocompatibility, so that it is used for fabrication of dense and porous bioceramics [7]. The strength of the scaffold is also associated with the level of porosity and high content of hydroxyapatite [8].

The design of the temporary scaffolds may include variety of conventional chemical or physical techniques, as solvent casting, particulate leaching, gas foaming, phase separation and freeze drying. However, these methods do not provide precise pore size, geometry, interconnectivity, and spatial distribution of the pores [9], which are all crucial factors for the scaffold creation and cells seeding.

Laser-induced surface texturing offers advantages over both the chemical and the other physical methods. It enables precise modification of certain surfaces that are difficult to treat with conventional chemical methods. As a result, the modified surfaces are free from contaminants [10]. The technique does not require special pre-treatment and improves cell adhesion, as it increases the roughness of the substrate. Control over the biomaterial's characteristics might be achieved, as the interaction of ultra-short pulses with biological tissues results in reduced crack formation and heat diffusion, absence of molten zones, and reduced ablation thresholds.

The laser-induced periodic surface structures (LIPSS) or ripples are of interest in what concerns the surface structuring. Ripples formation was first observed in 1965 by Birnbaum upon ruby-laser irradiation of various semiconductor surfaces [11]. LIPSS appear due the interference of the incident laser beam with the light reflected from the material surface and could be used to modify the functional characteristics of the surfaces, such as wetting, optical properties, or friction control [12]. It has been shown that LIPSS evolution and properties are influenced by the wavelength λ, the laser fluence F, the optical properties of the material and the number of applied pulses N. The LIPSS formation process is a non-contact phenomenon and provides possibilities of structuring the particular substance on a nanoscale. The process of ripples creation is being exhaustively investigated for many different biological applications due to the potential of improving the surface functionality and alter the wettability, both being important for the further cell behaviour on the material superficies [13].

2. Material and methods

2.1. LIPSS formation on Ch/HAp thin composites

Thin Ch/HAp composites were synthesized from a mixture of chitosan (medium molecular weight) and hydroxyapatite powder (with particle size 10 μm, $\geq 100\ m^2/g$) in a ratio 70:30, respectively, purchased from Sigma - Aldrich®. The solutions prepared were spread over 2×2 cm slides for further drying at room temperature for approximately 72 hours. The dried biopolymer/ceramic samples were fixed for further more precise experimental treatment. In order to obtain a clear visualization of the ripples (LIPSS) formation, a schematic representation of the basic steps of the process is given in Figure 1 below.

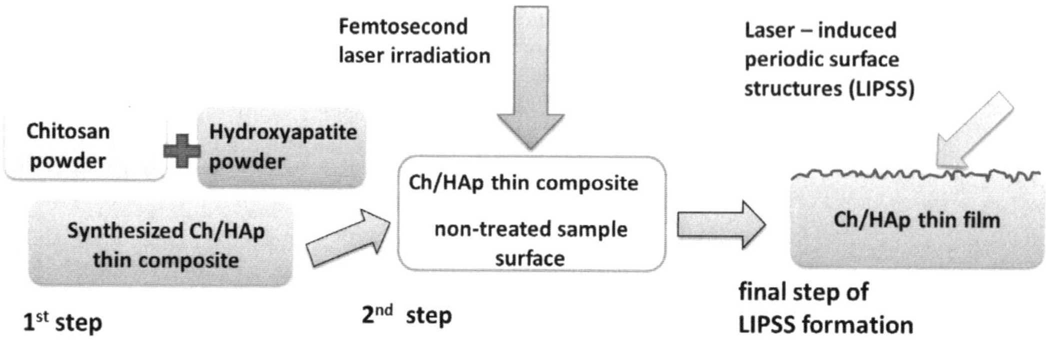

Figure 1. Schematic representation of the LIPSS (ripples) formation on synthesized Ch/HAp thin films.

In our experiments, we used a Ti:sapphire laser (Quantronix- Integra-C) delivering 130-fs pulses at a central wavelength of $\lambda = 800$ nm and a 50-Hz repetition rate. The beam profile has a Gaussian shape with $M^2 = 1.3$. The number of applied laser pulses (N) was controlled by a computer-driven fast mechanical shutter synchronized by controlling software. The experiments were performed in air with the laser beam focused to a focal spot with a diameter of approximately 40 μm using a lens with a 10 cm focal length.

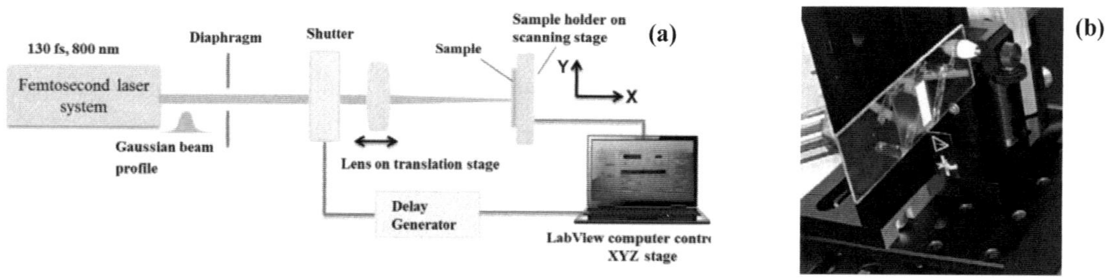

Figure 2. Experimental setup for femtosecond laser irradiation of Ch/HAp composites (a); image of the irradiation process of thin Ch/HAp layers (b).

The focusing lens was placed on a translation stage equipped with a micrometre screw for fine adjustment of the focus position on the specimen's working surface. The sample was positioned perpendicular to the focused beam on a high-precision XYZ translation stage. The experimental setup was controlled by specially written LabView software. The samples were processed by scanning the focused laser beam over the material surface at precisely-defined separation intervals d_x and d_y in order to avoid a spatial overlap between the separate laser spots.

3. Results and discussion

Generally, it has been shown that LIPSS evolution and properties are influenced by the laser beam parameters. A significant point is the laser polarization, which has a strong effect on the process of ripples formation and has been investigated in detail by various researchers using linear, elliptical and circular polarization [13,14]. For example, Gräf et al. [12] reported results concerning the formation process and properties of laser-induced periodic surface structures (LIPSS) on different technically relevant glasses, including fused silica, borosilicate glass, and soda-lime-silicate glass, which were irradiated by 300-fs laser pulses at the wavelength $\lambda = 1025$ nm. Another research group, Rebollar et al. [14], described LIPSS formation on spin-coated thin films of several model aromatic polymers, including poly(ethylene terephthalate), poly(trimethylene terephthalate) and polycarbonate bisphenol upon femtosecond irradiation at 795 nm and 265 nm and fluences well below the ablation threshold.

We followed the laser energy density influence on the created Ch/HAp surfaces via scanning electron microscopy (Hitachi S570). To characterize the ripples creation on the thin films, we explored the effect of varying the number of applied laser pulses. In order to avoid mechanical damage and formation of micro-cracks, femtosecond, rather than nanosecond, laser pulses were applied to the biomaterial composites [15]. The use of pulses of femtosecond duration, i.e., high peak laser power, results in a nonlinear absorption process and non-thermal ablation, thus providing the attractive benefits of the capability of treating a wide variety of materials [16].

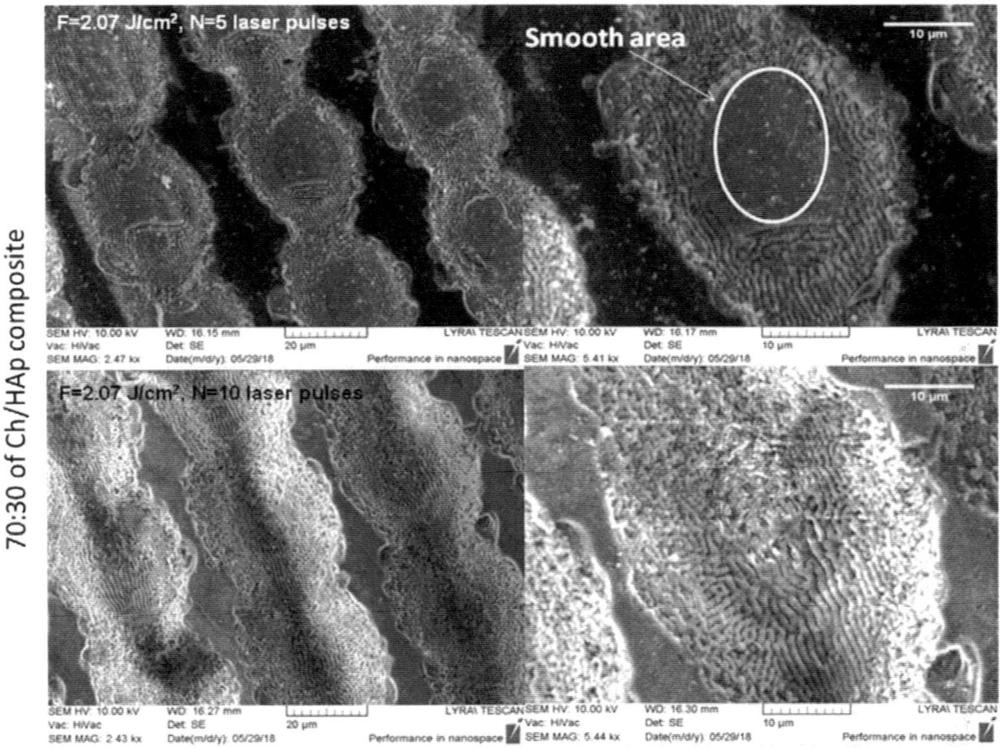

Figure 3. Scanning electron microscopy images of LIPSS formations on 70:30 Ch/HAp thin layers.

In our experiments, we observed formation of LIPSS when irradiating our samples by 5 and 10 laser pulses with duration 130 fs and $F = 2.07$ J/cm^2 (Figure 3). We monitored the changes in the substrate surface; namely for $N = 5$ laser pulses, ripples were formed only at the periphery of the craters and smooth zones of molten substrate were clearly seen at the bottom of the craters; whereas with increasing the applied laser pulses to $N = 10$, the process of formation LIPSS was completed and ripples appeared throughout the craters. LIPSS were induced upon irradiation by linearly polarized laser radiation.

Additionally, Fourier transform infrared spectroscopy (FTIR, IR Affinity-1 "Shimadzu") was used to obtain detailed information concerning the phase transformations and chemical bonds alterations in the Ch/HAp composite biomaterials before and after fs laser treatment (Figure 4). The measurements were conducted in a transmittance mode in the 4500 – 500 cm^{-1} range with a resolution of 4 cm^{-1} and 50 scan counts. The results illustrated significant differences in the spectra measured of the non-irradiated and laser-processed surfaces of the composite substrates. First, the FTIR spectra exhibited equal number and position of the various transmittance peaks, but with a major difference in the intensity of the peaks from the laser treated zones. Further, the shapes of the spectra from both processed areas (5 and 10 pulses) were similar, with the exception of a lower peak intensity of the saccharide structure of chitosan (around 900 cm^{-1}) in the matrix for $N = 10$ pulses.

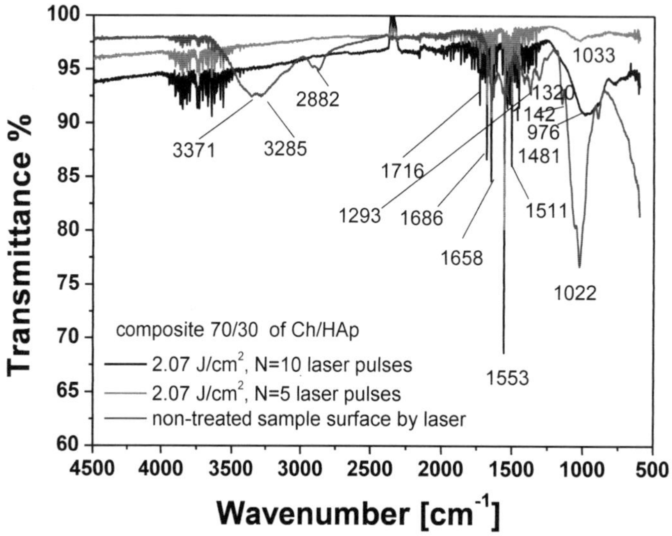

Figure 4. FTIR spectrum of synthesized Ch/HAp thin composite layers.

In the spectrum of the non-treated surface, between 3400 cm^{-1} and 3000 cm^{-1}, one characteristic group was observed, which corresponds to hydroxyl groups stretching in chitosan. The peaks near 1650 cm^{-1} correspond to = C = O stretching and = N-H vibrations of amide structures I and II. Another peak noticed at about 1267 cm^{-1} from the non-irradiated material surface is characteristic for the amide III structure. The band about 2880 cm^{-1} is attributed to -CH "backbone" vibrations, while a maximum at about 1400 cm^{-1} is assigned to -CH_3 and -CH_2 groups in plane deformation vibrations. The transmittance maxima identified at about 1030 cm^{-1} to 1020 cm^{-1} are commonly associated with the stretching of glucosamine in the chitosan structure. The groups observed in the range of about 1000 cm^{-1} and 600-500 cm^{-1} are attributed to the stretching and flexing vibrations of the PO_4^{3}- group.

The presence of chitosan in the non-irradiated Ch/HAp sample was clearly recorded, and, depending on the number of laser pulses, changes in the spectral shape occurred. This could be associated with the higher laser fluence and number of laser pulses applied in this case ($F = 2.07$ J/cm^2, $N = 5$, 10), which resulted in changes in the hydroxyapatite content in the matrix compared to our previous investigations of fs irradiation of pure chitosan thin films by 1 and 2 laser pulses with $F = 0.41$ J/cm^2, when discrepancies in the spectrum profile were not registered [17]. The results obtained confirmed the variation of the ablation thresholds due to the diversity of the materials considered.

4. Conclusion

The reported experimental results illustrated the generation of well-shaped LIPSS on chitosan/hydroxyapatite thin composites after femtosecond laser irradiation by laser pulses with wavelength 800 nm and duration 130 fs. The data was confirmed by SEM analysis, which clearly visualised structures throughout the laser-treated surface. FTIR was employed to acquire detailed information on the structural variations of the substrate content before and after laser treatment. We observed changes in the shapes of the transmittance spectra from the untreated and laser-processed surfaces depending on the number of laser pulses. No such changes occurred in our previous experiments on pure chitosan samples. The probable cause is the presence of hydroxyapatite for which the ablation threshold appears to be higher. The results obtained provide comprehensive evidence related to femtosecond laser processing of biomaterials and the boundary conditions for ripples formation as needed in view of improving the characteristics of the surface of scaffolds. The data confirm that fs laser processing is a technique suitable for material surfaces functionalization for further biomedical applications.

Acknowledgments
This work was financially supported by the Bulgarian National Science Fund (NSF) under Project no. DM18/6 - "Ultra-short laser surface modification and property analysis of synthesized ceramic scaffolds used for bone tissue engineering" (2017-2019) and Project no. DN08/5 - "Bioactivity improvement of biomimetic materials by texturing with ultrashort laser pulses" (2016-2018).

References
[1] Mistry S and Mikos A 2005 *Adv Biochem Engin/Biotechnol.* vol 94 pp 1-22
[2] Costantino P and Friedman C 1994 *OtolaryngolClin North Am.* vol 27 pp 1037-1074
[3] Wang X, Wang X, Tan Y, Zhang B, Gu Z and Li X 2009 *J. Biomed. Mater.* Res. A vol 89 pp 1079- 87
[4] Shrivats A, McDermott M and Hollinger J 2014 *Drug Discovery Today* vol 19 pp 781-786
[5] Enel S and McClure S J 2004 *Advanced Drug Delivery Reviews* vol 56 pp 1467-1480
[6] Glerentes P, Vachoud L, Doury J and Domard A 2002 *Biomaterials* vol 23 pp 1295-1302
[7] Suchanek W and Yoshimura *M J* 1998 *Mater Res* vol 13 pp 94-117
[8] Pallela R, Venkatesan J, Janapala V R and Kim S K 2012 *J. Biomed. Mater.* Res. A vol 100 pp 486-95
[9] Dhandayuthapani B, Yoshida Y, Maekawa T and Kumar D S 2011 *International Journal of Polymer Science* p 19
[10] Tiaw K, et al. 2005 *Biomaterials* vol 26 pp 763–769
[11] Barb R A, Hrelescu C, Dong L, Heitz J, Siegel J, Slepicka P, Vosmanska V, Svorcik V, Magnus B, Marksteiner R, Schernthaner M and Groschner K 2014 *Appl. Phys. A* (doi 10.1007/s00339013-8219-9)
[12] Gräf S, Kunz C and Müller F A 2017 *Formation and Properties of Laser-Induced Periodic Surface Structures on Different Glasses Materials* vol 10 p 933
[13] MüllerF A, Kunz C and Gräf S 2016 *Bio-Inspired Functional Surfaces Based on Laser-Induced Periodic Surface Structures Materials* vol 9 p 476
[14] Rebollar E, Va´zquez de AldanaJ R, Martı´n-Fabiani I, Herna´ndez M, Rueda D R, Ezquerra T A, Domingo C, Moreno P and Castillejo M 2013 *Phys.Chem. Chem. Phys.* vol 15 p 11287
[15] Bliedtner J, Schindler C, Seiler M, Wächter S, Friedrich M and Giesecke J 2016 *Laser Technik Journal* pp 46-50
[16] Friedrich M, Waechter S, Giesecke J, Bliedtner J and Bergmann J P 2017 *Journal of Laser Applications* vol 29 p 022502
[17] Daskalova A, Nathala C, Bliznakova I, Slavov D and Husinsky W 2015 *Proc. of SPIE* vol. 9447 pp 94470I-1

IOP Conf. Series: Journal of Physics: Conf. Series **1186** (2019) 012030 doi:10.1088/1742-6596/1186/1/012030

Crosslinked Chitosan/Casein Polyelectrolyte Multilayers for Drug Delivery

T Yovcheva[1], B Pilicheva[2], A Marinova[1], A Viraneva[1], I Bodurov[1], G Exner[1], S Sotirov[1], I Vlaeva[3], Y Uzunova[2], M Marudova[1]

[1]Faculty of Physics and Technology, University of Plovdiv "Paisii Hilendarski", 24 Tzar Assen str., 4000 Plovdiv, Bulgaria
[2]Faculty of Pharmacy, Medical University – Plovdiv, 15A Vassil Aprilov Blvd, 4002 Plovdiv, Bulgaria
[3]Faculty of Economics, University of Food Technologies - Plovdiv, 26 Maritsa Blvd, 4000 Plovdiv, Bulgaria

marudova@uni-plovdiv.net

Abstract. Polyelectrolyte multilayers (PEMs) are widely used as drug delivery systems, but still remain challenging for their small drug immobilizing capacity. One way to increase the immobilized drug amount may be crosslinking of the PEMs, which stabilize them and increase their porosity. The aim of the present study is fabrication and characterization of chitosan/casein PEMs, which are crosslinked with different crosslinking agents – glutaraldehyde, sodium tripolyphosphate, $CaCl_2$ and combinations of two of them. XPS method was used to prove the PEMs crosslinking. SEM was used to observe film morphology and its variation due to cross-linking. Water capacity of PEMs in 100 % relative humidity was investigated. Release of model drug Benzydamine Hydrochloride was monitored spectrophotometrically at 306 nm. The crosslinking improves the PEMs stability and causes formation of porous surface. After crosslinking the amount of the immobilized drug increased several times.

1. Introduction

Controlled drug release materials are widely used in biomedical implants, tissue engineering, and targeted delivery devices [1]. Over the past decade polyelectrolyte multilayer films (PEMs) produced by the layer-by-layer (LbL) assembly technique [2, 3] find an ever wider application as drug delivery systems [4].The layer-by-layer method is based on the electrostatic attraction between polyanions and polycations, hydrogen bonding or covalent bonding. The advantage of this is the possibility to structure PEMs with nanometer thickness and the ability to tune their internal properties [5] These films can contain a large variety of internal structures, depending on the used assembly conditions (pH and ionic strength) [6] during the deposition process or the type of used polyelectrolytes. LbL-assembly PEMs can be further chemically crosslinked to increase their stability or decrease the permeability [7].

The present research reports chemically crosslinked casein/chitosan PEMs as drug delivery systems. Some structural properties and drug upload capacity are commented.

2. Materials and methods

2.1. Materials and samples preparation

2.1.1. Pads formation. Biodegradable pads were prepared from polylactic acid (PLA) with ester end groups and intrinsic viscosity 0.55 - 0.75 dL/g purchased from Lactel Absorbable Polymers (USA). The PLA pads were cast from 2 % w/v PLA solution in chloroform and dried at 35 °C for 48 hours. Then the PLA pads were kept in a desiccator, at room temperature, and 54 % relative humidity (RH). Before the deposition process, the pads were charged in a corona discharge system, consisting of a corona electrode (needle), a grounded plate, and a metal grid placed between them. Positive 5 kV voltage was applied to the corona electrode. 1kV positive voltage was applied to the grid. The samples were charged under standard room conditions ($T = 21 \div 23$ °C and $RH = 50\text{-}60$ %) for 1 min.

2.1.2. Polyelectrolyte multilayers deposition. Casein (molecular mass 24000 Da) and low molecular mass chitosan (molecular mass 127000 Da) were purchased from Sigma-Aldrich. They were used without further purification. The layer-by-layer (LbL) deposition technique was applied for multilayer build-up. For the LbL assembly 1% w/v casein solution in phosphate buffer (containing sodium dihydrogen phosphate and disodium hydrogen phosphate) with pH 8 and ionic strength 100 mM, and 1% w/v chitosan solutions in acetate buffer (containing acetic acid and sodium acetate) with pH 4 and ionic strength 100 mM, were prepared. The deposition was done by the dip-coating process. The first built-up layer was casein, which possesses opposite to the pad electric charge. Each polyelectrolyte deposition step was followed by rinsing in distilled water.

Four different crosslinking procedures to crosslink the chitosan and/or the casein in the PEMs were used:

- crosslinking of chitosan by **glutaraldehyde** (GA) – after each chitosan layer the PEMs were immersed in 0.1 % w/v glutaraldehyde for 20 min;
- crosslinking of chitosan by two agents – **glutaraldehyde** and **sodium tripolyphosphate (GA + Na TPP)** - after each chitosan layer the PEMs were immersed consequently in 3% sodium tripolyphosphate for 20 min and 0.1 % w/v glutaraldehyde for 20 min.
- crosslinking of casein by **$CaCl_2$** ($CaCl_2$) – after each casein layer the PEMs were immersed in 1 % w/v $CaCl_2$ for 20 min;
- crosslinking of chitosan by **glutaraldehyde** and casein by **$CaCl_2$** (GA + $CaCl_2$) - after each chitosan layer the PEMs were immersed in 0.1 % w/v glutaraldehyde for 20 min; then after each casein layer the PEMs were immersed in 1 % w/v $CaCl_2$ for 20 min.

After each chitosan layer (after crosslinking) the PEMs were immersed in 1% solution of model drug Benzydamine hydrochloride (BH) and hold there for 30 min. The procedure was repeated until obtaining 8 layers casein/chitosan. After the deposition of the last layer the film was dried in hot air at temperature 60 °C for 30 min. The ready samples were stored in desiccator at room temperature and relative humidity 54 %.

2.2. Polyelectrolyte multilayers characterization

2.2.1. The X-ray photoelectron spectroscopy (XPS). XPS studies were performed in a VG Escalab II electronspectrometer using AlKα radiation with energy of 1486.6 eV under base pressure 10–7 Pa and a total instrumental resolution 1 eV. The binding energies (BE) were determined utilizing the C1s line (from an adventitious carbon) as a reference with energy of 285.0 eV. The accuracy of measuring the BE values was 0.2 eV. The C1s, Si2p, Na1s, P2p, Cl2p, Ca2p , N1s and O1s photoelectron lines were recorded and corrected by subtracting a Shirley-type of background and quantified using the peak area and Scofield's photoionization cross-sections. The XPSPEAK41 software was used for deconvolution of recorded spectra, where was necessary.

2.2.2. Scanning electron microscopy (SEM). The general morphology of the obtained polyelectrolyte multilayers films was revealed by means of SEM. A scanning electron microscope Lyra 3 XMU (Tescan) was employed. The working voltage was 8.1 kV. Prior to the measurements, the samples were covered with a thin film of gold (about 30 nm).

2.2.3. Moisture absorption. The moisture absorption test was carried out to assess the physical stability of the films at high humid conditions. In the present study the moisture absorption capacity of the films was determined gravimetrically. Before the test the samples were stored at 25 °C in desiccator with relative humidity 54 % for 72 hour. Then the films (2x2 cm, n=3) were weighed accurately and replaced into a desiccator containing distilled water (100 % RH) at 25 °C. The samples were left to equilibrate for 72 hours before new weight measurement was done. Moisture absorption was calculated as the increase in weight, expressed as a percentage.

2.2.4. Drug content. Benzydamine hydrochloride (BH) loaded films (3 samples of each type) were placed into 20 mL phosphate buffer saline (pH 6.8) and stirred continuously for 72 hours on a magnetic stirrer. Then, the samples were sonicated for 5 minutes and filtered using Chromafil® syringe filter (0.45 µm). The amount of BH was determined using UV/Vis spectrophotometer (Evolution 300, Thermo Fisher Scientific, USA), monitoring the band at λ_{max} 306 nm [8]. The drug concentration was calculated from a standard calibration curve of BH in phosphate buffer saline (pH 6.8).

3. Results and discussion

3.1. XPS investigations

Three different agents were used to achieve crosslinked PEMs: GA and Na TPP, which interact with the amine groups and $CaCl_2$, which forms ionic bonds with the carboxylate ions. To verify the crosslinking process XPS elemental analysis was employed. The XPS spectra of the non-cross-linked and cross-linked PEMs were compared.

The binding energy of C, O, and N atoms would change during the interaction between the protonated amine group and the Na TPP. The change of XPS spectra of N atom before and after cross-linking by Na TPP was clearly visible due to the high electrowithdrawing effect of O^- ion shifting the XPS signal to higher energies, as shown in figure 1. Before crosslinking only 1 peak shows up at 399.3 eV indicating the existence of amine groups connected to the carbon chain. After crosslinking, in addition to the peak at around 399 eV, a new peak appears at about 402 eV belonging to the N atom of amide group [9]. The similar behavior was observed by others [10].

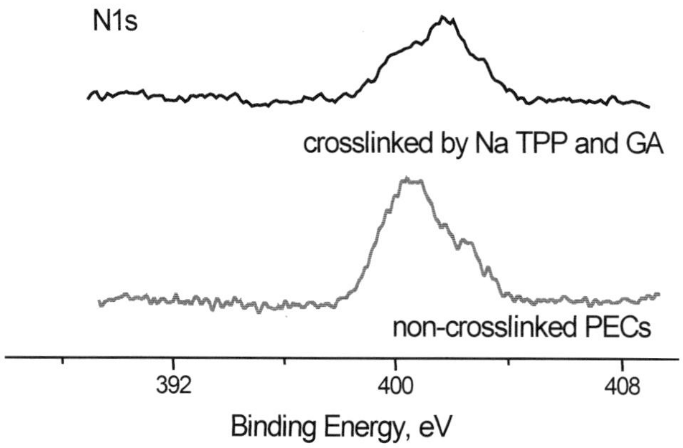

Figure 1. XPS spectra of the N atom (1 s) of the non-cross-linked and cross-linked by Na TPP and GA PEMs.

According to Öztop et al. [11] the crosslinking of chitosan with GA occurs by destruction of C=O group in the GA and formation of N=C bond between the GA and the N atom in the chitosan. The lack of the high energy binding at 289.259 eV in the XPS spectra of crosslinked by GA layers verifies the C=O destruction (figure 2).

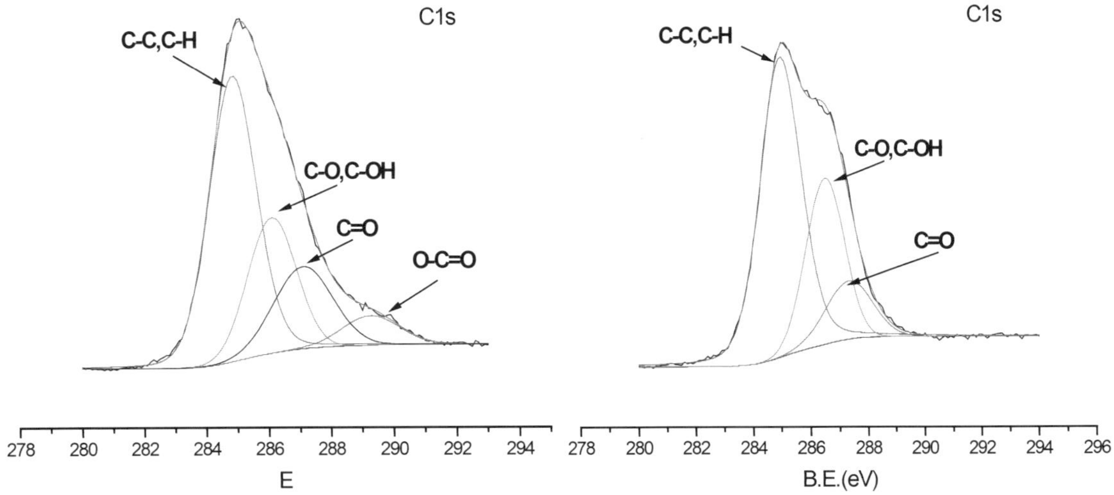

a) *Non-crosslinked layers* *b) Crosslinked layers by GA*

Figure 2. XPS spectra of the C atom (1 s) of the non-cross-linked and cross-linked by GA PEMs.

After crosslinking of PEMs by $CaCl_2$, a Ca 2p line in the XPS spectrum could be observed, while for the non-crosslinked samples this line was in the range of noise. Calcium was bound to the casein since no chlorine was detected in the XPS spectrum.

3.2. SEM morphology

The surface morphology of the PLA pads and casein/chitosan PEMs cross-linked with different agents, are presented in figure 3.

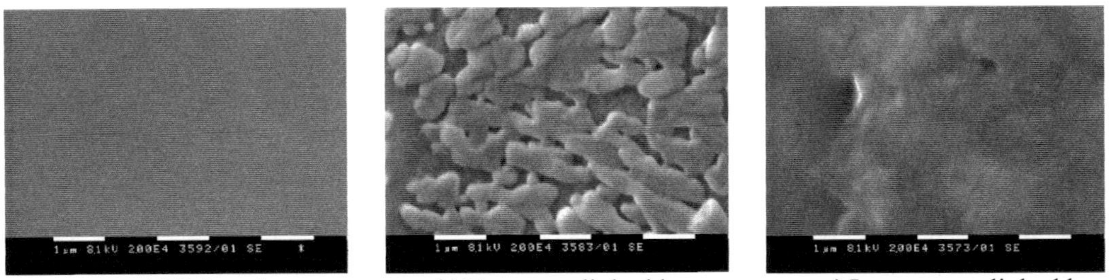

a) PLA pad b) Non-crosslinked layers c) Layers crosslinked by $CaCl_2$

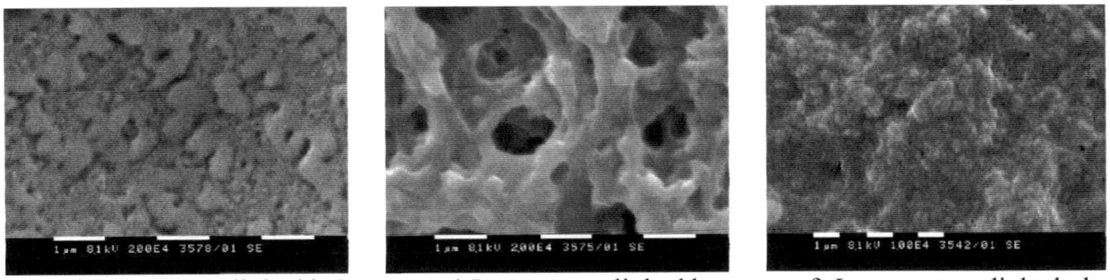

d) Layers crosslinked by GA e) Layers crosslinked by GA and $CaCl_2$ f) Layers crosslinked by GA and Na TPP

Figure 3. SEM micrographs of PLA pad (a), non-crosslinked layers (b), layers crosslinked by $CaCl_2$ (c), layers crosslinked by GA (d), layers crosslinked by GA and $CaCl_2$ (e), layers crosslinked by GA and Na TPP (f).

It is important initially to study the morphology of the pad, since it largely determines the mode and the type of the deposited thereon thin layers. As it is seen from figure 3 a) PLA pad is characterized by homogeneous structure.

LbL deposition of the PEMs let to a gradual change of the morphology of the surface (figure 3b). The polyelectrolytes were adsorbed in separate morphological elongated entities with size of about several microns. This structure might be a reason of partial chitosan desorption during the rinsing process. The structure significantly changed after the crosslinking. It became denser and smoother. Crosslinking with one agent (GA) caused smoothening of the surface, but the morphological entities were retained (figure 3d). When 2 agents were used, the structure became denser due to more stable binding of the chitosan molecules (figure 3f). When the casein layer before the last one is crosslinked the surface became more homogeneous (figure 3c). In case of double crosslinking of both casein and chitosan (figure 3e), the structure is porous with pore size of approximately 1 micron.

3.3. Water content

As it was shown the crosslinking process let to formation of denser structure, which resulted in different water uptakeabsorption properties of the PEMs. The water content of crosslinking samples decreased several times in comparison to the non-crosslinked ones. The lowest water content is observed in case of double crosslinking, where it is less than 10 % - figure 4.

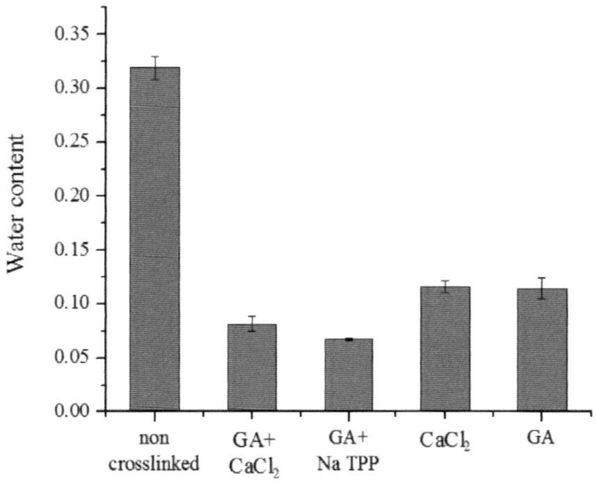

Figure 4. Water content of PEMs.

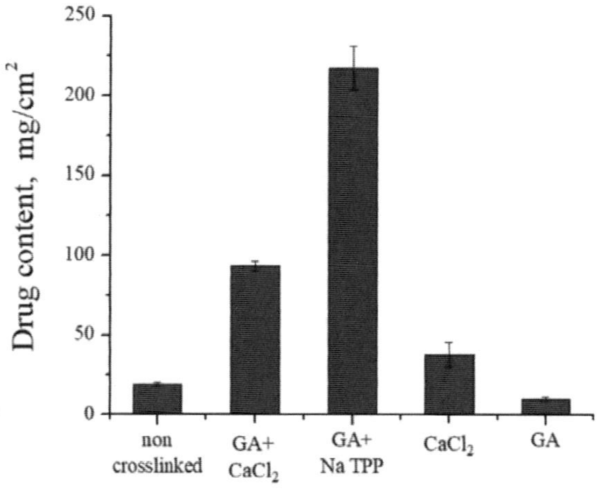

Figure 5. Drug content of PEMs.

3.4. Drug content investigations

The amount of uploaded model drug BH strongly depends on the PEMs structure and its ability for water uptake. If the layers are too dense there is no enough space for drug molecules entrapment. On the other hand, in case of very loose and swollen structure the drug release process dominates the process of uploading. Hence porous structure with an optimal pore size should be achieved for maximum drug content.

The drug content of the layers is presented in figure 5. The lowest drug is gripped in the non-crosslinked PEMs and in those crosslinked by means of glutaraldehyde. The growth of the layers in this case is insular type (see figure 3 b, d) and probably their thickness and density are not suitable for gripping a drug. The highest drug amount is uploaded in PEMs wherein the chitosan is crosslinked by double agents – glutaraldehyde and Na TPP. The PEMs in this case are strongly bound to the pad and to each other and the water uptake is small, i. e. the air pore size is not too big. Similar structure and water capacity possess PEMs crosslinked via glutaraldehyde and $CaCl_2$ where the drug amount is again high.

4. Conclusion

PEMs constructed from natural polymers casein and chitosan by layer-by-layer method on preliminary charged PLA pads were successfully crosslinked by glutaraldehyde, Na TPP, and $CaCl_2$. The PEMs crosslinking let to formation of stable and porous structure with less water capacity and higher potential for drug upload. The most effective crosslinking process is achieved by the usage of two agents - glutaraldehyde and Na TPP.

Acknowledgments

This study was funded by Bulgarian National Scientific Fund, Project No DFNI B-02/7.

References:

[1] Berg M, Zhai L, Cohen R and Rubner M 2006 *Biomacromolecules* **71** 357-64
[2] Decher G, Hong J and Schmitt J 1992 *Solid Films* **210-211** 831–35
[3] Decher G 1997 *Science* **277** 1232–37
[4] Anandhakumar S, Gokul P and Raichur A 2016 *Mat. Scie. and Engin: C* **58** 622-28
[5] Klitzing R 2006 *Phys. Chem.* **8** 5012–33
[6] Frueh J Gai M Halstead S and He Q 2014 *Polyelectrolyte Thermodyn. Rheol.* (Springer Berlin Heidelberg)
[7] Hu N, Frueh J, Zheng C, Zhang B and He Q 2015 *Colloids and Surfaces A: Physicochem. and Engin. Aspects* **482** 315-33
[8] Karavana S, Güneri P and Ertan G 2009 *Pharmaceut. Dev. Technol.* **14** 623-31
[9] Moulder J and Chastain J 1992 *Handbook of X-ray Photoelectron Spectroscopy: A Reference Book of Standard Spectra for Identification and Interpretation of XPS Data* (Perkin-Elmer Corporation Eden Prairie) p 259
[10] Zhu X, Jańczewski D, Lee S, Teo S and Vancso G 2013 *ACS appl. materials and interfaces* **5** 13 5961-68
[11] Öztop H, Saraydin D and Cetinus Ş 2002 *Appl. Biochem. and biotech.* **101** 3 239-49

IOP Conf. Series: Journal of Physics: Conf. Series **1186** (2019) 012031 doi:10.1088/1742-6596/1186/1/012031

Optical, electro-optical, electrical and dielectric characterization of nematic liquid crystal (E7) layers doped with graphene nanoparticles for electro-optics

Y G Marinov [1,*], G B Hadjichristov [1], P M Rafailov [1], S H Lin [2], V M Marinova [3] and A G Petrov [1]

[1] Georgi Nadjakov Institute of Solid State Physics, Bulgarian Academy of Sciences, 72 Tzarigradsko Chaussee Blvd., BG-1784 Sofia, Bulgaria
[2] Department of Electrophysics/Photonics, National Chiao Tung University, Hsinchu 30010, Taiwan
[3] Institute of Optical Materials and Technologies, Bulgarian Academy of Sciences, BG-1113 Sofia, Bulgaria

[*]Corresponding author's e-mail address: ymarinov@issp.bas.bg

Abstract. Thin (7 μm) layers of nanocomposites from graphene nanoflakes (GrFs) dispersed at concentrations of 10^{-3} wt.% into the nematic liquid crystal (NLC) E7 were characterized by various investigation techniques, such as Raman spectroscopy, impedance measurements and dielectric spectroscopy, as well as by electro-optical measurements (optical transmittance of the NLC layers versus the voltage of the applied external AC electric field). Conducting behaviour, dielectric permittivity and electric energy loss of our planar-aligned NLC layers at room temperature were analysed as a function of frequency in the range from 0.5 Hz to 1 MHz. The analysis of experimental data indicates that the molecular alignment through GrFs/NLC surface interactions is responsible for the reduction of the ionic conductivity of E7 NLC in the presence of GrFs. As compared to pure LC E7, this leads to improved characteristics for the studied nanocomposites, necessary in their practical applications in electro-optics.

1. Introduction

A new topic and currently trending approach to discover new interesting and practically useful properties of the liquid crystals (LCs) is their modification by addition of nanoparticles (NPs), such as graphene NPs (GrNPs) and NPs of graphene derivatives. The research on these advanced and multifunctional nanocomposite materials [1-8] demonstrated that one can broaden in this way the LCs usage in display technology area (notebooks, computer displays, and TVs), sensitive screens, in a variety of other electro-optical (EO) devices and technical applications, as well as in sensorics, optics and organic electronics (e.g., in the field of organic light-emitting devices).

Usually, the LC materials for such applications are multi-component mixtures of different compounds thus utilizing their optimal combination of physico-chemical parameters, properties and characteristics (phase transition, dielectric and optical anisotropy, viscosity, etc.) that determine their EO response relevant to device applications. The doping of LC materials with small amounts of GrNPs can improve their EO and other responses [1,3,5,8], in particular due to the controllable change in their complex refractive index, complex dielectric permittivity, complex electrical conductivity, etc.

In the last decade, such nano-induced modifications are object of growing scientific interest, because the doping with GrNPs brings new important features of the LCs.

In the presented work, we have added graphene nanoflakes (GrFs) to the room-temperature nematic LC (NLC) mixture E7 (widely used in LC devices) in order to produce a nanocomposite material with improved electrical, dielectric and EO properties. Our attention was focused on the effect of the presence of a diluted concentration (10^{-3} wt.%) of GrFs on the transport of ions in the E7 NLC, as well as on the dielectric characteristics of the produced NLC nanocomposites (nanocolloids). Recently, the investigations on alternating-current (AC) electrical conductivity and dielectric characteristics of such NLC material formed as dispersion of GrNPs into E7 showed that all dielectric parameters are strongly influenced by frequency and applied bias voltage [1]. Whereas in the latter study the concentration of GrNPs was 5×10^{-3} wt.%, we have further reduced the concentration of the functional carbon nanostructure materials, in our case – GrFs.

2. Experimental

The NLC used is the well known eutectic mixure E7 with a positive dielectric anisotropy ($\Delta\varepsilon > 0$). This cyano-biphenylic LC was purchased from Merck Eurolab (Darmstadt, Germany) and was used as received, without any further chemical purification. GrNPs (nanoflakes, GrFs, with an average thickness of 12 nm) were commercially obtained from Graphene Supermarket, Inc., in the form of nanopowder (Grade AO-3). The powder was dried in an oven to evaporate any residual moisture, then the GrFs were ball milled in a few drops of ethanol (wet milling) using a Fritsch pulverisette machine. The GrFs/ethanol suspension was steered in an ultrasonic bath for 30 min and added into E7 LC in its nematic state. To obtain a homogeneous solution, the mixture was sonicated for 1 h. Next, the ethanol was evaporated at an elevated temperature of 80°C for 8 hrs. The GrFs concentration in the LC E7 was 10^{-3} wt.% relative to the total mass of the LC.

We used commercially available empty glass cells KSRP-07/B107P6NTS05 Up/Low (E.H.C. Co., Ltd, Tokyo, Japan) with a cell gap of 7 μm, optically-transparent indium-tin oxide (ITO) contacts and alignment layers of polyimide, parallel-rubbed in order to achieve a planar orientation of the NLC content of the cells. The GrFs/E7LC dispersion was capillary filled (in the isotropic phase of E7) into the LC cells. For reference measurements, identical empty cells were filled with undoped NLC E7. Micro-Raman spectra were obtained in backscattering geometry by HORIBA Jobin Yvon Labram HR visible spectrometer equipped with a Peltier-cooled CCD detector. The 632.8-nm line of a He-Ne-laser was used for photo-excitation. Using microscope optics, the laser beam was focused on the sample surface to a spotsize of about 10 μm in diameter. The peak frequencies were determined using calibration with neon lines by least-squares fit to Voigt profiles. The absolute accuracy of the measurements was 0.5 cm^{-1}. The optical transmittance of the NLC layers versus the voltage of the applied external AC electric field (the Fréedericksz transition) was measured as described in [9]. The complex electrical impedance and dielectric spectroscopy measurements were performed as in [10].

3. Results and discussion

Raman spectroscopy has great potential in the studying of nanostructured LCs at molecular level. In our case, we used the Raman spectra to obtain rapid information on the sample content and to look for the changes of the host NLC E7 upon mixing with GrFs. We had to deal with intense Raman lines from the LC phenyl rings and much weaker Raman signal from GrFs. Figure 1(a) reports the micro-Raman spectra of our samples, as recorded under identical experimental conditions. The samples were studied as prepared, i.e., in their initial planar configuration. The E7 response can be recognized by the four sharp and intense Raman lines at 1183, 1286, 1608 and 2231 cm^{-1}. In the Raman spectrum of GrFs-doped E7 a clearly discernible shoulder at the low-frequency side ($\approx$ 1582 cm^{-1}) (shown with arrow in figure 1a) of the characteristic E7 peak marks the presence of GrFs. Due to the relatively low percentage of dispersed GrFs, the change of the Raman spectrum of E7 resulting from the addition of GrFs is a little, to negligible. That is why, no other spectral peculiarities were observed, as well as no change in the intensity, spectral positions and the widths of the E7 Raman lines.

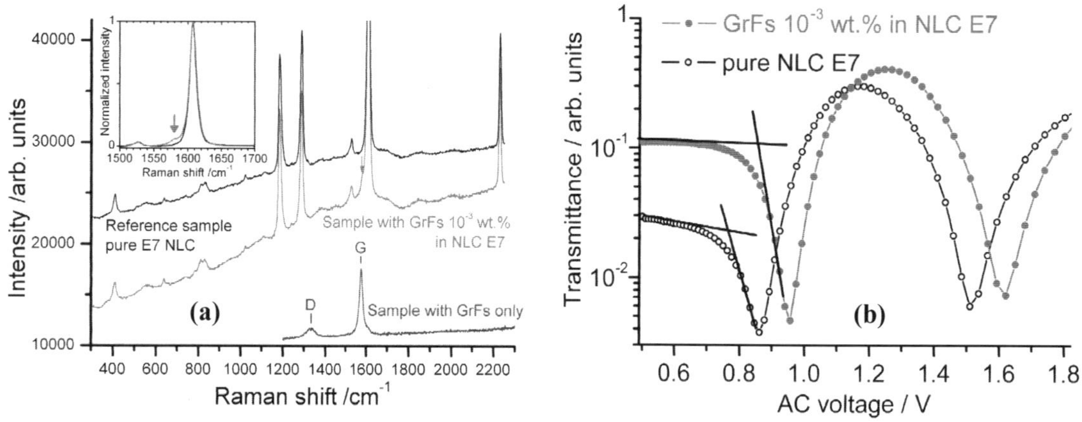

Figure 1. (a) Raman spectra of pure E7 and E7 doped with GrFs 10^{-3} wt.%, demonstrating spectroscopic evidence for the presence of the GrFs nanodopants. The Raman signal of the GrFs nanodopants (in pristine state) is also displayed for comparison. The insert: normalized spectra in expanded scale of wavenumbers; (b) Transmittance vs voltage curves of pure and GrFs-doped NLC E7. AC driving field of 1 kHz. In all cases, the samples were in identical cells, temperature was 25°C.

The EO transmittance characteristics of our samples of pure and GrFs-doped NLC E7 were also measured to be compared (figure 1b). To avoid the screening effect, the EO activation of the samples was driven by 1 kHz AC voltage when the voltage-dependent intensity of the transmitted light (He-Ne laser beam) through the LC cells between crossed polarizers was measured. Our EO measurements of the voltage-dependent transmittance of the examined GrFs-doped E7 samples upon AC electric field indicated a well pronounecd GrFs-produced change - the Fréedericksz transition for the GrFs-doped E7 samples was characterized with a higher threshold voltage (U_F) than that of the pure E7 measured under the same experimental conditions. The Fréedericksz transition is the molecular reorientation of the NLCs in the presence of an external electric field [11]. When the NLC posseses a positive dielectric anisotropy ($\Delta\varepsilon > 0$), the nematic director is forced to be oriented along the applied field. Fréedericksz thresholds of pure and nanostructured E7 are obtained by extrapolation of the steep parts of the transmittance curves to the initial, low-voltage transmittance. In the representative example in figure 1(b) one can adopt the U_F-values of 0.77 V and 0.86 V (RMS), for pure E7 and GrFs-doped E7, respectively. The observed change suggests a GrFs-induced over-alignment of the LC molecules and enhancement of their orientational order even at the very low concentration (10^{-3} wt.%) of GrFs nanodopants we have employed here.

The increase of threshold U_F could be associated with a possible modification of LC order parameter (S). U_F of a strongly anchored planar nematic layer depends mainly on two material parameters - the splay elastic constant K_{11} and dielectric anisotropy $\Delta\varepsilon$ [12]. Taking into account that $K \propto S^2$ and $\Delta\varepsilon \propto S$ [13], U_F can be expressed as $U_F \propto \sqrt{S}$. Thus, the increased U_F measured for GrFs-doped E7 might suggests an increased S in this nanocomposite, as compared to the undoped E7 NLC. But in view of the weak (square root of S) dependence of this effect on S, an additional enhancement in the planar anchoring due to adsorption of flakes over electrode surfaces could not be excluded.

Further, we compared the electrical responses of pure and GrFs-doped E7 NLC samples. In particular, by use of complex electrical impedance spectroscopy applied for characterization of both materials in a broad frequency range, we observed a clear effect from GrFs nanodopants in E7 NLC. Nyquist plots (the imaginary part, Z_i, vs the real part, Z_r, of the complex electrical impedance) display semicircles and almost vertical straight lines for both kinds of samples (figure 2). The studied nematic layers demonstrate a relatively high electrical impedance of the order of mega-ohms. A considerable shift between these curves is seen that displays the effect from GrFs nanodopants. Essentially, figure 2 shows that the resistance of the samples increases by doping with GrFs. Because the value of the bulk resistance (R_b) of the material is associated with the point of intersection of the semicircle with the Z_r-axis (the low-frequency end of the semicircle) [14,15], one can concluce from figure 3, that our GrFs-doped E7 samples exhibit a lower electrical conductivity (ionic conductivity) than the undoped reference E7 samples. This effect comes from the reduction of the concentration of the free ions in the NLC E7 – the known and complex ion-trapping process reported by other authors for GrNPs inclusions in NLCs [5,8,16,17]. Note that the ionic concentrations in both GrFs-doped and undoped E7 NLCs cells are the same due to the use of the same LC material and identical cells.

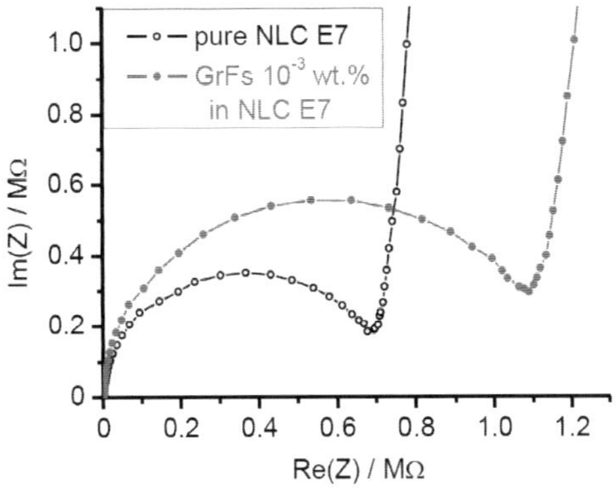

Figure 2. Nyquist complex impedance plots for temperature 25 °C.

In the considered NLC nanocomposites, by the GrFs is realized trapping of mobile ions, and thereby, a reduction of their diffusivity in the bulk. The corresponding physical mechanism is complex [5,8,16,17]. Most likely this occurs through π-π electron stacking and interfacial interactions (LC-graphene anchoring [17,18], similar to other such mechanisms of surface interactions known for carbonic NPs [19-23]). In this way, the introduction of GrFs nanodopants into NLCs shoud improve the EO performance of the doped NLCs. As known, the high concentration of free ion charges (excessive ionic impurities) in the NLCs is unwanted because it leads to problems such as slowing down reactions, retention of images, long-term effects of image tracking and short-term vibrations in LC displays and in other NLCs-based EO devices [24,25].

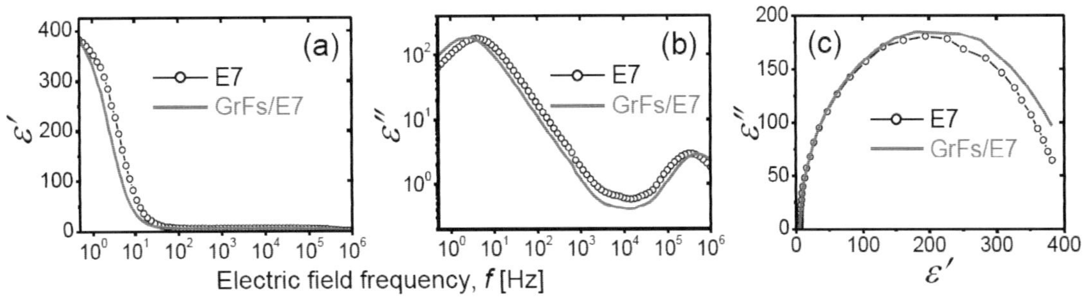

Figure 3. The Bode plots in log-log scale of the real (a) and imaginary (b) parts of the complex dielectric permittivity for planar layers of GrFs-doped NLC E7 and undoped E7, both in the nematic phase at 25 °C; (c) The Cole-Cole plots of the same dielectric spectra given in (a) and (b).

The analysis of dielectric spectra can be also useful for the consideration of the effect of GrFs inclusions in the NLC E7. The real (the storage component ε') and imaginary (the loss component ε'') parts of complex dielectric permittivity of the material in alternating electric fields can be evaluated from real (Z_r) and imaginary (Z_i) parts of the complex electrical impedance. Figure 3(a,b) presents the variation of the dielectric components ε' and ε'' of the measured NLC layers (perpendicular components, according to geometry of measurements) as a function of the frequency f of the AC electric field applied on the layers. As compared to ε' of the undoped E7 layers (figure 3a), the lower value of ε' for GrFs-doped E7 layers in the low-frequency region (in our case from 1 Hz to 100 Hz) corroborates the above conclusion that the planar alignment of the NLC molecules is improved by the doping with GrFs. Regarding the frequency behavior of ε'', one can note the shift of the relaxations peaks (figure 3b). Two dielectric relaxations were present. At the high-frequency region ($f > 100$ kHz) of the dielectric loss, the dielectric relaxation is referred to reorientation around the short molecular axis [6] and is not of importance for our present considerations. The low-frequency relaxations (a few Hz) is relevant to ionic contribution to the observed dielectric relaxations. It is seen that the GrFs nanodopants do shift the low-frequency relaxation peak towards the lower frequencies, which suggests a decrease of the number of free ions, and hence a decrease of cut-off frequency.

The Cole-Cole plots (the graphs of $\varepsilon''(f)$ vs $\varepsilon'(f)$ of both kinds of samples under study is shown in figure 3(c). From this diagram one can find the relaxation frequency (f_R) of ion polarization. The value of f_R corresponds to the top points of the semicircle curves of the Cole-Cole plots. At this frequency, the ions can no more follow the electric field and the dielectric heating is maximum. From figure 3(c) we determined that f_R (2.39 Hz) for GrFs-doped NLC E7 layer is lower than f_R (4.14 Hz) for undoped E7 layers, both at 25°C. The observed change of f_R is related to the ion mobility and the diffusion constant (D). Most likely, the decrease of f_R in GrFs-doped NLC is due to trapping of ions. These ions respectively exhibit larger inertia for the electric driven oscillations and are less mobile. The reduction of f_R corresponds to this.

Since the LCs are anisotropic media, one has to consider the LC alignment [2,26]. As might be expected, some of the GrFs should be localized at the aligning boundary interfaces, hence affecting the LC anchoring. This leads to improvement of the planar alignment [5,17,18,27,28]. The directional anchoring and LC alignment effect by GrNPs has been partly explained by π-π stacking interactions (π-π anchoring) [19,21,29-31]. On the other hand, the diffusion constant D is anisotropic, being related to the complex dielectric permittivity and electric conductivity of the medium. For instance, by impedance measurements in the range of 0.1 Hz – 1 MHz it was shown that impurity ions exhibit different transport behaviors along and perpendicular to the molecular axis, leading to anisotropy od D [32]. In fact, the diffusion constant along the molecular axis is greater than in perpendicular direction [12,32]. The anchoring modification and molecular orientation have important contributions to D. In considered GrFs-doped E7 NLC layers, the GrFs deposited on the boundary surfaces can quite possible decrease the pretilt angle, approaching to perfectly aligned planar cell. As a consequence, D in the direction perpendicular to the layer plane becomes smaller. This explanation is consistent with our dielectric and conductivity observations. Thus, the modification of the dielectric properties of the GrFs-NLC system is induced by the coupling of the molecular orientation of the NLCs with the GrFs properties. Accordingly, the dielectric response of the GrFs-NLC system can be modified and finely tuned by incorporating various amounts of GrFs.

4. Conclusion

We have studied the room-temperature EO, electrical and dielectric response of NLC nanocomposite materials (in 7 μm thin nematic layers) produced by dispersion of graphene nanoflakes (GrFs, graphene forms promising for nanoelectronics) into NLC E7. The results obtained by complex impedance and dielectric spectroscopy show that a sizable reduction of diffusivity of the free ions in the NLC E7 is possible even at the extremely low concentration of GrFs, equal to 10^{-3} wt.%. This happens by altering the molecular alignment through GrFs/LC surface interactions. The addition of this carbon nanomaterial to E7 NLC does influence the orientational order of the NLC, actually the graphene nanoflakes do increase the nematic ordering.

At the very low concentration of the carbonic nanodopants we have used, the material properties of the NLCs are not changed, the threshold voltage is slightly increased, but one can achieve a significant reduction of the transport of the mobile ions. Such positive effect is of great importance for LC displays and other NLCs-based EO devices, where the presence of ions of impurities is very problematic because it leads to an image sticking/flickering, changing the gray level, and slowing down of the switching response. Complementary measurements and further analyses are necessary to clarify all aspects of the GrFs-induced alignment and its relationship with the reduction of the ion diffusivity in the considered graphene-LC nanocomposites (work in progress).

Acknowledgments
Work supported by research project DFNI-TO2/18 "Nanostructured liquid crystals for tunable photonic devices" by the Ministry of Education and Science, through the National Science Fund of Bulgaria.

References

[1] Gökçen M, Yıldırım M and Köysal O 2012 *Eur. Phys. J. Appl. Phys.* **60** 30104
[2] Alam T M and Pearce C J 2014 *Chem. Phys. Lett.* **592** 7
[3] Wu P W and Lee W 2013 *Appl. Phys. Lett.* **102** 162904
[4] Kumar V, Kumar A, Biradar A M, Reddy G B, Sachdev D and Pasricha R 2014 *Liq. Cryst.* **41** 1719
[5] Basu R, Garvey A and Kinnamon D 2015 *J. Appl. Phys.* **117** 074301
[6] Al-Zangana S, Iliut M, Turner M, Vijayaraghavan A and Dierking I 2016 *Adv. Opt. Mater.* **4** 1541
[7] Lapanik V, Timofeev S and Haase W 2016 *Phase Transitions* **89** 133
[8] Garbovskiy Y and Glushchenko A 2017 *Nanomaterials (Basel)* **7** 361
[9] Hadjichristov G B, Marinov Y G, Petrov A G, Bruno E, Marino L and Scaramuzza N 2015 *Molec. Cryst. Liq. Cryst.* **610** 135
[10] Marinov Y G, Marinov M P, Hadjichristov G B, Petrov A G, Prasad S K, Marino L and Scaramuzza N 2017 *J. Phys. Conf. Ser.* **780** 012009
[11] Blinov L M and Chigrinov V G 1994 *Electrooptic Effects in Liquid Crystal Materials* (New York: Springer-Verlag)
[12] Blinov L M 1978 *Electro- and Magnetooptics of Liquid Crystals* (Moscow: Nauka) [in Russian]
[13] Maier W and Meier G 1961 *Z. Naturforsch. A* **163** 262
[14] Grant F A 1958 *J. Appl. Phys.* **29** 76
[15] Barsoukov E and Macdonald J R 2005 *Impedance Spectroscopy: Theory, Experiment, and Applications,* 2nd edn. (Hoboken: John Wiley & Sons)
[16] Garbovskiy Y and Glushchenko I 2015 *Crystals* **5** 501
[17] Basu R, Kinnamon D and Garvey A 2016 *Liq. Cryst.* **43** 2375
[18] Basu R 2017 *Phys. Rev. E* **96** 012707
[19] Park K A, Lee S M, Lee S H and Lee Y H 2007 *J. Phys. Chem. C* **111** 1620
[20] Basu R and Iannacchione G 2009 *Appl. Phys. Lett.* **95** 173113
[21] Kim D W, Kim Y H, Jeong H S and Jung H T 2012 *Nat. Nanotechnol* **7** 29
[22] Schymura S and Scalia G 2013 *Philos. T. Roy. Soc. A* **371** 20120261
[23] Shehzad M A, Tien D H, Iqbal M W, Eom J, Park J H, Hwang C and Seo Y 2015 *Sci. Rep.* **5** 13331
[24] De Vleeschouwer H, Bougrioua F and Pauwels H 2001 *Mol. Cryst. Liq. Cryst.* **360** 29
[25] Sasaki N 2001 *Mol. Cryst. Liq. Cryst.* **367** 671
[26] Lin F C,Wu P C, Jian B R and Lee W 2013 *Adv. Condens. Matter Phys.* **2013** 271574
[27] Basu R and Lee A 2017 *Appl .Phys Lett.* **111** 161905
[28] Basu R and Shalov S 2017 *Phys. Rev. E* **96** 012702
[29] Jeon S Y, Park K A, Baik I S, Jeong S J, Jeong S H, An K H, Lee S H and Lee Y H 2007 *Nano* **2** 41
[30] Lu S Y and Chien L C 2008 *Opt. Express* **16** 12777
[31] Jung Y U, Park K W, Hur S T, Choi S W and Kang S J 2013 *Liq. Cryst.* **41** 101
[32] Costa M R, Altafim R A C and Mammana A P 2001 *Liq. Cryst.* **28** 1779

IOP Conf. Series: Journal of Physics: Conf. Series **1186** (2019) 012032 doi:10.1088/1742-6596/1186/1/012032

Impedance and dielectric spectroscopy study of graphene-doped liquid crystal E7

G B Hadjichristov [1,*], T E Vlakhov [2] and Y G Marinov [2]

[1] Laboratory of Optics and Spectroscopy, Georgi Nadjakov Institute of Solid State Physics, Bulgarian Academy of Sciences, 72 Tzarigradsko Chaussee Blvd., BG-1784 Sofia, Bulgaria

[2] Laboratory of Biomolecular Layers, Georgi Nadjakov Institute of Solid State Physics, Bulgarian Academy of Sciences, 72 Tzarigradsko Chaussee Blvd., BG-1784 Sofia, Bulgaria

[*]Corresponding author's e-mail address: georgibh@issp.bas.bg

Abstract. The change in electrical conductivity and dielectric response of liquid crystal (LC) E7 by doping with graphene nanoparticles was experimentally studied by complex electrical impedance and dielectric spectroscopy over the frequency range from 0.1 Hz to 1 MHz of the applied electric field. In the measured graphene/E7 nanocomposite films (7 μm-thick, with planar orientation), graphene nanoflakes were dispersed in the nematic E7 at the concentration of 10^{-3} wt.%. Graphene nanodopants lead to a change of both electrical transport and dielectric permittivity of the LC E7 which should render improved electro-optic response of the considered room-temperature nanocomposite nematic films.

1. Introduction

Nanodoping with small amount of graphene is currently trending nanotechnology to largely improve the properties of liquid crystals LC (LCs), in particular their electrical and electro-optic responses. As found quite recently, graphene nanoparticles can considerably depress the ion transport [1-3] and reduce the ionic shielding and related ion-induced side effects in the practical LC devices, e.g., LC displays [4-6]. Such negative effects occur due to free ion carriers present in the LCs [7]. Although the initial density of mobile ions in the nematic LCs is relatively low, the ion-charge effects can strongly influence the performance of electro-optical (EO) devices based on nematic LCs [8,9]. In the study presented here, we deal with thin (7 μm) planar films of nematic LC E7 doped with 10^{-3} wt.% graphene nanoparticles. In order to assess the effect from this small amount of graphene doping, we compared the electrical and dielectric response of such nanocomposite nematic films with that of the same LC E7 but undoped. The study is in the area of physics/material nanoscience and nanotechnology, and is closely related to practical applications of nematic nanomaterials.

2. Experimental

The LC mixture E7, composed of 4-cyano-4′-*n*-pentyl-biphenyl (5CB), 4-cyano-4′-*n*-heptyl-biphenyl (7CB), 4-cyano-4′-*n*-octyloxy-biphenyl (8OCB), and 4-cyano-4"-*n*-pentyl-*p*-terphenyl (5CT), has been widely used in LC devices due to its large birefringence (~ 0.2 in the visible) and wide temperature range of the nematic phase (from −10 °C to 58 °C). E7 has relatively large positive dielectric anisotropy, $\Delta\varepsilon_{\|} = 19$ and $\Delta\varepsilon_{\perp} = 5.2$ at 20 °C and 1 kHz [10]. The graphene nanoparticles used in the present study (nanoflakes with an average thickness of 12 nm) were in the form of nanopowder – a commercial product from Graphene Supermarket Inc. Subsequently, they were ball milled. The graphene nanoflakes was added at the concentration of 10^{-3} wt.% to the nematic LC E7 (from Merck). To obtain a homogeneous dispersion, graphene-E7 mixture was tip-sonicated for 30 min at 4000 rpm by DW-SLD20-1200 (Dowell Ultrasonic Technology). Then this mixture was heated above 60 °C to reach isotropic phase of the LC and was capillary filled into glass cells with a gap of 7 ± 0.5 μm. The inner surfaces of the glass plates of the cells have transparent conductive layers of indium-tin oxide (ITO) (to serve as electrodes), being further overcoated with polyimide alignment layers, parallel-rubbed to achieve a planar orientation of the films. As reference cells, identical empty cells were filled with undoped LC E7.

Complex electrical impedances of the films were measured by a potentiostat/galvanostat (SP–200, Biologic) in the frequency range 0.1 Hz – 1 MHz of the external alternating-current (AC) electric field. The electrically active area (the contact area, the area of ITO) of the samples were $A = 1$ cm^2, as imposed from the ITO electrodes of the cells. Sinusoidal voltage of 10 mV$_{RMS}$ was applied to the ITO electrodes. At such voltage, the examined LC films keep their initial planar orientations. The temperature of the films was maintained by Mettler FP82 hot stage having an accuracy of ± 0.1°C. The complex impedance spectra were recorded at 5 °C intervals within the range from 15 °C to 60 °C.

3. Results and discussion

By the technique of complex electrical impedance spectroscopy, both real (Z_r) and imaginary (Z_i) parts of the complex electrical impedance (Z^*) were simultaneously measured as a function of the frequency (f) of the AC electric field applied on the sample. The Nyquist plots (Z_i versus Z_r) for the studied samples demonstrated semi-circular arcs (figure 1a). The deep valleys in these plots are associated with the bulk resistance (R_b) of the samples. The R_b values can be associated to the projections of the valleys on the abscissa (Z_r axis) [11,12]. As seen from figure 1(a), R_b of our thin planar films with graphene-doped E7 is approximately twice higher than R_b of such films with pure E7 and this applies to any temperature in the range 15 – 60 ° C (figure 1b).

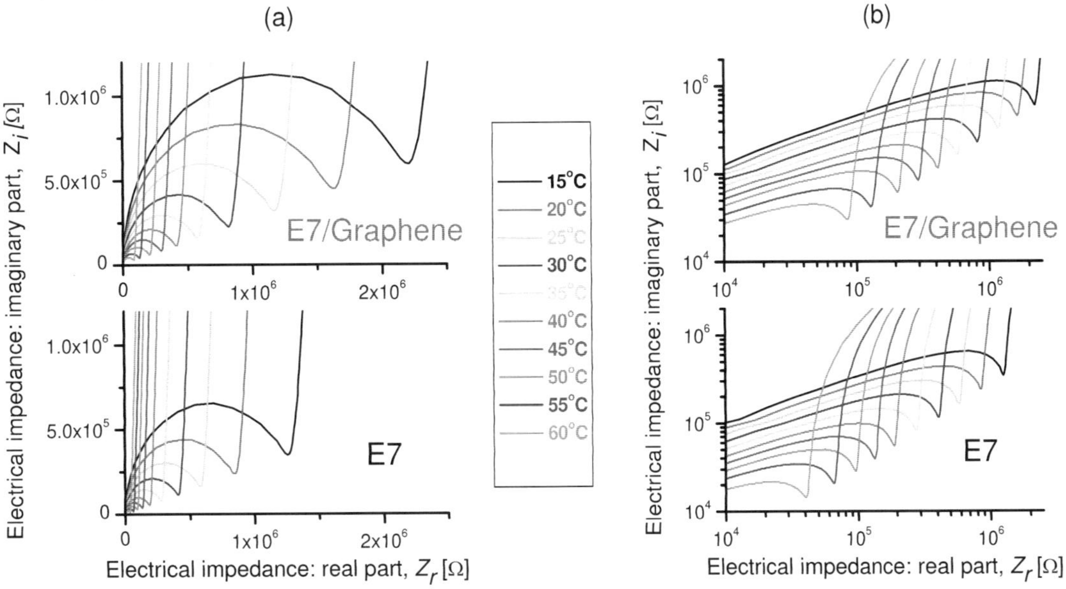

Figure 1. The complex Nyquist plots for planar films of graphene/E7 nanocomposite and undoped E7, measured upon identical experimental conditions, but at various temperatures; (b) The same as in (a) but in logarithmic scale

The corresponding values of the ionic conductivity (σ) of the samples (in direction that is perpendicular to the nematic director) were calculated by $\sigma = d/(R_b A)$, where d and A represent the film thickness and electrically-active area of the samples, respectively. This bulk conductivity is relevant to the low-frequency region ($f < 10^2$ Hz) where the dielectric behavior is attributable to the space-charge polarization. Hence, the lower σ resulting from the addition of graphene nanoparticles to the ion-rich nematic host E7 implies that they effectively capture mobile ions and reduce the space-charge polarization.

Indeed, the reducing of the ionic conductivity of the nematic LC E7 due to the presence of graphene nanoparticles can be explained by ion trapping effect, well known for nematic LCs doped by relatively small amount of graphene nanoparticles [5]. Basically, the LCs are weakly conducting dielectrics. Their ionic conduction comes from impurities in these materials (uncontrollable contamination during their synthesis). The impurity ion concentration in cyanobiphenilic E7 is relatively low and corresponds to electrical conductivity ~ $10^{-8} - 10^{-7}$ S/m [13]. However, a charge injection from the cell electrodes may induce migration of free ions from the electrodes into the bulk of the LC medium. The presence of alignment layers and planar boundary conditions should increase both the concentration and the mobility of the free ions in the bulk of the nemtic LCs in upon electric field. In the case of the nematic LC E7, this fact is well established, e.g. in the frequency range 0.5 Hz – 200 kHz [14]. As known, the high density of mobile ions in the nematic LCs leads to several negative effects that restrict their EO applications [6,8,9,15-17]. In this context, it should be noted that an improvement of EO, optical and non-linear optical responses has been reported for a number of nematic LCs doped with graphene and graphene derivatives [1,18-21]. Most probably this happens owing to ion trapping due to interaction (π-π electron stacking and interfacial interactions) between graphene nanodopants and nematic LC molecules [5,22]. We have to point out that the result for the depression of the ion-charge effects observed here for the considered graphene/E7 nanocomposite is consistent with the scientific reports published previously.

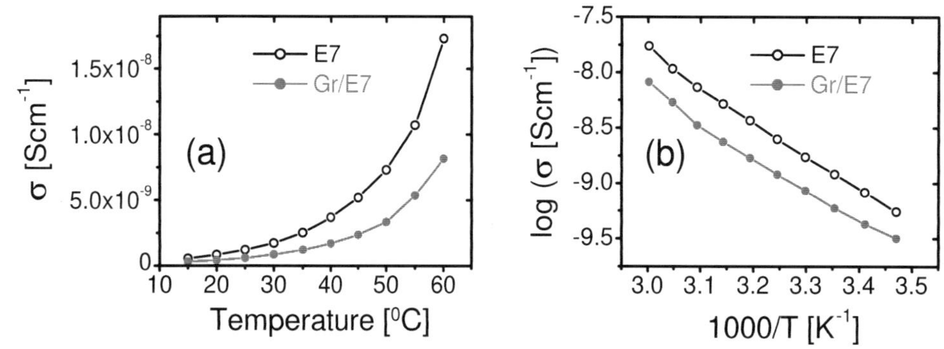

Figure 2. Variation of ionic conductivity (σ) of E7 and graphene/E7 samples vs temperature.

Figure 2 shows the temperature dependence of the ionic conductivities of E7 and graphene/E7 samples. The σ values for both E7 and graphene/E7 increase by elevating temperature (figure 2a) that is reasonable. Further, the plots for both materials demonstrated a linear relation between the logarithmic conductivity ($\log \sigma$) and inverse temperature (figure 2b), i.e., their temperature behaviours follow the Arrhenius relation $\sigma = \sigma_0 \exp(-E_a/kT)$, where T, k, E_a and σ_0 are the absolute temperature, the Boltzmann constant, activation energy and pre-exponential factor, respectively. The linear slopes (s) in figure 2(b) suggest a hopping mechanism of ion conduction (hopping of charge carriers between localized states). Determined from linear fits to the data in figure 2(b) according to the Arrhenius equation, the linear slope is related to the values of E_a by $E_a = - s \times R$, where $R = 8.314$ J.mol^{-1}.K^{-1} is the universal gas constant. Within the uncertainty limits by the measurements, E_a values calculated in this way for both graphene/E7 and undoped E7 are nearly equal, hence a very little change in the threshold energy for the ions to hop from one site to another takes place due to doping with graphene nanoparticles. This can be attributed to negligible structural changes in the nematic LC E7 that are relevant to and influence the ion transport, e.g., eventual change in viscosity that could have impact on the diffusivity of the mobile ions. The lack of such structural modifications corroborates the ion-trapping mechanism as a main reason for lowering of the ionic conductivity of E7 by graphene doping.

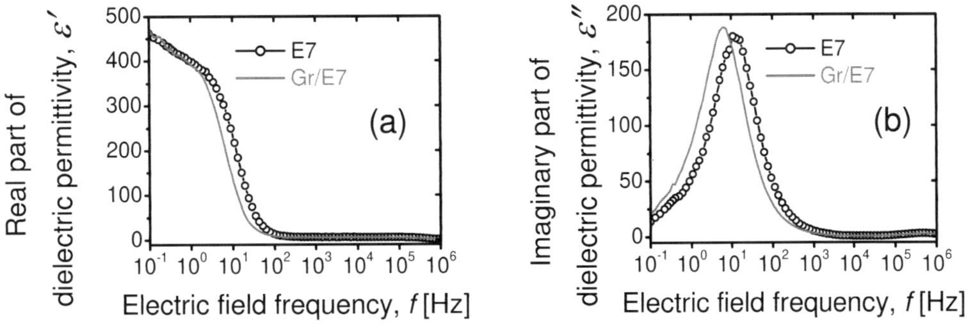

Figure 3. (a) The frequency spectra of the real (a) and imaginary (b) parts of the complex dielectric permittivity function ε* for planar films of graphene/E7 nanocomposite and undoped E7 measured at 26° C and identical other experimental conditions.

In fact, a clear difference between the graphene/E7 nanocomposite and the undoped E7 was also present in their dielectric spectra (figure 3). The real (ε') and imaginary (ε'') components of the complex dielectric permittivity were calculated from the measured frequency spectra $Z_r(f)$ and $Z_i(f)$ by means of the well known relations [12]:

$$\varepsilon'(f) = -Z_i d / [2\pi f \varepsilon_0 A (Z_r^2 + Z_i^2)] \tag{1}$$

and

$$\varepsilon''(f) = Z_r d / [2\pi f \varepsilon_0 A (Z_r^2 + Z_i^2)] \tag{2}$$

where $\varepsilon_o = 8.85 \times 10^{-12}$ F.m^{-1} is the permittivity of free space. ε' has the same significance as that of the ordinary dielectric constant of the material. It is a measure of the energy stored in the material during each cycle, to be returned to the electric field at the end of the cycle. The high values of ε' in the low-frequency region ($f < 10^2$ Hz)) can be attributed to a build-up of space charge near electrode/LC interface, which partially blocks the charge transport. With increasing f, ε' decreases up to a nearly constant value, because at such frequencies the ions will hardly able to follow the periodic reversal of the electric field. Significantly, graphene-doped E7 exhibits a lower value of dielectric constant ε' than the undoped E7. This difference is most pronounced in the low-frequency region, in our case from 1 Hz to 100 Hz. As known, in this range the dielectric behavior is attributable to the space-charge polarization, i.e. the addition of graphene nanoparticles to the ion-rich nematic host E7 effectively leads to capturing of mobile ions and reducing of the space-charge polarization. The decrease in the dielectric constant represents a fractional decrease in charge density due to the addition of graphene nanoparticles in the LC E7, resulting in a lower ion conductivity.

Using the frequency spectrum of ε'', one can calculate the dielectric loss tangent, tan $\delta = \varepsilon''/\varepsilon'$ [11], and the real part of the AC conductivity $\sigma'_{AC} = 2\pi f \varepsilon_o \varepsilon''$ [23]. Figure 4 shows the corresponding frequency-dependent characteristics. Regarding tanδ, the effect from graphene nanodopants was evident in the frequency range 1 Hz – 10 kHz (figure 4a). In the half of this range ($f < 10^2$ Hz), the dielectric loss is enhanced by graphene nanodopants, and in the other part is reduced. The shift of the frequency dependence of tanδ by doping with graphene suggests that the relaxation frequency of the graphene-doped E7 is different (slightly lower) from that of pure E7 sample, i.e. the graphene nanodopants do affect at detectable level the molecular dynamics of the nematic LC molecules. As a ratio of dielectric loss factor to the dielectric constant, the loss tangent is a measure of the ratio of the electrical energy lost to the energy stored in a periodic field. The shift of the frequency dependence of tanδ towards the lower frequencies by doping with graphene nanoparticles may be a sign that the nanodopants effectively suppress the space-charge polarization and ion-charge effects.

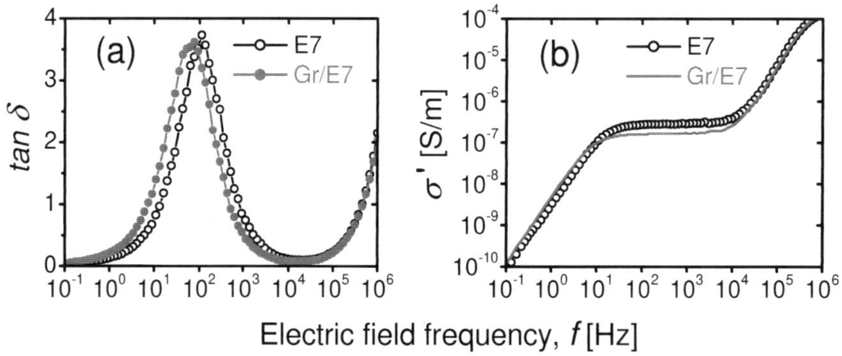

Figure 4. Frequency spectra of dielectric loss tangent (a) and the real part of the AC conductivity (b) for pure E7 (open circles) and graphene/E7 nanocomposite (colored with red), at 26° C.

The frequency-dependent behaviour of σ'_{AC} (in our case, measured perpendicular to the nematic LC director) is shown in figure 4(b). It is apparent that σ'_{AC} increases with f, but a nearly frequency-independent plateau at intermediate frequency region takes place. The increasing conductivity behaviour is relevant to the electrode polarization effects, while saturation-like behaviour results from the failure of movement of ions. As seen from figure 4(b), in the frequency range 10 Hz – 10 kHz graphene/E7 nematic nanocomposite displays a lower σ'_{AC} than that of the undoped nematic LC E7. Both reduced AC conductivity and dielectric loses in the above mentioned frequency range can be connected with the immobilization of the ions in the presence of graphene. In particular, these effects are well pronounced by graphene-doped ferroelectric LCs [5, 20,24,25].

4. Conclusion

Our study shows that the electrical and dielectric behaviours of planarly-aligned relatively thin films of nematic LC (E7 in the present case) are influenced by its doping with graphene nanoflakes, even at very low concentration of 10^{-3} wt.%. Essentially, the results obtained here confirm the concept that the doping with graphene nanoparticles reduces the ion diffusivity, and thereby, lowers the room-temperature electrical conductivity of the doped nematic LCs. This applies to the ionic conductivity (relevant to the low frequencies) (and direct-current conductivity), as well as to AC conductivity in certain frequency range (in our case, 10 Hz – 10 kHz). Slight amounts of graphene nanodopants reduce the concentration of impurity and injected ions and their diffusivity, thus depressing the unwanted ion-charge effects in the nematic LCs. The ion immobilization effect through graphene nanoparticles is well pronounced. As such, it will be useful in view of practical applications of nematic nanocomposites produced from graphene-doped nematic LCs (e.g., as EO materials in high-quality LC display technology). Providing an efficient reducing of the negative side effects, the purification by graphene nanodopants is of paramount importance for the functionality of nematic LCs EO devices and for applicability of nematic LCs in organic electronics. Further, the considered important modification is useful for development of new advanced and multifunctional nanocomposite nematic materials working at ambient temperatures, as well as over a wide temperature range of the nematic phase.

Acknowledgments
Work supported by research project DFNI-TO2/18 "Nanostructured liquid crystals for tunable photonic devices" by the Ministry of Education and Science, through the National Science Fund of Bulgaria. One of us (T E V) acknowledges the project DPMNF 01/7–23.07.2018 by the Ministry of Education and Science, Bulgarian National Science Fund.

References

[1] Gökçen M, Yıldırım M and Köysal O 2012 *Eur. Phys. J. Appl. Phys.* **60** 30104
[2] Wu P W and Lee W 2013 *Appl. Phys. Lett.* **102** 162904
[3] Garbovskiy Y and Glushchenko A 2017 *Nanomaterials (Basel)* **7** 361
[4] Urbanski M 2015 *Liq. Cryst. Today* **24** 102
[5] Garbovskiy Y and Glushchenko I 2015 *Crystals* **5** 501
[6] De Vleeschouwer H, Verweire B, D'Have K and Zhang H 1999 *Mol. Cryst. Liq. Cryst.* **331** 567
[7] Blinov L M 1983 *Electro-Optical and Magneto-Optical Properties of Liquid Crystals* (Chichester: J. Wiley)
[8] De Vleeschouwer H, Bougrioua F and Pauwels H 2001 *Mol. Cryst. Liq. Cryst.* **360** 29
[9] De Vleeschouwer H, Verschueren A, Bougrioua F, Van Asselt R, Alexander E, Vermael S, Neyts K and Pauwels H 2001 *Jpn. J. Appl. Phys.* **40** Part 1 3272
[10] Drzaic P S and Muller A 1989 *Liq. Cryst.* **5** 1467
[11] Grant F A 1958 *J. Appl. Phys.* **29** 76
[12] Barsoukov E and Macdonald J R 2005 *Impedance Spectroscopy: Theory, Experiment, and Applications,* 2nd edn. (Hoboken: John Wiley & Sons)
[13] Chigrinov V G 1999 *Liquid Crystal Devices: Physics and Applications* (Boston: Artech House) p 360
[14] Murakami S and Naito H 1997 *Jpn. J. Appl. Phys.* **36** 2222
[15] Yang K H 1990 *J. Appl. Phys.* **67** 36
[16] Takahashi S 1991 *J. Appl. Phys.* **70** 5346
[17] Sasaki N 2001 *Mol. Cryst. Liq. Cryst.* **367** 671
[18] Kumar V, Kumar A, Biradar A M, Reddy G B, Sachdev D and Pasricha R 2014 *Liq. Cryst.* **41** 1719
[19] Basu R, Garvey A and Kinnamon D 2015 *J. Appl. Phys.* **117** 074301
[20] Lapanik V, Timofeev S and Haase W 2016 *Phase Transitions* **89** 133
[21] Al-Zangana S, Iliut M, Turner M, Vijayaraghavan A and Dierking I 2016 *Adv. Opt. Mater.* **4** 1541
[22] Basu R, Kinnamon D and Garvey A 2016 *Liq. Cryst.* **43** 2375
[23] Smyth C P 1955 *Dielectric Behavior and Structure: Dielectric Constant and Loss, Dipole Moment and Molecular Structure* (New York: McGraw-Hill)
[24] Singh D P, Gupta S K, Vimal T and Manohar R 2014 *Phys. Rev. E* **90** 022501
[25] Basu R 2014 *Appl. Phys. Lett.* **105** 112905

IOP Conf. Series: Journal of Physics: Conf. Series **1186** (2019) 012033 doi:10.1088/1742-6596/1186/1/012033

Development of low-cost technology to obtain carbon based supercapacitors

N Tyutyundziev[1,2], M Petrov [1], K Lovchinov [3], H Nichev [1], H Angelov[4], P Petrov[1]

[1]Institute of Electronics, 72 Tzarigradsko chausse, Blvd. Sofia, Bulgaria
[2]Central Laboratory of Solar Energy and New Energy Sources, 72 Tzarigradsko chausse, Blvd. Sofia, Bulgaria
[3]Institute of Optical Materials and Technologies, Bulgarian Academy of Sciences, Acad. G. Bonchev Str., bl. 109, 1113 Sofia, Bulgaria
[4]Moussala BEO, Institute for Nuclear Research and Nuclear Energy, 72 Tzarigradsko chausse Blvd. Sofia, Bulgaria

E-mail: pv-jet@phys.bas.bg

Abstract. Recent advances in supercapacitor technology reveals new opportunities for energy storage devices for application in electro-transportation and renewable energy sectors. In this paper results on device fabrication are reported. A low-cost screen-printing technology to obtain electrodes of supercapacitor has been applied. The device structure is based on Cu plate/hydrocarbon ink electrodes and gel electrolyte. Carbon ink is prepared from naturally occurring not full carbonized particles in the chimney's soot. The cyclic charge/discharge measurements confirm supercap behaviour and specific capacitance around 5 F/g at discharge current of 0.62 A/g. The flat configuration of the supercaps allows easy connection in series and adjustment to DC loads. A pack of four supercaps driving conventional LED has been thoroughly tested for more than 100 cycles.

1. Introduction

Nowadays supercapacitors are used frequently in autonomous power supply systems. Large capacitance allows generating a small value current for a long period time or alternatively to provide high value current for a short period.

The electrode material is a key component that determines the electrical capacity of the device [1]. Supercapacitors based on activated carbon electrodes have a wide pore size distribution [2], consisting of micropores (< 2 nm), mesopores (2 – 50 nm), and macropores (> 50 nm). Mesopores contribute the most to the capacitance in an electrical double layer capacitor (EDLC) [3]. In principle, pores greater than 0.5 nm are electrochemically accessible for aqueous electrolytes [4].

A typical supercapacitor cell is comprised of two microporous electrodes that are isolated from electrical contact by a separator. Metallic current collectors are used to conduct electrical current from each electrode to outside circuitry. The separator and the electrodes are impregnated with an electrolyte, which allows ionic current to flow reversibly between the electrodes [5]. Active electrode material could be produced from many sources [6 - 9] but in this investigation the emphasis is placed on low-cost resource - domestic soot.

Soot is a particulate carbonaceous matter with a characteristic lattice structure. It is formed during high temperature pyrolysis or combustion of hydrocarbons. It consists mainly of carbon and small amounts of other elements, such as hydrogen and oxygen [10]. Assuming that soot is formed as is described in [11] they consist of internal graphite core and nascent poly-aromatic hydrocarbons (PAHs) with different functional groups [12] which increase absorption capability of the material [13].

This paper presents results of identical supercapacitors connected in series. Each element is operating at (1.0-1.1 V) and utilizes stable inorganic electrolyte.

2. Experiment

2.1. Ink preparing

In this experiment we test ink produced from particles derived from soot. The recipe is as next: The soot was collected from a chimney of domestic fire system, after pine tree burning. The collected soot is flooded in ethanol 96% and threated in ultrasonic wane for 10 minutes. After that the mixture is sieved through primary 150μm mesh and washed with additional amount of ethanol. All contents which pass through the mesh are collected and sieved again with a secondary 47 μm mesh and washed with additional amount of ethanol. The mixture is left for 48 hours to allow the particles to precipitate, after that the liquid is removed and the grouts are dried in air at 150°C.

To prepare the ink dried grouts are added to PVA / DI water solution. PVA acts as a binder. By screen-printing technology or brush the as prepared ink is deposited on the Cu foil. The electrode with approx size 4 x 5 cm is ready after careful drying to avoid cracking and defoliation. The deposited ink mass is measured at 0.06g using microbalance

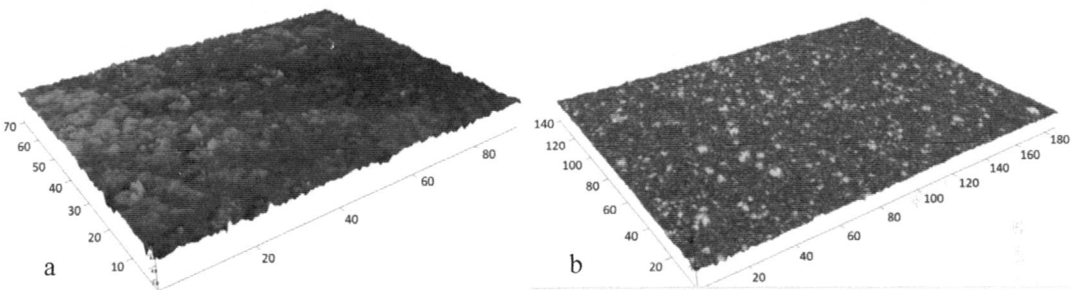

a b

Figure1.An electrode surface with as prepared ink, 3D surface roughness a); optical image b)

2.2. Device assembling

Symmetric capacitor is prepared in sandwich structure by pressing together two electrodes and separator film in between. The separator and electrolyte are added together. Filter paper (2-3 μm pore size) soaked with the electrolyte (3ml, 5M KOH) solution is used as separator. The composite is dried in the air at 90°C for about 15 min. The volume and concentration defines the exact amount of KOH (~0.86g). The as prepared capacitor is encapsulated in PE plastic foil pack using hot lamination technique. Electrical measurements are then carried out.

2.3. Electrical measurements

Measurement procedures include electrochemical measurements and device parameters such as electrode mass and thickness. Cell capacitance is best determined from galvanostatic or constant current (CC) discharge curves. In our case self-made CC charge/discharge testing tool has been developed. The charging current is controlled by a simple C++ code, current source and an open-source programmable microcontroller. In order to increase the activation of micro pores charge/dicharge cycle in a ratio of 10:1 is appplied. At the initial stage, the voltage is starting from 0V, the supercapacitor testing structure is charged at 20 mA for 8 s up to 1.0V, held at open circuit for 2 s, and subsequently discharged for 10 s at 2 mA. As can be seen at Figure 2, the sequences are repeated with slow drift up to the limit of 1.2V to prevent splitting of the water content of the electrolyte. The slow increase of trapped charge is an indication of supercapacitor behaviour.

The voltage versus time profile during the constant current discharging process can be approximated with a straight line after a correction of the initial voltage drop (IR drop), and the slope of the line can be used to calculate the cell capacitance from the relation $C = I/(dV/dt)$.

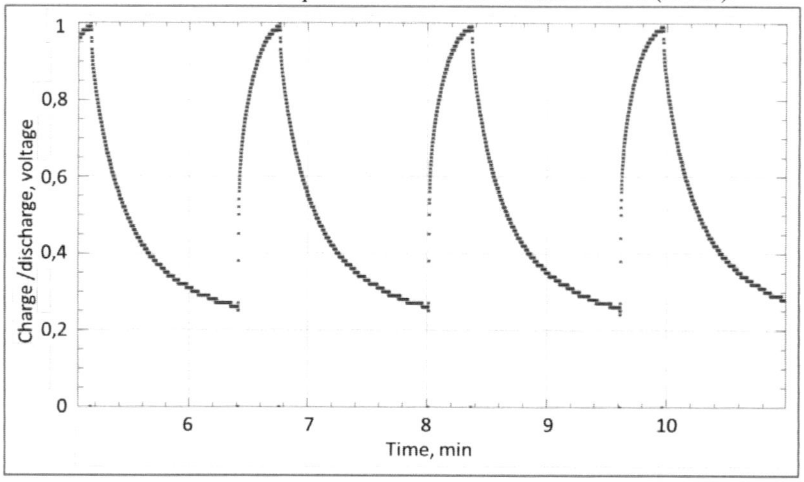

a

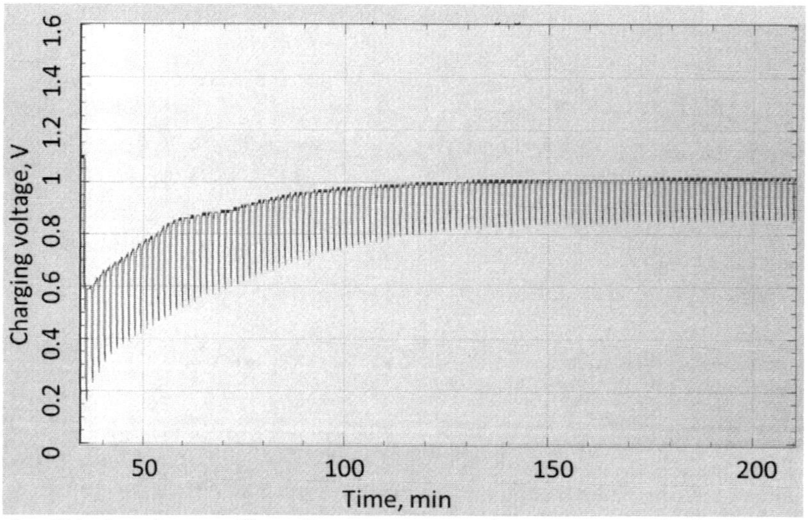

b

Figure 2 . Charging/Discharging profiles of test structures a) narrow gap - 5min, b) nearly 4 hour charge/discharge.

Portable electronics needs operating voltages in the range of 1.5V -3 V which determines the voltage window of energy storage devices. To testify the real potential application of the devices, we connected four capacitors in a pack. The useful voltage window of the device is extended to 4.2V. The charge-discharge duration of the developed device is nearly the same as for individual element suggesting consistent capacitance performance of each supercapacitor.

Finally, we tried to charge the serial pack of capacitors and succeed to light up a blue light-emitting diode (LED) which has operating potential at about 3.0V (illustrated in figure 3). The LED can light up over 300 seconds thus it proves the potential of LED driving application.

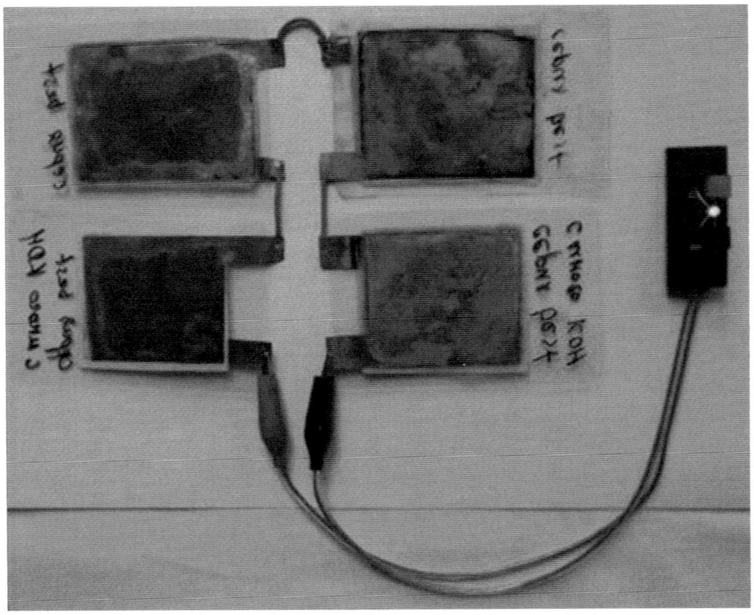

Figure 3. SuperCAPs charging pack

3. Conclusions

A supercapacitor based on waste material without expensive intentionally created carbonaceous materials has been fabricated, tested and demonstrated in an application. The analysis of the results revealed a potential for soot implementation in supercapacitor devices with a suitable treatment for increasing the parameters of the capacitors. Connection in series is possible and blue LED may be operated for 300s with a prototype. The specific capacitance (F/g) of whole capacitor in our case simply could increase with increasing quantity of KOH, by using thinner separator and by charging at a maximal voltage.

References

[1] Vasile V N, Obreja 2014 *AIP Conference Proceedings* **98** 1597
[2] Wu F, Tseng R L, Hu C C, Wang C C, 2004 *Journal of Power Sources* **138** 351–59
[3] Gryglewicz G, Machnikowski J, Lorenc-Grabowska E, Lota G, Frackowiak E, 2005 *Electrochimica Acta,* **5** 1197-1206
[4] Frackowiak E, Beguin F 2001*Carbon* **39** 937–50
[5] Sakka M A, Gualous H, Omar N, Mierlo J Van 2012, *Batteries and Supercapacitords for Electric Vechicles* **5** http://dx.doi.org/10.5772/53490
[6] Pérez-Madrigal M M, Edo M G, Alemán C, 2016 *Green Chem.* **18** 5930
[7] Adinaveen T, Vijaya J J, Sivakumar R, Kennedy L J 2016 *Materials Science Poland* **34(2)** 302-14
[8] Farma R, Deraman M, Awitdrus, Talib I A, Omar R, Manjunatha J G, Ishak M M, Basri N H, Dolah BN M 2013 *Int. J. Electrochem. Sci.* **8** 257 – 73
[9] Rufford T E, Hulicova-Jurcakova D, Fiset E, Zhu Z, Lu G Q 2009 *Electrochemistry Communications* **11** 974 - 77
[10] Xi J, Zhong B J 2006 *Chem Eng Technol* **29** 65–73
[11] Kholghy M R, Veshkini A, Thomson M J 2016 *Carbon* **100** 508-36
[12] Raj A, Robert da Silva G, Chung S H 2012 *Combustion and Flame* **159** 3423–36
[13] Ternero-Hidalgo J J, Rosas J M, Palomo J, Valero-Romero M J, Rodríguez-Mirasol J, Cordero T 2016 *Carbon* **101** 409-19

IOP Conf. Series: Journal of Physics: Conf. Series **1186** (2019) 012034 doi:10.1088/1742-6596/1186/1/012034

Graphene/Polyaniline flexible supercapacitors using non-metalic electrodes

N Tyutyundzhiev[1,2], K Lovchinov[3], M Petrov[1] and Hr Nichev[1]

[1]Institute of Electronics, 72 Tzarigradsko chausse, Blvd. Sofia, Bulgaria
[2]Central Laboratory of Solar Energy and New Energy Sources, 72 Tzarigradsko chausse, Blvd. Sofia, Bulgaria
[3]Institute of Optical Materials and Technologies ''Acad. J. Malinowski'', Bulgarian Academy of Sciences, Acad. G. Bonchev str., bl. 109, 1113 Sofia, Bulgaria

E-mail: pv-jet@phys.bas.bg

Abstract. Flexible supercapacitors are attractive devices for portable applications including those integrated in clothes. Highly conducting flexible electrodes can easily be fabricated by graphene powder and PE polymer foil but reliable fabrication of supercapacitor structure is still a challenge. A promising solution could be the incorporation of conducting polymer as polyaniline (PANI) intended to stabilize the mesoporous carbon and graphene plates, to form a conductive porous composite and to increase the capacitance. Low-cost aqueous electrolyte (1V) supercapacitor devices based on graphene/PANI composite were developed, tested in series configurations and compared in electrical performance. Preliminary applications in portable solar chargers are discussed.

1. Introduction

In Standalone-powering systems with small renewable energy generator (series of solar cells, small wind-generators or Peltier elements) are popular in smart energy microsystems as a reliable, low-cost and environmentally independent source of energy. Nowadays, the bottleneck in energy production pipeline has been transferred from energy generation to short-time energy storage devices capable of withstanding the increasing need for electricity. By this reason storage unit (for example, series of supercapacitor devices) have recently attracted more and more attention for use in novel applications especially for flexible electronics [1-3].

The principle of energy storage in a supercapacitor can be either electrostatic surface charge accumulation in double layers capacitors (EDLC) or charge transfer, via reversible (Faradaic) redox reactions known as pseudo capacitance. The high surface area and high cycle life, makes these novel devices preferred energy storage banks for many applications based on non-constant renewable energy sources and also regenerative braking in vehicles.

Carbon nanomaterials, especially graphene and carbon nanotubes (CNTs), are very promising candidates to replace activated carbons due to their large surface area, high mesoporosity and electrolyte accessibility, and excellent electrical conductivity. The mesoporous structure of nano-carbon materials have been introduced by many authors [4, 5] due to its low cost and easy fabrication. In various papers [6], research efforts have been focused on design of 3-D structures based on graphene oxide (GO), reduced graphene oxide (rGO), CNTs materials as well as novel hybrid composites based on graphene and conducting polymer or metal oxides.

Graphene sheet is a two-dimensional (2-D) honeycomb lattice of one-atom-thick layer of sp^2-hybridized carbon with good mechanical properties and additional flexibility. Of particular interest, graphene sheets with tunable through-thickness p–p stacking interactions and/or edge functionalities have been shown [7]. Within a reasonable range of oxygen functional groups content in graphene sheets, a large specific area and high conductivity could be obtained.

In this paper recent results are reported on low-cost technologies for fabrication of flexible flat all-carbon supercapacitors using graphite conducting electrodes prepared on PE polymer substrate.

2. Experimental

2.1. Graphene exfoliation

Nano graphene plates has been produced by many authors [6] using oxidation of graphite. The carbon sheets can be easily "unfolded" by applying liquid-assisted exfoliation procedures [8]. In our study, electrochemical procedure of graphite electrodes exfoliation in $(NH_3)_4SO_4$ is applied. Electrochemical potential of 10 V DC between graphite cathode and anode electrodes for 2 hours and electrolyte solution of 0.1M $(NH_3)_4SO_4$ are effective low-cost method for graphene plate extraction. Next step is cleaning of SO_4 radicals by rinsing 3 times in DI water at 80°C for 30 min followed by carbon microcrystal separation by filter paper. Graphene-based powder has been dried and prepared for electrodes fabrication. Microscopic photos of dispersed graphene plates on glass surface are presented in figure 1.

Homogeneous conducting graphene layers has been further obtained by rubbing the powder carefully in between two 125 μm Polyethylene sheets (PE) and by hot laminating the sandwich structure at 70°C together. After lamination the PE sheets was exfoliated mechanically. Double-exfoliated graphene layers have been obtained on the internal surfaces of laminated PE foils. Graphene oxide platelets form highly developed surface as can be seen at figure 2.

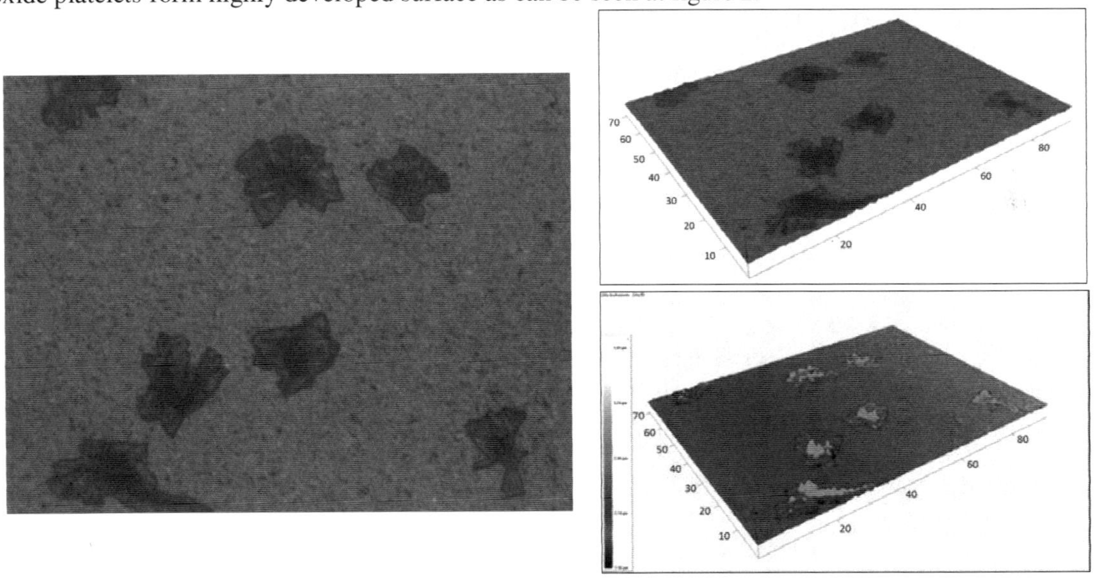

Figure 1. Graphene plates.

2.2. Supercapacitor based on Graphene/PANI composite

The problem with aggregation and re-stacking of graphene sheets caused by Van der Waals force tends to result in a loss of surface area and reduced capacitance. Therefore, intermediate external particles or structures between graphene sheets could improve mesoporosity. Conducting polymers are being explored as supplementary materials for supercapacitors owing to their flexibility, low cost and high conductivity in doped states with little environmental hazards. Graphene layer can serve as a underlying conductive network, while polyaniline (PANI) can serve as a spacer to separate neighboring graphene sheets. Capacitance values higher than EDLC capacitors are possible to achieve due to fast pseudo-faradaic charge-transfer in the polymer. The GO/PANI composite material is expected to improve the overall capacitance values and cycling stability, as conducting polymers enhance the capacitance of carbon nano-materials by introducing voltage-dependent pseudo capacitance.

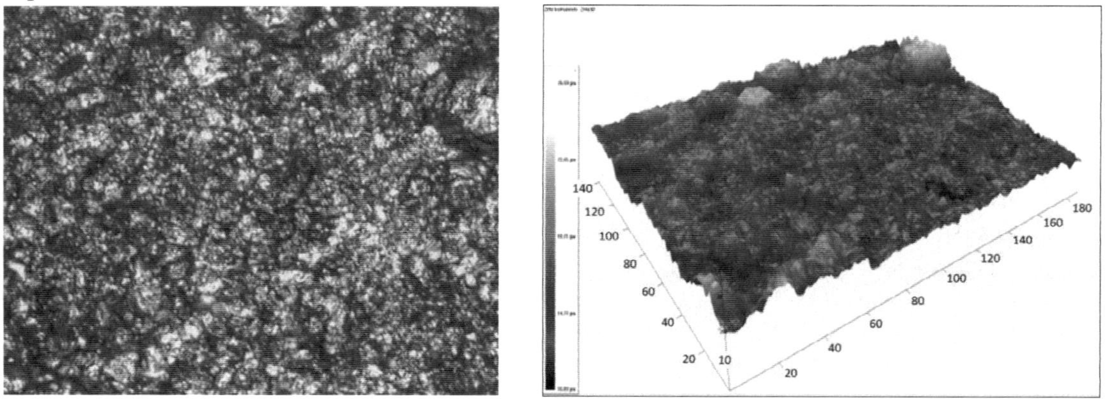

Figure 2. Graphene layer on PE polymer substrate.

Moreover, the GO/PANI composite displays an excellent electrochemical performance due to an extra synergistic effect between PANI and GO. The availability of functional groups on GO, such as hydroxyl, epoxide, carbonyl and carboxyl groups on its basal planes and edges, can be utilized for chemical functionalization and doping of conducting polymers. GO with oxygen-contenting groups can provide a large number of active sites for the nucleation of PANI and for reversible modulation of conductivity.

It seems that smart engineering of advanced 3-D structures is a promising option for next generation of energy storage devices. The performance can be further enhanced by coating of graphene underlayered matrix with PANI polymer caps or preparation of multi-layered GO/PANI construction.

2.3. Electropolymerization of PANI

It is well known that PANI exists in three states: leucoemeraldine, emeraldine, and pernigraniline. In general, conducting polymer is a semicrystalline material and the crystallinity can be attributed to the systematic alignment of the polymer chains by chain folding or the formation of single or multiple helices. In this study, emeraldine salt of PANI has been prepared by electrochemical deposition of aniline monomer in $HClO_4$ solution on Palladium working electrode and on PE polymer/graphene electrode. The electrochemical potential of 0.75 V is sufficient to extract electron and to start polymerization of green conducting state of PANI.

Figure 3 presents optical reflection spectra of suitable emeraldine PANI film. The appearance of absorption bands at 330, 430 and 750nm for PANI is an indication of the emeraldine salt form. The characteristic bands were assigned as follows: 1) The lowest wavelength band around 330 nm was often related to the presence of cation radicals and corresponded to a π–π^* transition in the benzenoid structure; 2). The mediate band was related to the n–π^* transition and on behalf of the doping degree on polymer chains; 3) The band around 750nm was attributed to π–n transition and it is also ascribed to exciton formation in the quinonoid rings.

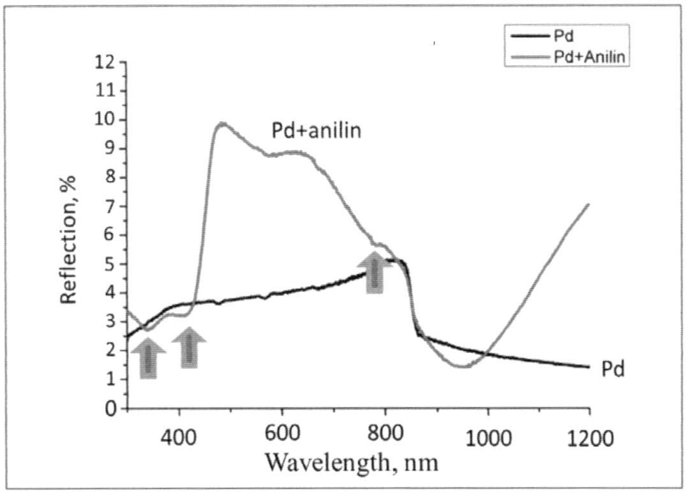

Figure 3. Optical reflection spectra of conducting emeraldine salt of polyaniline (dark green in color) deposited electrochemically on Pd electrode.

Figure 4. Emeraldine film of polyaniline (PANI) electrodeposited on the top of graphene layer.

3. Fabrication of flexible supercapacitors

3.1. Electrode preparation

Fine powder of graphene sheets (particles size < 50 um) were diluted in DMF binder to form graphene ink. The electrode is fabricated by sequential deposition of dry graphene powder on PE substrate to form current collecting electrode, electrodeposition of PANI film and screen-printing of grapheme ink. The resultant multilayer composite has conducting graphene/PANI structure. Top surface and 3-D pictures of as prepared double layered structure are shown at figure 4.

3.2. Gel electrolyte separator:

The electrolyte used is gel polymer electrolyte. The host polymer is PVA, the solution is water and the electrolytic salt is KCl. Filter paper (2-3 μm pore size) was dipped in the KCl/PVA electrolyte solution and dryed in air at 90°C for 15 min.

3.3. Device assembling

Symmetric rectangular capacitor with 25 x 70 mm area is prepared in sandwich structure by pressing together two electrodes and separator film in between. As prepared EDLC-pseudo Cap is encapsulated in PE plastics foil pack using hot lamination, as can be seen at Figure 5.

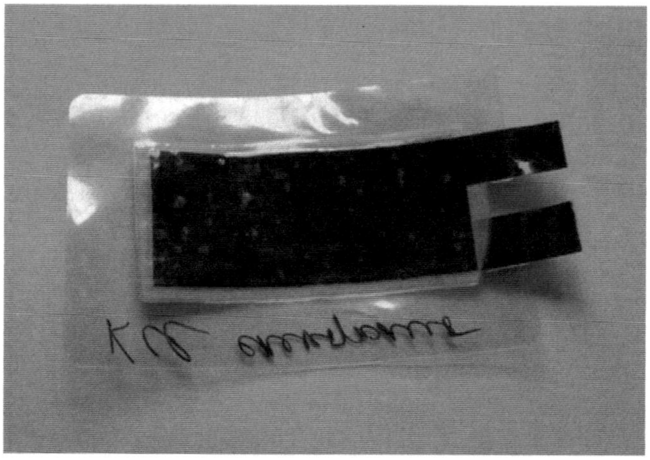

Figure 5. Encapsulated graphene/PANI structure.

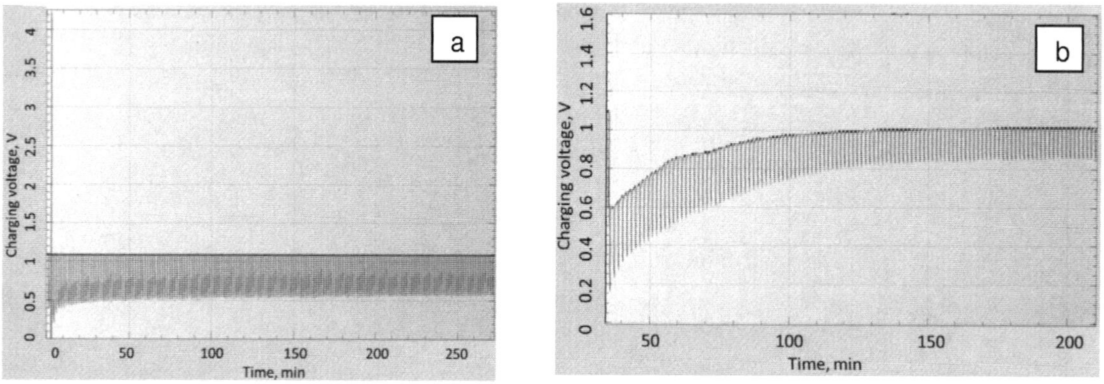

Figure 6. Comperative charge/discharge charts of two test structures a)Non-metalic graphene/PANI supercapacitor b) Cu sheet/Activated carbon supercapacitor

4. Electrical testing

The electrochemical properties of the supercapacitors are investigated by means of galvanostatic charge-discharge techniques. Pseudocapacitive and hybrid systems can exhibit large deviations in linearity based upon varying capacitance with voltage compared to pure EDLC. After preparation of graphene/PANI composite the charge-discharge curves show that the discharge time has increased. The initial full charge to 1V and bending of the discharge curve only shows extended charge trapping. Redox reaction in polyaniline has appeared during charge-discharge process together with the electrostatic charging in graphene micropores which indicates the synergistic effect of the two phenomena and improved device performance.

5. Discussions

Figure 6 shows the galvanostatic initial charging curves at charge-discharge currents of 20mA/2mA in 20s for two individual supercapacitors with the same area and gel-electrolyte but prepared on different plates -metallic and non-metallic. The non-metallic graphene/PANI supercapacitor during initial charging period has a different electrical behavior compared to Cu/activated carbon capacitor produced by the same technology. It is assumed that activated carbon needs longer time to charge pores of the electrodes while graphene/PANi structure has bigger pores or even cavities able to capture ions. The second reason is the additional source of charge - not only electrolyte ions but also the extracted electrons from polymer chains of PANI. Obviously, graphene/PANi structure is more attractive for fast charging devices, for example in portable applications.

IOP Conf. Series: Journal of Physics: Conf. Series **1186** (2019) 012034 doi:10.1088/1742-6596/1186/1/012034

A key challenge is to design and assemble flexible electrode materials which can harvest/store human body energy through actions such as breathing, arm pressing, jogging, etc. The new flexible energy storage devices need to maintain high-quality performance under continuous mechanical deformation, such as bending, folding, twisting and stretching, for long cycles. For this reason, the importance of non-metallic supercapacitor has been increased due to the extended flexibility of the incorporated materials. Flexible electrodes can be assembled in thin, light and smart designs of any shape and size. Portable electronics needs operating voltages in the range of (1.5V -3V) which determines the voltage window of flexible energy storage devices. One reasonable approach is a single element to be used based on new organic electrolytes or ionic liquids operating at higher voltages (2.5V - 4V) instead of aqueous solutions. However, recent publications reveal effects of aging and fast degradation in organic electrolyte as for example acetonitrile [9, 10], which slows down the interest. On the other hand, ionic liquids are complex toxic metal-organic compositions at an early stage of development with high prices for mass applications. Obviously, a useful low-cost "battery" approach can be formed from connected in series identical supercapacitors each of them operating at (1.0-1.1 V) and based on human-friendly stable inorganic electrolyte. Our experiments reveal no degradation in performance of up to 4 identical supercapacitors in a 2D single plane configuration.

Possible light-weight application as a portable solar charger/powerbank could be graphene/PANI supercaps connected in series and suitable energy management circuitry assembled together on the back surface a small PV module. Such solar energy configuration could be feasible for daily in-house lighting applications.

Conclusions

All-carbon based thin film supercapacitors were successfully manufactured. These structures consist of conductive carbon/plastic films as current collectors and high surface area graphene-polyaniline (PANI) mixture films as electrodes. We developed a technique to increase the surface area as well as the capacitance of the electrode layer by electrochemical –exfoliation of graphene and preparation of graphene/PANI multi-layered structure. The flexible rectangular shaped graphene/PANI supercapacitor (25 x 70 mm) were encapsulated by hot lamination. The manufacturing process has a good reproducibility. The rectangular supercapacitors can be used in various combinations of series to obtain desired energy or power. The in-plane assembly of supercapacitors in series will lead to space-saving integration of supercapacitors pack. The main advantages of the developed supercapacitors are the scalable low cost production, the inexpensive materials and the large variety of possible practical applications.

References

[1] Xu Y, Schwab M G, Strudwick A J, Hennig I, Feng X, Wu Z and Müllen K 2013 *Adv. Energy Mater* 201300184
[2] Zhang R, Xu Y, Harrison D, Fyson J, Qiu F and Southee D 2015 *International Journal of Automation and Computing* **12(1)** 43-49
[3] Dubal D P, Chodankar N R, Kim D H, Gomez-Romero P 2018 *Chem. Soc. Rev* **47** 2065-2129
[4] Markoulidis F, Lei C, Lekakou C, Figgemeier E, Duff D, Khalil S, Martorana B and Cannavaro I 2012 *IOP Conf. Ser. Mater. Sci. Eng.* **40** 012021
[5] Li B, Dai F, Xiao Q, Yang L, Shen J, Zhang C and Cai M 2016 *Energy Environ. Sci.* **9** 102
[6] Liu J, Mirri F, Notarianni M, Pasquali M, Motta N 2015 *J. of Power Sources* **274** 823-30
[7] Singh R and Tripathi C C 2016 *Materials Today: Proceedings* **16**
[8] Andres B, Forsberg S, Vilches A P, Zhang R, Andersson H, Hummelgård M, Bäckström J and Olin H 2012 *Nordic Pulp and Paper Research Journal* **27** 481-85
[9] Ruch P W, Cericola D, Foelske-Schmitz A, Kötz R and Wokaun A 2010 *Electrochimica Acta*, **55** 4412–20
[10] M Zhu, C J Weber, Y Yang, M Konuma, U Starke, K Kern, A M Bittner, 2008, *CARBON*, **46**, 1829 –1840.

IOP Conf. Series: Journal of Physics: Conf. Series **1186** (2019) 012035 doi:10.1088/1742-6596/1186/1/012035

Carrier transport in methylammonium lead iodide perovskite single crystals studied by dark current, steady state and transient photocurrent measurements

S Reynolds[1], A R Houghton[1,2] and D J Keeble[1]

[1] School of Science and Engineering, University of Dundee, Dundee DD1 4HN, UK.
[2] Present address: Department of Chemical Engineering, Imperial College, South Kensington Campus, London SW7 2AZ, UK.

E-mail: s.z.reynolds@dundee.ac.uk

Abstract. Electronic transport properties of methylammonium lead iodide perovskite single crystals prepared by inverse temperature crystallisation have been investigated using dark current, steady state and transient photocurrent measurements. Above 250 K the dark conductivity at electric fields below 100 V/cm is thermally activated, with single activation energy 0.7 eV and magnitude 4×10^{-8} S/cm at 300 K. At higher electric fields, conductivity increases as a power law suggesting space charge limitation. Steady state photoconductivity exhibits a roughly linear dependence on photon flux below 10^{15} $cm^{-2}s^{-1}$ and square root dependence above this, consistent with trap-limited and band-to-band recombination mechanisms respectively. Photocurrent overshoot and undershoot on a timescale of 10-100 s are observed on application and removal of a steady light source. Transient photocurrent measurements with temperature reveal a peak in the localised density of states 0.26 eV into the band gap, with an attempt to escape frequency of 10^{11} s^{-1}. Deeper states are distributed exponentially with characteristic energy 54 meV, decreasing towards mid-gap.

1. Introduction

Over the past decade, methylammonium lead tri-halide perovskite solar cells have maintained a spectacular year-on-year increase in headline photovoltaic conversion efficiency, now surpassing 22% [1, 2], but their tendency to degrade and become unstable on exposure to light, heat and ambient atmosphere is also well-known [3]. The member of this family studied most extensively is *methylammonium lead tri-iodide* $CH_3NH_3PbI_3$, commonly referred to by the acronym MAPI. MAPI has a direct bandgap of 1.5 eV, high optical absorption coefficient, steep absorption edge and good carrier transport properties, and may be solution-processed into solar cell absorber layers.

While thin films are of greater interest for large-area solar cell applications, single crystals generally have the advantage of being less affected by surface phenomena and structural disorder, and enable the intrinsic material properties to be studied more directly. Simple methods of growing single MAPI crystals several cm^2 on a face, of good electronic quality, from solution have been developed, including *inverse temperature crystallisation* (ITC) [4, 5]. This relies upon the fact that crystallisation of product from solutions of precursor compounds is favoured, in certain solvents, at higher temperatures. The choice of solvent also depends upon the solubility of the precursors.

We have carried out a range of measurements on MAPI single crystals grown using ITC. These include thermally-activated dark conductivity, space-charge limited currents, steady-state and switch-

IOP Conf. Series: Journal of Physics: Conf. Series **1186** (2019) 012035 doi:10.1088/1742-6596/1186/1/012035

on/switch-off photoconductivity, and short pulse transient photoconductivity (TPC). From this we seek to quantify their electronic transport properties, in particular intrinsic electronic defect density of states (DOS) and its energetic distribution, and to gain insight into why these semiconductors are such promising photovoltaic (PV) absorbers but also show a strong tendency towards instability.

2. Experimental

2.1. Growth of MAPI single crystals

1 mole quantities of PbI_2 (4.61 g) and CH_3NH_3I (1.59 g) precursors were dissolved in 10 ml γ-butyrolactone solvent by magnetic stirring at approximately 60 °C for 30 minutes. A 2 ml portion was then pipetted into a flat bottom screw-top glass vial and placed in a silicone oil bath at 110 °C. Over a period of some three hours, MAPI crystals 1-2 mm in size were formed, and were recovered by passing the liquor through filter paper. To increase their size, the process was repeated using selected crystals with good morphology as seeds, placed in fresh solution. The above procedures were carried out in a fume cupboard under ambient conditions and 20-30% relative humidity.

MAPI crystals exhibit rhombo-hexagonal dodecahedral morphology; examples are shown in figure 1(a). Planar electrical contacts were formed on the upper facet using silver conductive paste, leaving a gap of approximately 0.5 mm, as shown in figure 1(b). This rather crude gap-cell arrangement enables photocurrent measurements to be made by directing light between the contacts. As the thickness of the crystal (in the direction orthogonal to the upper facet) was typically 2 mm, this sample geometry does not conform particularly well to any of the usual approximations. To estimate dark conductivity, the active sample thickness was assumed equal to the contact separation. The crystal was mounted on a ceramic substrate, and placed in a cryostat pumped to 10^{-2} torr over the entire measurement period to minimise contamination. A 6 mm diameter quartz rod was used to transfer light on to the sample.

While MAPI crystals were normally mounted and measured shortly after growth, they could be stored in a dessicator for periods of weeks without significant degradation being visible or electrically detectable. Discarded samples may develop regions of yellow colouration after several days at ambient, associated with an undesirable hydrated form of the material.

2.2. Electrical measurements

Dark current and steady-state photocurrent measurements were made using a Keithley 485 auto-ranging picoammeter and low-noise DC supply. Light sources were LEDs of the required wavelength, coupled to the sample via the quartz rod. The light output from the rod was calibrated versus LED current prior to use. For switch-on/switch-off measurements, the LED was driven from the 50 ohm output of a Thurlby Thandar TG1304 function generator, via a series resistor to set the desired current. For TPC measurements, shorter current pulses (0.1 to 1 μs) at up to 2 A were provided using a Renesas EL7104 laser driver IC. Photocurrents were pre-amplified, recorded on a Tektronix 3052 DSO and transferred to a PC for analysis.

In the TPC experiment, 4 to 6 time segments spanning from microseconds to seconds were edited into one logarithmically-spaced photocurrent decay curve, used to calculate the density of localised states (DOS) versus energy via a Fourier-transform technique [6, 7].

3. Results and discussion

3.1. Dark current activation

An Arrhenius plot of dark conductivity vs. reciprocal absolute temperature is shown in figure 2. The graph divides broadly into: (i) a higher temperature region (260-350 K) of activation energy 0.7 eV, (ii) a lower temperature region (160-180 K) of activation energy 0.2 eV, (iii) a curved intermediate region (180-260 K), with no single activation energy. Region (i) may be attributed unambiguously to thermal band-to-band (intrinsic) kinetics. The activation energy is roughly half the bandgap energy of 1.5 eV, as expected [7]. Region (ii) suggests extrinsic behaviour, with conduction

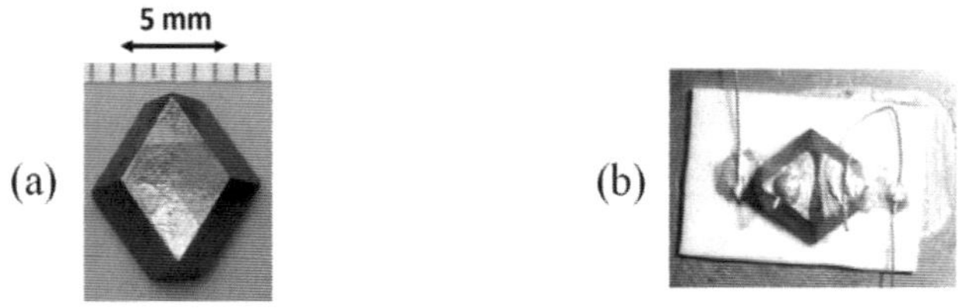

Figure 1. (a) MAPI crystal approximately 2 mm thick following three growth stages, (b) MAPI crystal with silver paste contacts, mounted on ceramic substrate.

taking place via a different thermally activated process. This might involve transport via surface channels or interlinked structural defects within the crystal bulk, which we have yet to investigate. The intermediate region (iii) would then represent a transition from (inhomogeneous) current flow via extrinsic processes at low temperatures, where the bulk crystal is largely insulating, through a mixed phase in which both intrinsic and extrinsic processes may be conjoined, to the intrinsic phase where the bulk crystal predominates. The measured dark conductivity of 4×10^{-8} S/cm at 300 K, with a variation between samples of around a factor of 2, lies within the range reported in the literature [8, 9].

Discontinuities in transport measurements with temperature, attributed to phase transitions in the perovskite crystal structure, have been reported [10]. These are known to occur at 165 K (orthorhombic to tetragonal) and 327 K (tetragonal to cubic). However, we observed no specific effects over the temperature range studied here. We thus conclude that the crystal remained in its tetragonal phase. We note however that small hysteresis [9] effects, dependent upon whether the sample current is measured moving from high to low temperature or vice-versa, are observed but do not substantially change the above reasoning.

3.2. Dark current-voltage curves

Dark current-voltage curves measured at temperatures of 298, 250 and 210 K are shown in figure 3. Below 3 V the I-V curve is substantially linear and shows the thermal activation discussed in section 3.1 above. However in the range 10-50 volts the curve deviates significantly from linear towards a power-law, of index 2 to 3. This behaviour has been reported previously for these materials [4, 5] and linked to the onset of space-charge limited currents (SCLC). The conditions for SCLC involve trap-filling, with the 'knee' representing the trap-filled voltage:

$$V_{TFL} = \frac{qn_{trap}L^2}{2\varepsilon_0\varepsilon_r} \quad (1)$$

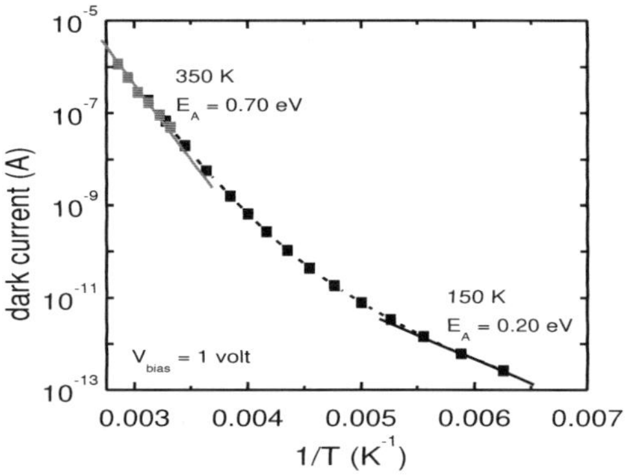

Figure 2. Arrhenius plot of dark current vs. reciprocal temperature.

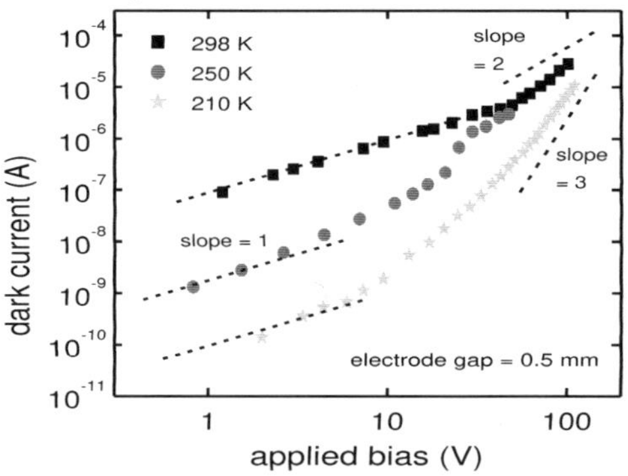

Figure 3. DC current-voltage curve measured at three different temperatures.

where n_{trap} is the trap density, L the conduction path length, ε_0 the permittivity of free space, ε_r the relative permittivity of the material and q the electronic charge. Setting $L = 0.1$ cm, $n_{trap} = 10^{10}$ cm^{-3} and $\varepsilon_r = 6$ (the high frequency limit based on the refractive index) yields $V_{TFL} = 15$ V, in accordance with our observations. A trap density of 10^{10} cm^{-3} is remarkably low, on a par with the best inorganic semiconductor samples. However, recent investigations of the dielectric properties of organic-inorganic lead halide perovskites [11] has revealed their remarkably high value of low-frequency relative permittivity, of order 10^4 in the range 10 mHz-1 Hz corresponding to the timescale of the above experiment. Adopting this value, the trap density would be of order 10^{14} cm^{-3}, still a low figure, more in keeping with multicrystalline and thin-film semiconductors.

3.3. Steady-state and switch-on/switch-off photocurrent

The room-temperature steady-state photocurrent versus photon flux at 650 nm and 830 nm plotted on a log-log scale is shown in figure 4. Both curves bear a similar form, but differ somewhat in detail. Below 10^{15} cm^{-2} s^{-1}, photocurrent is roughly linearly dependent on photon flux. Above this value, the photocurrent increases less steeply and follows a square-root dependence. These two regions strongly suggest two different recombination mechanisms. A close to linear dependence suggests recombination via bandgap states, while square-root dependence at higher generation rates in direct-gap materials is characteristic of bimolecular recombination [12]. Although the behaviour at both wavelengths is quite similar, there are minor differences which may be attributed to the smaller optical absorption depth of around 1 µm at 650 nm, compared with 10 to 100 µm at 830 nm which lies on the MAPI optical absorption edge [13]. This will give rise to a greater carrier density close to the surface in the former case, resulting in earlier onset of bimolecular recombination.

The photocurrent overshoot (and undershoot) observed on application and removal of light shown in figure 5 is typical of a variety of longer time-scale phenomena reported for bulk MAPI samples and solar cells, which may involve ionic redistribution [14]. This exemplifies the challenges encountered when measuring and interpreting carrier transport in these materials. It should be noted that the data in figure 4 were recorded after allowing the current to stabilise.

3.4. Transient photocurrent spectroscopy

Transient photocurrent decays over a range of temperatures from 140 K to 270 K are shown in figure 6. There are a number of features (changes in gradient) in the decays, which are likely associated with carrier trapping and emission kinetics between band (transport) states and localised gap states. In order to identify these states we have applied DOS spectroscopy [6, 7], with the results shown in figure 7 on a log scale and in figure 8 on a linear scale. These plots emphasise different aspects of the DOS. With reference to figure 7, a reasonably consistent overlap of the DOS curves at deeper energies is obtained

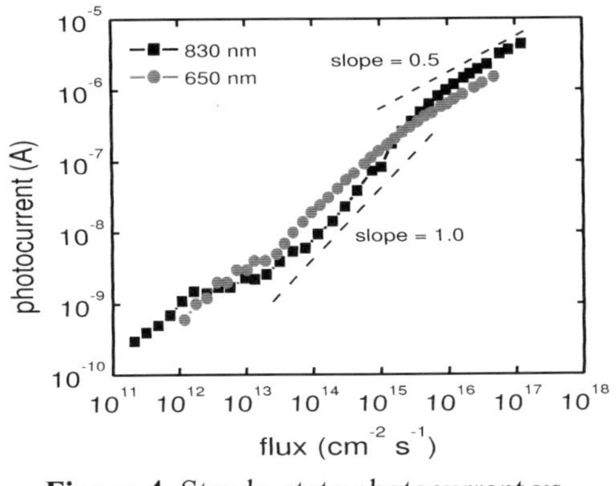

Figure 4. Steady-state photocurrent vs. photon flux at 650 nm and 830 nm.

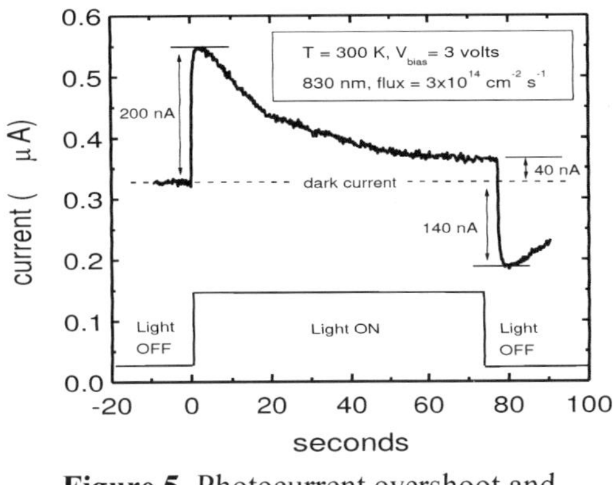

Figure 5. Photocurrent overshoot and undershoot relative to dark current.

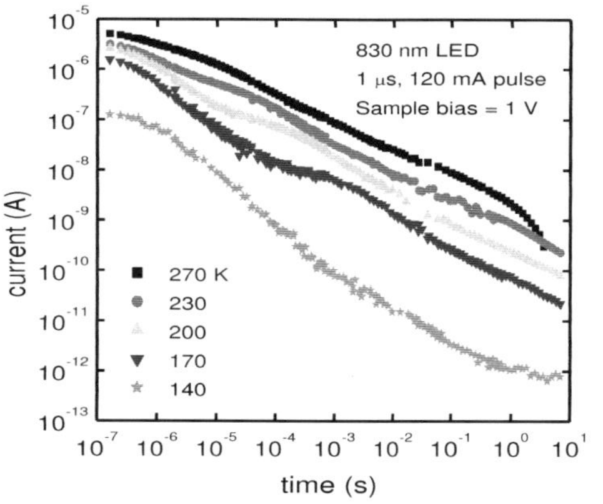

Figure 6. Transient photocurrent decays between 140 K and 270 K.

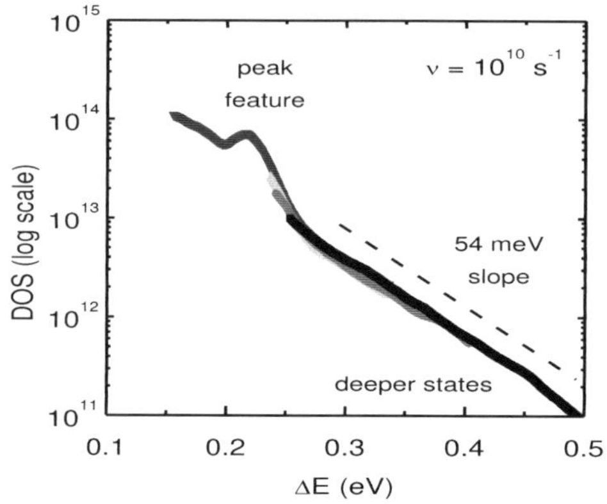

Figure 7. DOS on a log scale vs. energy from band edge ΔE. An attempt to escape frequency of 10^{10} s^{-1} gives optimal fit for deeper states.

by adopting an attempt to escape frequency of 10^{10} s^{-1}. This parameter essentially defines the energy scale of the DOS relative to the band edge. The DOS follows a broadly exponential reduction with increasing energy, and a characteristic slope of 54 meV. However the absolute DOS scale (y-axis) is more difficult to quantify as it depends on several factors, such as carrier mobility, which has not been determined here. For simplicity, the DOS scale in figure 7 has been aligned broadly with our estimate for the localised state density obtained from SCLC (section 3.2). This figure is in keeping with more recent estimates of defect density in single crystal MAPI [15]. Figure 8 reveals the peak 'feature' evident in figure 7 more clearly. It can be observed that the attempt to escape frequency shifts the peak along the energy axis by an amount dependent on the experimental temperature – ideally there will be a single frequency at which all coincide. The optimum overlap occurs at 10^{11} s^{-1}, with a corresponding peak energy $\Delta E_P = 0.26$ eV.

It should be noted that the TPC method applied to a planar sample is unable to distinguish *a priori* between electron and hole transport, and thus the above procedure may be sensitive to (empty) states above the Fermi level, or to (filled) states below it, frequently depending upon whether electrons or holes, respectively, are the majority carrier. As carrier mobilities in MAPI single crystals [16] do not consistently favour either electrons or holes, we cannot state here with any authority which half of the bandgap is being probed by TPC.

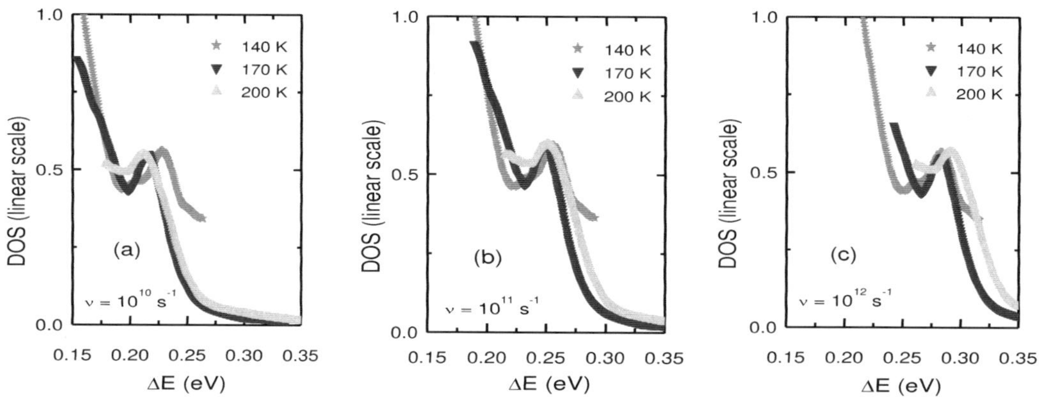

Figure 8. DOS on a linear scale vs. energy from band edge ΔE. An attempt to escape frequency of 10^{11} s^{-1} (graph b) gives optimal fit for the peak feature, at energy $\Delta E_P = 0.26$ eV.

4. Conclusions
We have investigated the electronic transport properties of methylammonium lead tri-iodide single crystals grown by the inverse temperature crystallisation method. Our dark current measurements show similar room-temperature conductivity (4×10^{-8} S/cm), thermal activation energy (0.70 eV) and space-charge limited current behaviour as reported previously in the literature. Steady-state and transient photoconductivity measurements have been applied to investigate carrier interactions with localised states. The steady-state measurements are consistent with trap-mediated recombination below a flux of 10^{15} $cm^{-2}s^{-1}$, and bimolecular band-to-band recombination above this level. Some small wavelength-dependent variations occur, which probably reflect the higher volume generation rates that occur with more strongly absorbed light. Transient photoconductivity decays are consistent with multiple trapping in a narrow band of states 0.26 eV below the band edge with an attempt to escape frequency of 10^{11} s^{-1}, plus a broader exponential distribution of deeper states, of slope 54 meV. We are presently unable to say whether these states lie in the upper or the lower half of the bandgap. More generally, photoconductivity measurements display a rich variety of behaviour, unlikely to be adequately explained in terms of a 'stationary' model. In common with other more comprehensive studies, we conclude that (reversible) ionic migration or redistribution takes place on application and removal of electric fields or carrier concentration gradients. The resulting modified potential landscape will in turn influence electronic carrier transport; some outcomes may be beneficial, for example increasing effective carrier diffusion lengths, but others may be less desirable, since transport properties will be synchronised to the long timescales dictated by the landscape response.

Acknowledgments
The authors are grateful to S Anthony and C Moore for technical assistance, and to C Main for the legacy of his DOS spectroscopy software. The SC-Simul simulation software (University of Oldenburg) was used to visualise and explore certain density of states distributions.

References
[1] Green M A, Hishikawa Y, Dunlop E D, Levi D H, Hohl-Ebinger J and Ho-Baillie A W Y 2018 *Prog. Photovolt. Res. Appl.* **26** 427
[2] Green M A and Ho-Baillie A 2017 *ACS Energy Lett.* **2** 822
[3] Wang D, Wright M, Elumalai N K and Uddim A 2016 *Sol. Energy Mater. Sol. Cells* **147** 255
[4] Saidaminov M I *et al* 2015 *Nat. Commun.* **6** 7586
[5] Liu Y *et al* 2015 *Adv. Mater.* **27** 5176
[6] Main C, Brüggemann R, Webb D P and Reynolds S 1992 *Sol. St. Commun.* **83** 40
[7] Main C 1997 *MRS Symp. Proc.* **467** 167
[8] Sveinbjörnsson K, Aitola K, Zhang X, Pazoki M, Hagfeldt A, Boschloo G and Johansson E M J 2015 *J. Phys. Chem. Lett.* **6** 4259
[9] Khenkin M V, Amasev D V, Kozyukhin S A, Sadovnikov A V, Katz E A and Kazanskii A G 2017 *Appl. Phys. Lett.* **110** 222107
[10] Li D, Wang G, Cheng H-C, Chen C-Y, Wu H, Liu Y, Huang Y and Duan X 2016 *Nat. Commun.* **7** 11330
[11] Slonopas A, Kaur B and Norris P 2017 *Appl. Phys. Lett.* **110** 222905
[12] Bube R H 1992 *Photoelectronic Properties of Semiconductors* (Cambridge: Cambridge University Press)
[13] De Wolf S, Holovsky J, Moon S-J, Löper P, Niesen B, Ledinsky M, Haug F-J, Yum J-H and Ballif C 2014 *J. Phys. Chem. Lett.* **5** 1035
[14] Eames C, Frost J M, Barnes P R F, O'Regan B C, Walsh A and Islam M S 2015 *Nat. Commun.* **6** 7497
[15] Wenger B, Nayak P B, Wen X, Kesava S V, Noel N K and Snaith H 2017 *Nat. Commun.* 8 590
[16] Herz L 2017 *ACS Energy Lett.* **2** 1539